Introduction to Genetic Algorithms

S.N.Sivanandam · S.N.Deepa

Introduction to Genetic Algorithms

With 193 Figures and 13 Tables

 Springer

Authors

S.N.Sivanandam
Professor and Head
Dept. of Computer Science and Engineering
PSG College of Technology
Coimbatore - 641 004
TN, India

S.N.Deepa
Ph.D Scholar Dept. of Computer Science
and Engineering
PSG College of Technology
Coimbatore - 641 004
TN, India

Library of Congress Control Number: 2007930221

ISBN 978-3-540-73189-4 Springer Berlin Heidelberg New York

This work is subject to copyright. All rights are reserved, whether the whole or part of the material is concerned, specifically the rights of translation, reprinting, reuse of illustrations, recitation, broadcasting, reproduction on microfilm or in any other way, and storage in data banks. Duplication of this publication or parts thereof is permitted only under the provisions of the German Copyright Law of September 9, 1965, in its current version, and permission for use must always be obtained from Springer. Violations are liable for prosecution under the German Copyright Law.

Springer is a part of Springer Science+Business Media
springer.com
© Springer-Verlag Berlin Heidelberg 2008

The use of general descriptive names, registered names, trademarks, etc. in this publication does not imply, even in the absence of a specific statement, that such names are exempt from the relevant protective laws and regulations and therefore free for general use.

Typesetting: Integra Software Services Pvt. Ltd., India

Cover design: Erich Kirchner, Heidelberg

Printed on acid-free paper   SPIN: 12053230   89/3180/Integra   5 4 3 2 1 0

Preface

The origin of evolutionary algorithms was an attempt to mimic some of the processes taking place in natural evolution. Although the details of biological evolution are not completely understood (even nowadays), there exist some points supported by strong experimental evidence:

- Evolution is a process operating over chromosomes rather than over organisms. The former are organic tools encoding the structure of a living being, i.e., a creature is "built" decoding a set of chromosomes.
- Natural selection is the mechanism that relates chromosomes with the efficiency of the entity they represent, thus allowing that efficient organism which is well-adapted to the environment to reproduce more often than those which are not.
- The evolutionary process takes place during the reproduction stage. There exists a large number of reproductive mechanisms in Nature. Most common ones are mutation (that causes the chromosomes of offspring to be different to those of the parents) and recombination (that combines the chromosomes of the parents to produce the offspring).

Based upon the features above, the three mentioned models of evolutionary computing were independently (and almost simultaneously) developed.

An Evolutionary Algorithm (EA) is an iterative and stochastic process that operates on a set of individuals (population). Each individual represents a potential solution to the problem being solved. This solution is obtained by means of a encoding/decoding mechanism. Initially, the population is randomly generated (perhaps with the help of a construction heuristic). Every individual in the population is assigned, by means of a fitness function, a measure of its goodness with respect to the problem under consideration. This value is the quantitative information the algorithm uses to guide the search.

Among the evolutionary techniques, the genetic algorithms (GAs) are the most extended group of methods representing the application of evolutionary tools. They rely on the use of a selection, crossover and mutation operators. Replacement is usually by generations of new individuals.

Intuitively a GA proceeds by creating successive generations of better and better individuals by applying very simple operations. The search is only guided by the fitness value associated to every individual in the population. This value is used to rank individuals depending on their relative suitability for the problem being solved. The problem is the fitness function that for every individual is encharged of assigning the fitness value.

The location of this kind of techniques with respect to other deterministic and non-deterministic procedures is shown in the following tree. This figure below outlines the situation of natural techniques among other well-known search procedures.

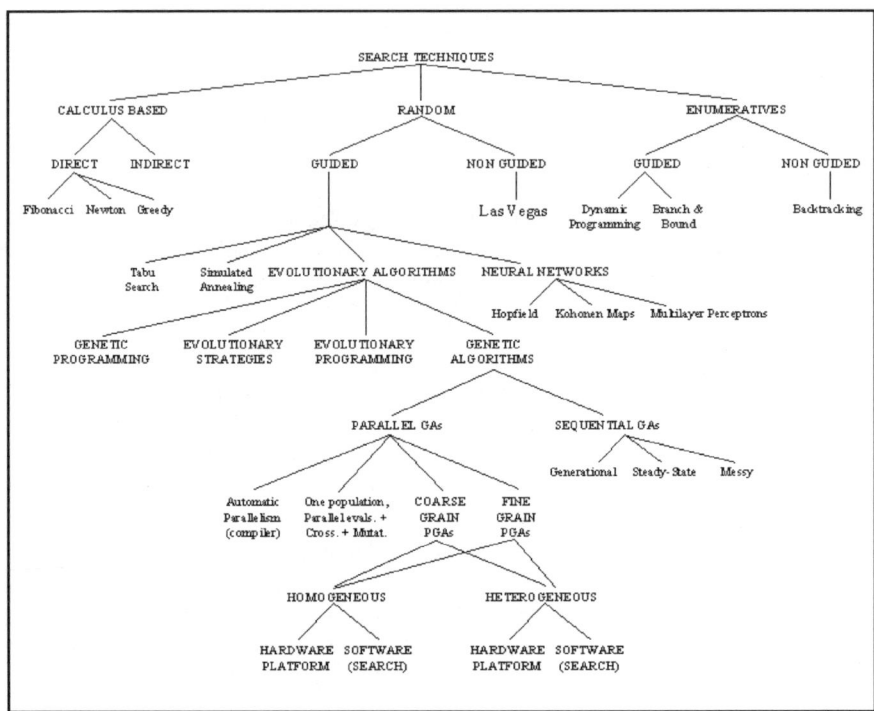

Combinations of EAs with Hill-Climbing algorithms are very powerful. Genetic algorithms intensively using such local search mechanism are termed *Memetic Algorithms*. Also parallel models increase the extension and quality of the search. The EAs exploration compares quite well against the rest of search techniques for a similar search effort. Exploitation is a more difficult goal in EAs but nowadays many solutions exist for EAs to refine solutions.

Genetic algorithms are currently the most prominent and widely used computational models of evolution in *artificial-life* systems. These decentralized models provide a basis for understanding many other systems and phenomena in the world. Researches on GAs in alife give illustrative *examples* in which the genetic algorithm is used to study how learning and evolution interact, and to model ecosystems, immune system, cognitive systems, and social systems.

About the Book

This book is meant for a wide range of readers, who wishes to learn the basic concepts of Genetic Algorithms. It can also be meant for programmers, researchers and management experts whose work is based on optimization techniques. The basic concepts of Genetic Algorithms are dealt in detail with the relevant information and knowledge available for understanding the optimization process. The various operators involved for Genetic Algorithm operation are explained with examples. The advanced operators and the various classifications have been discussed in lucid manner, so that a starter can understand the concepts with a minimal effort.

The solutions to specific problems are solved using MATLAB 7.0 and the solutions are given. The MATLAB GA toolbox has also been included for easy reference of the readers so that they can have hands on working with various GA functions. Apart from MATLAB solutions, certain problems are also solved using C and C++ and the solutions are given.

The book is designed to give a broad in-depth knowledge on Genetic Algorithm. This book can be used as a handbook and a guide for students of all engineering disciplines, management sector, operational research area, computer applications, and for various professionals who work in Optimization area.

Genetic Algorithms, at present, is a hot topic among academicians, researchers and program developers. Due to which, this book is not only for students, but also for a wide range of researchers and developers who work in this field. This book can be used as a ready reference guide for Genetic Algorithm research scholars. Most of the operators, classifications and applications for a wide variety of areas covered here fulfills as an advanced academic textbook.

To conclude, we hope that the reader will find this book a helpful guide and a valuable source of information about Genetic Algorithm concepts for their several practical applications.

1 Organization of the Book

The book contains 11 chapters altogether. It starts with the introduction to Evolutionary Computing. The various application case studies are also discussed.

The chapters are organized as follows:

- Chapter 1 gives an introduction to Evolutionary computing, its development and its features.
- Chapter 2 enhances the growth of Genetic Algorithms and its comparison with other conventional optimization techniques. Also the basic simple genetic algorithm with its advantages and limitations are discussed.
- The various terminologies and the basic operators involved in genetic algorithm are dealt in Chap. 3. Few example problems, enabling the readers to understand the basic genetic algorithm operation are also included.
- Chapter 4 discusses the advanced operators and techniques involved in genetic algorithm.
- The different classifications of genetic algorithm are provided in Chap. 5. Each of the classifications is discussed with their operators and mode of operation to achieve optimized solution.
- Chapter 6 gives a brief introduction to genetic programming. The steps involved and characteristics of genetic programming with its applications are described here.
- Chapter 7 discusses on various genetic algorithm optimization problems which includes fuzzy optimization, multi objective optimization, combinatorial optimization, scheduling problems and so on.
- The implementation of genetic algorithm using MATLAB is discussed in Chap. 8. The toolbox functions and simulated results to specific problems are provided in this chapter.
- Chapter 9 gives the implementation of genetic algorithm concept using C and C++. The implementation is performed for few benchmark problems.
- The application of genetic algorithm in various emerging fields along with case studies is given in Chapter 10.
- Chapter 11 gives a brief introduction to particle swarm optimization and ant colony optimization.

The Bibliography is given at the end for the ready reference of readers.

2 Salient Features of the Book

The salient features of the book include:

- Detailed explanation of Genetic Algorithm concepts
- Numerous Genetic Algorithm Optimization Problems
- Study on various types of Genetic Algorithms
- Implementation of Optimization problem using C and C++
- Simulated solutions for Genetic Algorithm problems using MATLAB 7.0
- Brief description on the basics of Genetic Programming
- Application case studies on Genetic Algorithm on emerging fields

S.N. Sivanandam completed his B.E (Electrical and Electronics Engineering) in 1964 from Government College of Technology, Coimbatore and M.Sc (Engineering) in Power System in 1966 from PSG College of Technology, Coimbatore. He acquired PhD in Control Systems in 1982 from Madras University. He has received Best Teacher Award in the year 2001 and *Dhakshina Murthy Award* for Teaching Excellence from PSG College of Technology. He received The CITATION for best teaching and technical contribution in the Year 2002, Government College of Technology, Coimbatore. He has a total teaching experience (UG and PG) of 41 years. The total number of undergraduate and postgraduate projects guided by him for both Computer Science and Engineering and Electrical and Electronics Engineering is around 600. He is currently working as a *Professor and Head Computer Science and Engineering Department*, PSG College of Technology, Coimbatore [from June 2000]. He has been identified as an outstanding person in the field of Computer Science and Engineering in MARQUIS "Who's Who", October 2003 issue, New providence, New Jersey, USA. He has also been identified as an outstanding person in the field of Computational Science and Engineering in "Who's Who", December 2005 issue, Saxe-Coburg Publications, United Kingdom. He has been placed as a VIP member in the continental WHO's WHO Registry of national Business Leaders, Inc. 33 West Hawthorne Avenue Valley Stream, NY 11580, Aug 24, 2006.

S.N. Sivanandam has published 12 books. He has delivered around 150 special lectures of different specialization in Summer/Winter school and also in various Engineering colleges. He has guided and coguided 30 Ph.D research works and at present 9 Ph.D research scholars are working under him. The total number of technical publications in International/National journals/Conferences is around 700. He has also received Certificate of Merit 2005–2006 for his paper from The Institution of Engineers (India). He has chaired 7 International conferences and 30 National conferences. He is a member of various professional bodies like IE (India), ISTE, CSI, ACS and SSI. He is a technical advisor for various reputed industries and Engineering Institutions. His research areas include Modeling and Simulation, Neural networks , Fuzzy Systems and Genetic Algorithm, Pattern Recognition, Multi dimensional system analysis, Linear and Non linear control system, Signal and Image processing, Control System, Power system, Numerical methods, Parallel Computing, Data Mining and Database Security.

S.N. Deepa has completed her B.E Degree from Government College of Technology, Coimbatore, 1999 and M.E Degree from PSG College of Technology, Coimbatore, 2004. She was a gold medalist in her B.E Degree Programme. She has received G.D Memorial Award in the year 1997 and Best Outgoing Student Award from PSG College of Technology, 2004. Her M.E Thesis won National Award from the Indian Society of Technical Education and L&T, 2004. She has published 5 books and papers in International and National Journals. Her research areas include Neural Network, Fuzzy Logic, Genetic Algorithm, Digital Control, Adaptive and Non-linear Control.

Acknowledgement

The authors are always thankful to the Almighty for perseverance and achievements. They wish to thank Shri G. Rangaswamy, Managing Trustee, PSG Institutions, Shri C.R. Swaminathan, Chief Executive; and Dr. R. Rudramoorthy, Principal, PSG College of Technology, Coimbatore, for their whole-hearted cooperation and great encouragement given in this successful endeavor. They also wish to thank the staff members of computer science and engineering for their cooperation. Deepa wishes to thank her husband Anand, daughter Nivethitha and parents for their support.

Contents

1 Evolutionary Computation 1
1.1 Introduction 1
1.2 The Historical Development of EC 2
    1.2.1 Genetic Algorithms 2
    1.2.2 Genetic Programming 3
    1.2.3 Evolutionary Strategies 4
    1.2.4 Evolutionary Programming 5
1.3 Features of Evolutionary Computation 5
    1.3.1 Particulate Genes and Population Genetics 6
    1.3.2 The Adaptive Code Book 7
    1.3.3 The Genotype/Phenotype Dichotomy 8
1.4 Advantages of Evolutionary Computation 9
    1.4.1 Conceptual Simplicity 10
    1.4.2 Broad Applicability 10
    1.4.3 Hybridization with Other Methods 11
    1.4.4 Parallelism 11
    1.4.5 Robust to Dynamic Changes 11
    1.4.6 Solves Problems that have no Solutions 12
1.5 Applications of Evolutionary Computation 12
1.6 Summary 13

2 Genetic Algorithms 15
2.1 Introduction 15
2.2 Biological Background 16
    2.2.1 The Cell 16
    2.2.2 Chromosomes 16
    2.2.3 Genetics 17
    2.2.4 Reproduction 17
    2.2.5 Natural Selection 19
2.3 What is Genetic Algorithm? 20
    2.3.1 Search Space 20
    2.3.2 Genetic Algorithms World 20
    2.3.3 Evolution and Optimization 22
    2.3.4 Evolution and Genetic Algorithms 23

2.4	Conventional Optimization and Search Techniques	24
	2.4.1 Gradient-Based Local Optimization Method	25
	2.4.2 Random Search	26
	2.4.3 Stochastic Hill Climbing	27
	2.4.4 Simulated Annealing	27
	2.4.5 Symbolic Artificial Intelligence (AI)	29
2.5	A Simple Genetic Algorithm	29
2.6	Comparison of Genetic Algorithm with Other Optimization Techniques	33
2.7	Advantages and Limitations of Genetic Algorithm	34
2.8	Applications of Genetic Algorithm	35
2.9	Summary	36

3 Terminologies and Operators of GA ... 39

3.1	Introduction	39
3.2	Key Elements	39
3.3	Individuals	39
3.4	Genes	40
3.5	Fitness	41
3.6	Populations	41
3.7	Data Structures	42
3.8	Search Strategies	43
3.9	Encoding	43
	3.9.1 Binary Encoding	43
	3.9.2 Octal Encoding	44
	3.9.3 Hexadecimal Encoding	44
	3.9.4 Permutation Encoding (Real Number Coding)	44
	3.9.5 Value Encoding	45
	3.9.6 Tree Encoding	45
3.10	Breeding	46
	3.10.1 Selection	46
	3.10.2 Crossover (Recombination)	50
	3.10.3 Mutation	56
	3.10.4 Replacement	57
3.11	Search Termination (Convergence Criteria)	59
	3.11.1 Best Individual	59
	3.11.2 Worst individual	60
	3.11.3 Sum of Fitness	60
	3.11.4 Median Fitness	60
3.12	Why do Genetic Algorithms Work?	60
	3.12.1 Building Block Hypothesis	61
	3.12.2 A Macro-Mutation Hypothesis	62
	3.12.3 An Adaptive Mutation Hypothesis	62
	3.12.4 The Schema Theorem	63
	3.12.5 Optimal Allocation of Trials	65

	3.12.6 Implicit Parallelism	66
	3.12.7 The No Free Lunch Theorem	68
3.13	Solution Evaluation	68
3.14	Search Refinement	69
3.15	Constraints	69
3.16	Fitness Scaling	70
	3.16.1 Linear Scaling	70
	3.16.2 Sigma Truncation	71
	3.16.3 Power Law Scaling	72
3.17	Example Problems	72
	3.17.1 Maximizing a Function	72
	3.17.2 Traveling Salesman Problem	76
3.18	Summary	78
	Exercise Problems	81

4 Advanced Operators and Techniques in Genetic Algorithm 83

4.1	Introduction	83
4.2	Diploidy, Dominance and Abeyance	83
4.3	Multiploid	85
4.4	Inversion and Reordering	86
	4.4.1 Partially Matched Crossover (PMX)	88
	4.4.2 Order Crossover (OX)	88
	4.4.3 Cycle Crossover (CX)	89
4.5	Niche and Speciation	89
	4.5.1 Niche and Speciation in Multimodal Problems	90
	4.5.2 Niche and Speciation in Unimodal Problems	93
	4.5.3 Restricted Mating	96
4.6	Few Micro-operators	97
	4.6.1 Segregation and Translocation	97
	4.6.2 Duplication and Deletion	97
	4.6.3 Sexual Determination	98
4.7	Non-binary Representation	98
4.8	Multi-Objective Optimization	99
4.9	Combinatorial Optimizations	100
4.10	Knowledge Based Techniques	100
4.11	Summary	102
	Exercise Problems	103

5 Classification of Genetic Algorithm 105

5.1	Introduction	105
5.2	Simple Genetic Algorithm (SGA)	105
5.3	Parallel and Distributed Genetic Algorithm (PGA and DGA)	106
	5.3.1 Master-Slave Parallelization	109
	5.3.2 Fine Grained Parallel GAs (Cellular GAs)	110
	5.3.3 Multiple-Deme Parallel GAs (Distributed GAs or Coarse Grained GAs)	111

		5.3.4	Hierarchical Parallel Algorithms	113

- 5.4 Hybrid Genetic Algorithm (HGA) 115
  - 5.4.1 Crossover 116
  - 5.4.2 Initialization Heuristics 117
  - 5.4.3 The RemoveSharp Algorithm 117
  - 5.4.4 The LocalOpt Algorithm 119
- 5.5 Adaptive Genetic Algorithm (AGA) 119
  - 5.5.1 Initialization 120
  - 5.5.2 Evaluation Function 120
  - 5.5.3 Selection operator 121
  - 5.5.4 Crossover operator 121
  - 5.5.5 Mutation operator 122
- 5.6 Fast Messy Genetic Algorithm (FmGA) 122
  - 5.6.1 Competitive Template (CT) Generation 123
- 5.7 Independent Sampling Genetic Algorithm (ISGA) 124
  - 5.7.1 Independent Sampling Phase 125
  - 5.7.2 Breeding Phase 126
- 5.8 Summary 127
  - Exercise Problems 129

6 Genetic Programming 131
- 6.1 Introduction 131
- 6.2 Comparison of GP with Other Approaches 131
- 6.3 Primitives of Genetic Programming 135
  - 6.3.1 Genetic Operators 136
  - 6.3.2 Generational Genetic Programming 136
  - 6.3.3 Tree Based Genetic Programming 136
  - 6.3.4 Representation of Genetic Programming 137
- 6.4 Attributes in Genetic Programming 141
- 6.5 Steps of Genetic Programming 143
  - 6.5.1 Preparatory Steps of Genetic Programming 143
  - 6.5.2 Executional Steps of Genetic Programming 146
- 6.6 Characteristics of Genetic Programming 149
  - 6.6.1 What We Mean by "Human-Competitive" 149
  - 6.6.2 What We Mean by "High-Return" 152
  - 6.6.3 What We Mean by "Routine" 154
  - 6.6.4 What We Mean by "Machine Intelligence" 154
- 6.7 Applications of Genetic Programming 156
  - 6.7.1 Applications of Genetic Programming in Civil Engineering 156
- 6.8 Haploid Genetic Programming with Dominance 159
  - 6.8.1 Single-Node Dominance Crossover 161
  - 6.8.2 Sub-Tree Dominance Crossover 161
- 6.9 Summary 161
  - Exercise Problems 163

7 Genetic Algorithm Optimization Problems 165
7.1 Introduction 165
7.2 Fuzzy Optimization Problems 165
    7.2.1 Fuzzy Multiobjective Optimization....................... 166
    7.2.2 Interactive Fuzzy Optimization Method 168
    7.2.3 Genetic Fuzzy Systems 168
7.3 Multiobjective Reliability Design Problem 170
    7.3.1 Network Reliability Design 170
    7.3.2 Bicriteria Reliability Design 174
7.4 Combinatorial Optimization Problem 176
    7.4.1 Linear Integer Model 178
    7.4.2 Applications of Combinatorial Optimization 179
    7.4.3 Methods 182
7.5 Scheduling Problems 187
    7.5.1 Genetic Algorithm for Job Shop Scheduling Problems (JSSP) .. 187
7.6 Transportation Problems 190
    7.6.1 Genetic Algorithm in Solving Transportation
          Location-Allocation Problems with Euclidean Distances....... 191
    7.6.2 Real-Coded Genetic Algorithm (RCGA) for Integer Linear
          Programming in Production-Transportation Problems
          with Flexible Transportation Cost 194
7.7 Network Design and Routing Problems 199
    7.7.1 Planning of Passive Optical Networks 199
    7.7.2 Planning of Packet Switched Networks 202
    7.7.3 Optimal Topological Design of All Terminal Networks 203
7.8 Summary 208
    Exercise Problems 209

8 Genetic Algorithm Implementation Using Matlab 211
8.1 Introduction 211
8.2 Data Structures 211
    8.2.1 Chromosomes 212
    8.2.2 Phenotypes 212
    8.2.3 Objective Function Values 213
    8.2.4 Fitness Values 213
    8.2.5 Multiple Subpopulations 213
8.3 Toolbox Functions 214
8.4 Genetic Algorithm Graphical User Interface Toolbox 219
8.5 Solved Problems using MATLAB 224
8.6 Summary 260
    Review Questions 261
    Exercise Problems 261

9 Genetic Algorithm Optimization in C/C++ 263
9.1 Introduction 263
9.2 Traveling Salesman Problem (TSP) 263

9.3	Word Matching Problem	271
9.4	Prisoner's Dilemma	280
9.5	Maximize $f(x) = x^2$	286
9.6	Minimization a Sine Function with Constraints	292
	9.6.1 Problem Description	293
9.7	Maximizing the Function $f(x) = x^*\sin(10^*\Pi^*x) + 10$	302
9.8	Quadratic Equation Solving	310
9.9	Summary	315
	9.9.1 Projects	315

10 Applications of Genetic Algorithms 317

- 10.1 Introduction 317
- 10.2 Mechanical Sector 317
  - 10.2.1 Optimizing Cyclic-Steam Oil Production with Genetic Algorithms 317
  - 10.2.2 Genetic Programming and Genetic Algorithms for Auto-tuning Mobile Robot Motion Control 320
- 10.3 Electrical Engineering 324
  - 10.3.1 Genetic Algorithms in Network Synthesis 324
  - 10.3.2 Genetic Algorithm Tools for Control Systems Engineering 328
  - 10.3.3 Genetic Algorithm Based Fuzzy Controller for Speed Control of Brushless DC Motor 334
- 10.4 Machine Learning 341
  - 10.4.1 Feature Selection in Machine learning using GA 341
- 10.5 Civil Engineering 345
  - 10.5.1 Genetic Algorithm as Automatic Structural Design Tool 345
  - 10.5.2 Genetic Algorithm for Solving Site Layout Problem 350
- 10.6 Image Processing 352
  - 10.6.1 Designing Texture Filters with Genetic Algorithms 352
  - 10.6.2 Genetic Algorithm Based Knowledge Acquisition on Image Processing 357
  - 10.6.3 Object Localization in Images Using Genetic Algorithm 362
  - 10.6.4 Problem Description 363
  - 10.6.5 Image Preprocessing 364
  - 10.6.6 The Proposed Genetic Algorithm Approach 365
- 10.7 Data Mining 367
  - 10.7.1 A Genetic Algorithm for Feature Selection in Data-Mining 367
  - 10.7.2 Genetic Algorithm Based Fuzzy Data Mining to Intrusion Detection 370
  - 10.7.3 Selection and Partitioning of Attributes in Large-Scale Data Mining Problems Using Genetic Algorithm 379
- 10.8 Wireless Networks 386
  - 10.8.1 Genetic Algorithms for Topology Planning in Wireless Networks 386
  - 10.8.2 Genetic Algorithm for Wireless ATM Network 387
- 10.9 Very Large Scale Integration (VLSI) 395

10.9.1	Development of a Genetic Algorithm Technique for VLSI Testing 395
10.9.2	VLSI Macro Cell Layout Using Hybrid GA 397
10.9.3	Problem Description 398
10.9.4	Genetic Layout Optimization 399
10.10	Summary 402

11 Introduction to Particle Swarm Optimization and Ant Colony Optimization 403
11.1 Introduction 403
11.2 Particle Swarm Optimization 403
  11.2.1 Background of Particle Swarm Optimization 404
  11.2.2 Operation of Particle Swarm Optimization 405
  11.2.3 Basic Flow of Particle Swarm Optimization 407
  11.2.4 Comparison Between PSO and GA 408
  11.2.5 Applications of PSO 410
11.3 Ant Colony Optimization 410
  11.3.1 Biological Inspiration 410
  11.3.2 Similarities and Differences Between Real Ants and Artificial Ants 414
  11.3.3 Characteristics of Ant Colony Optimization 415
  11.3.4 Ant Colony Optimization Algorithms 416
  11.3.5 Applications of Ant Colony Optimization 422
11.4 Summary 424
  Exercise Problems 424

Bibliography 425

Chapter 1
Evolutionary Computation

1.1 Introduction

Charles Darwinian evolution in 1859 is intrinsically a so bust search and optimization mechanism. Darwin's principle "Survival of the fittest" captured the popular imagination. This principle can be used as a starting point in introducing evolutionary computation. Evolved biota demonstrates optimized complex behavior at each level: the cell, the organ, the individual and the population. Biological species have solved the problems of chaos, chance, nonlinear interactivities and temporality. These problems proved to be in equivalence with the classic methods of optimization. The evolutionary concept can be applied to problems where heuristic solutions are not present or which leads to unsatisfactory results. As a result, evolutionary algorithms are of recent interest, particularly for practical problems solving.

The theory of natural selection proposes that the plants and animals that exist today are the result of millions of years of adaptation to the demands of the environment. At any given time, a number of different organisms may co-exist and compete for the same resources in an ecosystem. The organisms that are most capable of acquiring resources and successfully procreating are the ones whose descendants will tend to be numerous in the future. Organisms that are less capable, for whatever reason, will tend to have few or no descendants in the future. The former are said to be more *fit* than the latter, and the distinguishing characteristics that caused the former to be fit are said to be *selected for* over the characteristics of the latter. Over time, the entire population of the ecosystem is said to *evolve* to contain organisms that, on average, are more fit than those of previous generations of the population because they exhibit more of those characteristics that tend to promote survival.

Evolutionary computation (EC) techniques abstract these evolutionary principles into algorithms that may be used to search for optimal solutions to a problem. In a search algorithm, a number of possible solutions to a problem are available and the task is to find the best solution possible in a fixed amount of time. For a search space with only a small number of possible solutions, all the solutions can be examined in a reasonable amount of time and the optimal one found. This *exhaustive search*, however, quickly becomes impractical as the search space grows in size. Traditional search algorithms randomly sample (e.g., *random walk*) or heuristically sample (e.g., *gradient descent*) the search space one solution at a time in the hopes of finding the optimal solution. The key aspect distinguishing an evolutionary search algorithm from such traditional algorithms is that it is *population-based*. Through the adaptation of successive generations of a large number of individuals, an evolutionary algorithm performs an efficient directed search. Evolutionary search is generally better than random search and is not susceptible to the hill-climbing behaviors of gradient-based search.

Evolutionary computing began by lifting ideas from biological evolutionary theory into computer science, and continues to look toward new biological research findings for inspiration. However, an over enthusiastic "biology envy" can only be to the detriment of both disciplines by masking the broader potential for two-way intellectual traffic of shared insights and analogizing from one another. Three fundamental features of biological evolution illustrate the range of potential intellectual flow between the two communities: particulate genes carry some subtle consequences for biological evolution that have not yet translated mainstream EC; the adaptive properties of the genetic code illustrate how both communities can contribute to a common understanding of appropriate evolutionary abstractions; finally, EC exploration of representational language seems pre-adapted to help biologists understand why life evolved a dichotomy of genotype and phenotype.

1.2 The Historical Development of EC

In the case of evolutionary computation, there are four historical paradigms that have served as the basis for much of the activity of the field: genetic algorithms (Holland, 1975), genetic programming (Koza, 1992, 1994), evolutionary strategies (Recheuberg, 1973), and evolutionary programming (Forgel et al., 1966). The basic differences between the paradigms lie in the nature of the representation schemes, the reproduction operators and selection methods.

1.2.1 Genetic Algorithms

The most popular technique in evolutionary computation research has been the *genetic algorithm*. In the traditional genetic algorithm, the representation used is a *fixed-length bit string*. Each position in the string is assumed to represent a particular feature of an individual, and the value stored in that position represents how that feature is expressed in the solution. Usually, the string is "evaluated as a collection of *structural* features of a solution that have little or no interactions". The analogy may be drawn directly to genes in biological organisms. Each gene represents an entity that is structurally independent of other genes.

The main reproduction operator used is *bit-string crossover*, in which two strings are used as parents and new individuals are formed by swapping a sub-sequence between the two strings (see Fig. 1.1). Another popular operator is *bit-flipping mutation*, in which a single bit in the string is flipped to form a new offspring string

1.2 The Historical Development of EC

Fig. 1.1 Bit-string crossover of parents a & b to form offspring c & d

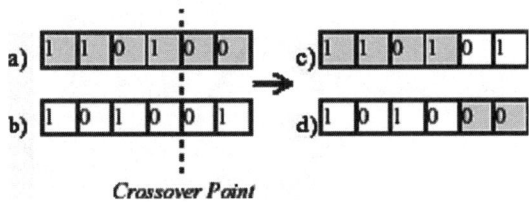

(see Fig. 1.2). A variety of other operators have also been developed, but are used less frequently (e.g., *inversion*, in which a subsequence in the bit string is reversed). A primary distinction that may be made between the various operators is whether or not they introduce any new information into the population. Crossover, for example, does not while mutation does. All operators are also constrained to manipulate the string in a manner consistent with the structural interpretation of genes. For example, two genes at the same location on two strings may be swapped between parents, but not combined based on their values. Traditionally, individuals are selected to be parents *probabilistically* based upon their fitness values, and the offspring that are created replace the parents. For example, if N parents are selected, then N offspring are generated which replace the parents in the next generation.

1.2.2 Genetic Programming

An increasingly popular technique is that of *genetic programming*. In a standard genetic program, the representation used is a variable-sized tree of functions and values. Each leaf in the tree is a label from an available set of value labels. Each internal node in the tree is label from an available set of function labels.

The entire tree corresponds to a single function that may be evaluated. Typically, the tree is evaluated in a leftmost depth-first manner. A leaf is evaluated as the corresponding value. A function is evaluated using arguments that is the result of the evaluation of its children. Genetic algorithms and genetic programming are similar in most other respects, except that the reproduction operators are tailored to a tree representation. The most commonly used operator is *subtree crossover*, in which an entire subtree is swapped between two parents (see Fig. 1.3). In a standard genetic program, all values and functions are assumed to return the same type, although functions may vary in the number of arguments they take. This *closure* principle (Koza, 1994) allows any subtree to be considered structurally on par with any other subtree, and ensures that operators such as sub-tree crossover will always produce legal offspring.

Fig. 1.2 Bit-flipping mutation of parent a to form offspring b

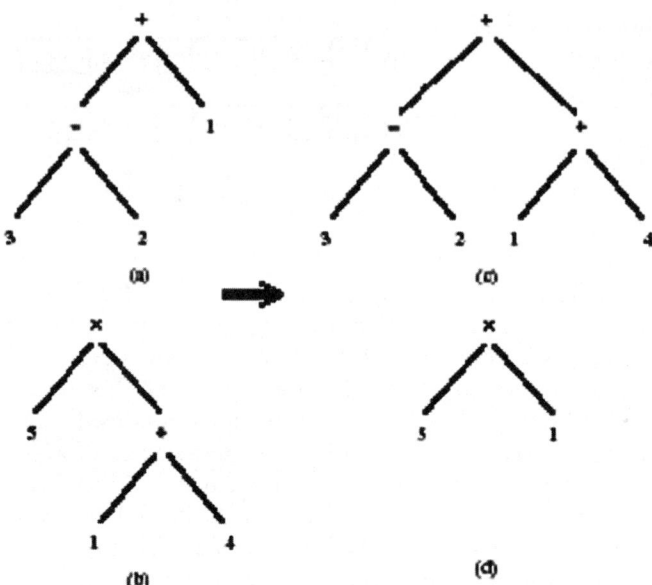

Fig. 1.3 Subtree crossover of parents a & b to form offspring c & d

1.2.3 Evolutionary Strategies

In evolutionary strategies, the representation used is a fixed-length real-valued vector. As with the bitstrings of genetic algorithms, each position in the vector corresponds to a feature of the individual. However, the features are considered to be behavioral rather than structural. "Consequently, arbitrary non-linear interactions between features during evaluation are expected which forces a more holistic approach to evolving solutions" (Angeline, 1996).

The main reproduction operator in evolutionary strategies is *Gaussian mutation*, in which a random value from a Gaussian distribution is added to each element of an individual's vector to create a new offspring (see Fig. 1.4). Another operator that is used is *intermediate recombination*, in which the vectors of two parents are averaged together, element by element, to form a new offspring (see Fig. 1.5). The effects of these operators reflect the behavioral as opposed to structural interpretation of the representation since knowledge of the values of vector elements is used to derive new vector elements.

The selection of parents to form offspring is less constrained than it is in genetic algorithms and genetic programming. For instance, due to the nature of the representation, it is easy to average vectors from many individuals to form a single offspring. In a typical evolutionary strategy, N parents are selected uniformly randomly

Fig. 1.4 Gaussian mutation of parent a to form offspring b

1.3 Features of Evolutionary Computation

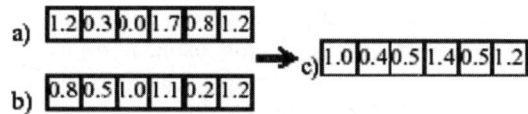

Fig. 1.5 Intermediate recombination of parents a & b to form offspring c

(i.e., not based upon fitness), more than N offspring are generated through the use of recombination, and then N survivors are selected deterministically. The survivors are chosen either from the best N offspring (i.e., no parents survive) or from the best N parents and offspring.

1.2.4 Evolutionary Programming

Evolutionary programming took the idea of representing individuals' phenotypic ally as finite state machines capable of responding to environmental stimuli and developing operators for effecting structural and behavioral change over time. This idea was applied to a wide range of problems including prediction problems, optimization and machine learning.

The above characterizations, leads one to the following observations. GA practitioners are seldom constrained to fixed-length binary implementations. GP enables the use of variable sized tree of functions and values. ES practitioners have incorporated recombination operators into their systems. EP is used for the evolution of finite state machines.

The representations used in evolutionary programming are typically tailored to the problem domain. One representation commonly used is a fixed-length real-valued vector. The primary difference between evolutionary programming and the previous approaches is that no exchange of material between individuals in the population is made. Thus, only mutation operators are used. For real-valued vector representations, evolutionary programming is very similar to evolutionary strategies without recombination.

A typical selection method is to select all the individuals in the population to be the N parents, to mutate each parent to form N offspring, and to probabilistically select, based upon fitness, N survivors from the total 2N individuals to form the next generation.

1.3 Features of Evolutionary Computation

In an evolutionary algorithm, a *representation scheme* is chosen by the researcher to define the set of solutions that form the search space for the algorithm. A number of individual solutions are created to form an *initial population*. The following steps are then repeated iteratively until a solution has been found which satisfies a pre-defined *termination criterion*. Each individual is evaluated using a *fitness function* that is specific to the problem being solved. Based upon their fitness values, a number of individuals are chosen to be *parents*. New individuals, or *offspring*, are produced from those parents using *reproduction operators*. The fitness values of those offspring are determined. Finally, survivors are selected from the old population and the offspring to form the new population of the next *generation*.

The mechanisms determining which and how many parents to select, how many offspring to create, and which individuals will survive into the next generation together represent a *selection method*. Many different selection methods have been proposed in the literature, and they vary in complexity. Typically, though, most selection methods ensure that the population of each generation is the same size.

EC techniques continue to grow in complexity and desirability, as biological research continues to change our perception of the evolutionary process.

In this context, we introduce three fundamental features of biological evolution:

1. particulate genes and population genetics
2. the adaptive genetic code
3. the dichotomy of genotype and phenotype

Each phenomenon is chosen to represent a different point in the spectrum of possible relationships between computing and biological evolutionary theory. The first is chosen to ask whether current EC has fully transferred the basics of biological evolution. The second demonstrates how both biological and computational evolutionary theorists can contribute to common understanding of evolutionary abstractions. The third is chosen to illustrate a question of biological evolution that EC seems better suited to tackle than biology.

1.3.1 Particulate Genes and Population Genetics

Mainstream thinking of the time viewed the genetic essence of phenotype as a liquid that blended whenever male and female met to reproduce. It took the world's first professor of engineering, Fleming Jenkin (1867), to point out the mathematical consequence of blending inheritance: a novel advantageous mutation arising in a sexually reproducing organism would dilute itself out of existence during the early stages of its spread through any population comprising more than a few individuals. This is a simple consequence of biparental inheritance. Mendels' theory of particulate genes (Mendel, 1866) replaced this flawed, analogue concept of blending inheritance with a digital system in which the advantageous version (allele) of a gene is either present or absent and biparental inheritance produces diploidy. Thus natural selection merely alters the proportions of alleles in a population, and an advantageous mutation can be selected into fixation (presence within 100% of individuals) without any loss in its fitness. Though much has been written about the Neo-Darwinian Synthesis that ensured from combining Mendelian genetics with Darwinian theory, it largely amounts to biologists' gradual acceptance that the particulate nature of genes alone provided a solid foundation to build detailed, quantitative predictions about evolution.

Indeed, decision of mathematical models of genes in populations as "bean bag genetics" overlooks the scope of logical deductions that follow from particulate genetics. They extend far beyond testable explanations for adaptive phenomena and into deeper, abstract concepts of biological evolution. For example, particulate genes introduce stochasticity into evolution. Because genes are either present or absent from any given genome, the genetic makeup of each new individual in a sexually reproducing population is a probabilistic outcome of which particular alleles it inherits from each parent. Unless offspring are infinite in number, their allele frequencies will not accurately mirror those of the parental generation, but instead will show some sampling error (genetic drift).

The magnitude of this sampling error is inversely proportional to the size of a population. Wright (1932) noted that because real populations fluctuate in size, temporary reductions can briefly relax selection, potentially allowing gene pools to diffuse far enough away from local optima to find new destinations when population size recovers and selection reasserts itself. In effect, particulate genes in finite populations improve the evolutionary heuristic from a simple hill climbing algorithm to something closer to simulated annealing under a fluctuating temperature. One final property of particulate genes operating in sexual populations is worthy of mention. In the large populations where natural selection works most effectively, any novel advantageous mutation that arises will only reach fixation over the course of multiple generations. During this spread, recombination and diploidy together ensure that the allele will temporarily find itself in many different genetic contexts. Classical population genetics (e.g., Fisher, 1930) and experimental EC systems (e.g., O'Reilly, 1999) have focused on whether and how this context promotes selective pressure for gene linkage into "co-adapted gene complexes". A simpler observation is that a novel, advantageous allele's potential for negative epistatic effects is integral to its micro-evolutionary success. Probability will favor the fixation of alleles that are good "team players" (i.e., reliably imbue their advantage regardless of genetic background). Many mainstream EC methods simplify the population genetics of new mutations (e.g., into tournaments), to expedite the adaptive process. This preserves non-blending inheritance and even genetic drift, but it is not clear that it incorporates population genetics' implicit filter for "prima donna" alleles that only offer their adaptive advantage when their genetic context is just so. Does this basic difference between biology and EC contribute anything to our understanding of why recombination seems to play such different roles in the two systems?

1.3.2 The Adaptive Code Book

Molecular biology's Central Dogma connects genes to phenotype by stating that DNA is transcribed into RNA, which is then translated into protein.

The terms transcription and translation are quite literal: RNA is a chemical sister language to DNA. Both are polymers formed from an alphabet of four chemical letters (nucleotides), and transcription is nothing more than a process of complementing DNA, letter by letter, into RNA. It is the next step, translation that profoundly influences biological evolution. Proteins are also linear polymers of chemical letters, but they are drawn from a qualitatively different alphabet (amino acids) comprising 20 elements. Clearly no one-to-one mapping could produce a genetic code for translating nucleotides unambiguously into amino acids, and by 1966 it was known that the combinatorial set of possible nucleotide triplets forms a dictionary of "codons" that each translate into a single amino acid meaning. The initial surprise for evolutionary theory was to discover that something as fundamental as the code-book for life would exhibit a high degree of redundancy (an alphabet of 4 RNA letters permits $4 \times 4 \times 4 = 64$ possible codons that map to one of only 20 amino acid meanings). Early interpretation fuelled arguments for Non-Darwinian evolution: genetic variations that make no difference to the protein they encode must be invisible to selection and therefore governed solely by drift. More recently, both computing and biological evolutionary theory have started to place this coding neutrality in the bigger picture of the adaptive heuristic. Essentially, findings appear to mirror Wright's early arguments on the importance of genetic drift: redundancy in the code adds selectively neutral dimensions to the fitness landscape that renders adaptive algorithms more effective by increasing the connectedness of local optima.

At present, an analogous reinterpretation is underway for a different adaptive feature of the genetic code: the observation that biochemically similar amino acids are assigned to codons that differ by only a single nucleotide. Early speculations that natural selection organized the genetic code so as to minimize the phenotypic impact of mutations have gained considerable evidential support as computer simulation enables exploration of theoretical codes that nature passed over. However, it seems likely that once again this phenomenon has more subtle effects in the broader context of the adaptive heuristic. An "error minimizing code" may in fact maximize the probability that a random effects on both traits defines a circle of radius around the organism.

The probability that this mutation will improve fitness (i.e., that the organism will move within the white area) is inversely proportional to its magnitude, mutation produces an increase in fitness according to Geometric Theory of gradualism (Fig. 1.6). Preliminary tests for this phenomenon reveal an even simpler influence: the error minimizing code smoothes the fitness landscape where a random genetic code would render it rugged. By clustering biochemically similar amino acids within mutational reach of one another it ensures that any selection towards a specific amino acid property (e.g., hydrophobicity) will be towards an interconnected region of the fitness landscape rather than to an isolated local optimum.

1.3.3 The Genotype/Phenotype Dichotomy

Implicit to the concept of an adaptive genetic code is a deeper question that remains largely unanswered by biology: why does all known life use two qualitatively different polymers, nucleic acids and proteins, with the associated need for translation? Current theories for the origin of this dichotomy focus on the discovery that RNA can act both as a genetic storage medium, and as a catalytic molecule. Within the

1.4 Advantages of Evolutionary Computation

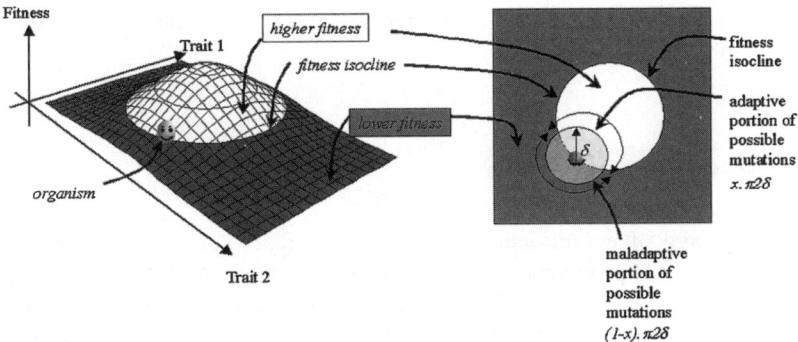

Fig. 1.6 The fitness landscape for an organism of 2 phenotypic traits: (**a**) for any organism, we may define an isocline that connects all trait combinations of equal fitness; (**b**) (the fitness landscape from above): a random mutation of magnitude that has tradeoff

most highly conserved core of metabolism, all known organisms are found to use RNA molecules in roles we normally attribute to proteins (White, 1976).

However, the answer to *how* the dichotomy evolved has largely eclipsed the question of *why* RNA evolved a qualitatively different representation for phenotype. A typical biological answer would be that the larger alphabet size of amino acids unleashed a greater catalytic diversity for the replicators, with an associated increase in metabolic sophistication that optimized self-replication. Interestingly, we know that nucleic acids are not limited to the 4 chemical letters we see today: natural metabolically active RNA's utilize a vast repertoire of posttranscriptional modifications and synthetic chemistry has demonstrated that multiple additional nucleotide letters can be added to the genetic alphabet even with today's cellular machinery. Furthermore, an increasing body of indirect evidence suggests that the protein alphabet itself underwent exactly the sort of evolutionary expansion early in life's history.

Given the ubiquity of nucleic acid genotype and protein phenotype within life, biology is hard-pressed to assess the significance of evolving this "representational language". The choice of phrase is deliberate: clearly the EC community is far ahead of biology in formalizing the concept of representational language, and exploring what it means. Biology will gain when evolutionary programmers place our system within their findings, illustrating the potential for biological inspiration *from* EC.

1.4 Advantages of Evolutionary Computation

Evolutionary computation, describes the field of investigation that concerns all evolutionary algorithms and offers practical advantages to several optimization problems. The advantages include the simplicity of the approach, its robust response to changing circumstances, and its flexibility and so on. This section briefs some of these advantages and offers suggestions in designing evolutionary algorithms for real-world problem solving.

1.4.1 Conceptual Simplicity

A key advantage of evolutionary computation is that it is conceptually simple. Figure 1.7 shows a flowchart of an evolutionary algorithm applied for function optimization. The algorithm consists of initialization, iterative variation and selection in light of a performance index. In particular, no gradient information needs to be presented to the algorithm. Over iterations of random variation and selection, the population can be made to converge to optimal solutions. The effectiveness of an evolutionary algorithm depends on the variation and selection operators as applied to a chosen representation and initialization.

1.4.2 Broad Applicability

Evolutionary algorithms can be applied to any problems that can be formulated as function optimization problems. To solve these problems, it requires a data structure to represent solutions, to evaluate solutions from old solutions. Representations can be chosen by human designer based on his intuition. Representation should allow for variation operators that maintain a behavioral link between parent and offspring. Small changes in structure of parent will lead to small changes in offspring, and similarly large changes in parent will lead to drastic alterations in offspring. In this case, evolutionary algorithms are developed, so that they are tuned in self adaptive

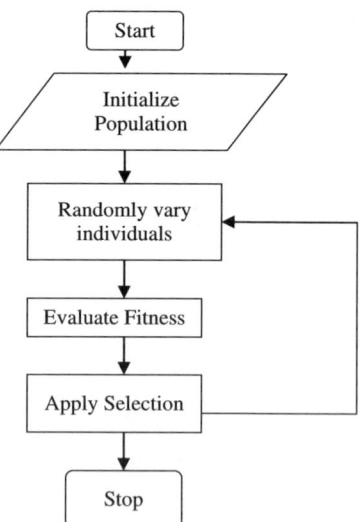

Fig. 1.7 Flowchart of an evolutionary algorithm

manner. This makes the evolutionary computation to be applied to broad areas which includes, discrete combinatorial problems, mixed-integer problems and so on.

1.4.3 Hybridization with Other Methods

Evolutionary algorithms can be combined with more traditional optimization techniques. This is as simple as the use of a conjugate-gradient minimization used after primary search with an evolutionary algorithm. It may also involve simultaneous application of algorithms like the use of evolutionary search for the structure of a model coupled with gradient search for parameter values. Further, evolutionary computation can be used to optimize the performance of neural networks, fuzzy systems, production systems, wireless systems and other program structures.

1.4.4 Parallelism

Evolution is a highly parallel process. When distributed processing computers become more popular are readily available, there will be increased potential for applying evolutionary computation to more complex problems. Generally the individual solutions are evaluated independently of the evaluations assigned to competing solutions. The evaluation of each solution can be handled in parallel and selection only requires some serial operation. In effect, the running time required for an application may be inversely proportional to the number of processors. Also, the current computing machines provide sufficient computational speed to generate solutions to difficult problems in reasonable time.

1.4.5 Robust to Dynamic Changes

Traditional methods of optimization are not robust to dynamic changes in the environment and they require a complete restart for providing a solution. In contrary, evolutionary computation can be used to adapt solutions to changing circumstances. The generated population of evolved solutions provides a basis for further improvement and in many cases, it is not necessary to reinitialize the population at random. This method of adapting in the face of a dynamic environment is a key advantage. For example, Wielaud (1990) applied genetic algorithm to evolve recurrent neural networks to control a cart-pole system consisting of two poles as shown in Fig. 1.2.

In the above Fig. 1.8, the objective is to maintain the cart between the limits of the track while not allowing either pole to exceed a specified maximum angle of deflection. The control available here is the force, with which pull and push action on the cart is performed. The difficulty here is the similarity in pole lengths. Few researchers used evolutionary algorithms to optimize neural networks to control this plant for different pole lengths.

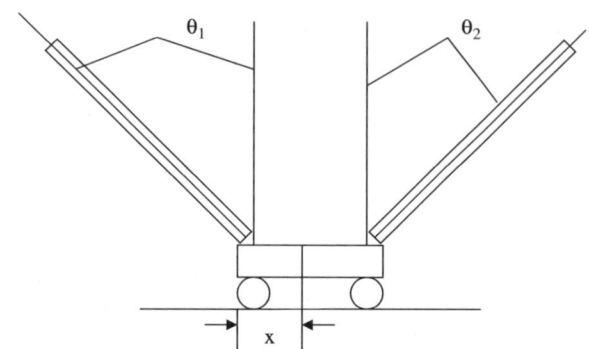

Fig. 1.8 A cart with two poles

1.4.6 Solves Problems that have no Solutions

The advantage of evolutionary algorithms includes its ability to address problems for which there is no human expertise. Even though human expertise should be used when it is needed and available; it often proves less adequate for automated problem-solving routines. Certain problems exist with expert system: the experts may not agree, may not be qualified, may not be self-consistent or may simply cause error. Artificial intelligence may be applied to several difficult problems requiring high computational speed, but they cannot compete with the human intelligence, Fogel (1995) declared artificial intelligence as "They solve problems, but they do not solve the problem of how to solve problems." In contrast, evolutionary computation provides a method for solving the problem of how to solve problems.

1.5 Applications of Evolutionary Computation

Evolutionary computation techniques have drawn much attention as optimization methods in the last two decades. From the optimization point of view, the main advantage of evolutionary computation techniques is that they do not have much mathematical requirements about the optimization problems. All they need is an evaluation of the objective function. As a result, they are applied to non-linear problems, defined on discrete, continuous or mixed search spaces, constrained or unconstrained.

The applications of evolutionary computation include the following fields:

- Medicine (for example in breast cancer detection).
- Engineering application (including electrical, mechanical, civil, production, aeronautical and robotics).
- Traveling salesman problem.
- Machine intelligence.
- Expert system

- Network design and routing
- Wired and wireless communication networks and so on.

Many activities involve unstructured, real life problems that are difficult to model, since they require several unusual factors. Certain engineering problems are complex in nature: job shop scheduling problems, timetabling, traveling salesman or facility layout problems. For all these applications, evolutionary computation provides a near-optimal solution at the end of an optimization run. Evolutionary algorithms are thus made efficient because they are flexible, and relatively easy to hybridize with domain-dependent heuristics.

1.6 Summary

The basics of evolutionary computation with its historical development were discussed in this chapter. Although the history of evolutionary computation dates back to the 1950s and 1960s, only within the last decade have evolutionary algorithms became practicable for solving real-world problems on desktop computers. The three basic features of the biological evolutionary algorithms were also discussed. For practical genes, we ask whether Evolutionary computation can gain from biology by considering the detailed dynamics by which an advantageous allele invades a wild-type population. The adaptive genetic code illustrates how Evolutionary computation and biological evolutionary research can contribute to a common understanding of general evolutionary dynamic. For the dichotomy of genotype and phenotype, biology is hard-pressed to assess the significance of representational language. The various advantages and applications of evolutionary computation are also discussed in this chapter.

Review Questions

1. Define Evolutionary computation.
2. Briefly describe the historical developments of evolutionary computation.
3. State three fundamental features of biological evolutionary computation.
4. Draw a flowchart and explain an evolutionary algorithm.
5. Define genotype and phenotype.
6. Mention the various advantages of evolutionary computation.
7. List a few applications of evolutionary computation.
8. How are evolutionary computational methods hybridized with other methods?
9. Differentiate: Genetic algorithm and Genetic Programming.
10. Give a description of how evolutionary computation is applied to engineering applications.

Chapter 2
Genetic Algorithms

2.1 Introduction

Charles Darwin stated the theory of natural evolution in the origin of species. Over several generations, biological organisms evolve based on the principle of natural selection "survival of the fittest" to reach certain remarkable tasks. The perfect shapes of the albatross wring the efficiency and the similarity between sharks and dolphins and so on, are best examples of achievement of random evolution over intelligence. Thus, it works so well in nature, as a result it should be interesting to simulate natural evolution and to develop a method, which solves concrete, and search optimization problems.

In nature, an individual in population competes with each other for virtual resources like food, shelter and so on. Also in the same species, individuals compete to attract mates for reproduction. Due to this selection, poorly performing individuals have less chance to survive, and the most adapted or "fit" individuals produce a relatively large number of offspring's. It can also be noted that during reproduction, a recombination of the good characteristics of each ancestor can produce "best fit" offspring whose fitness is greater than that of a parent. After a few generations, species evolve spontaneously to become more and more adapted to their environment.

In 1975, Holland developed this idea in his book "Adaptation in natural and artificial systems". He described how to apply the principles of natural evolution to optimization problems and built the first Genetic Algorithms. Holland's theory has been further developed and now Genetic Algorithms (GAs) stand up as a powerful tool for solving search and optimization problems. Genetic algorithms are based on the principle of genetics and evolution.

The power of mathematics lies in technology transfer: there exist certain models and methods, which describe many different phenomena and solve wide variety of problems. GAs are an example of mathematical technology transfer: by simulating evolution one can solve optimization problems from a variety of sources. Today, GAs are used to resolve complicated optimization problems, like, timetabling, job-shop scheduling, games playing.

2.2 Biological Background

The science that deals with the mechanisms responsible for similarities and differences in a species is called Genetics. The word "genetics" is derived from the Greek word "genesis" meaning "to grow" or "to become". The science of genetics helps us to differentiate between heredity and variations and seeks to account for the resemblances and differences due to the concepts of Genetic Algorithms and directly derived from natural heredity, their source and development. The concepts of Genetic Algorithms are directly derived from natural evolution. The main terminologies involved in the biological background of species are as follows:

2.2.1 The Cell

Every animal/human cell is a complex of many "small" factories that work together. The center of all this is the cell nucleus. The genetic information is contained in the cell nucleus. Figure 2.1 shows anatomy of the animal cell and cell nucleus.

2.2.2 Chromosomes

All the genetic information gets stored in the chromosomes. Each chromosome is build of Dioxy Ribo Nucleic Acid (DNA). In humans, a chromosome exists in the form of pairs (23 pairs found). The chromosomes are divided into several parts called genes. Genes code the properties of species i.e., the characteristics of an individual. The possibilities of the genes for one property are called allele and a gene can take different alleles. For example, there is a gene for eye color, and all the different possible alleles are black, brown, blue and green (since no one has red or violet eyes). The set of all possible alleles present in a particular population forms a gene tool. This gene pool can determine all the different possible variations for

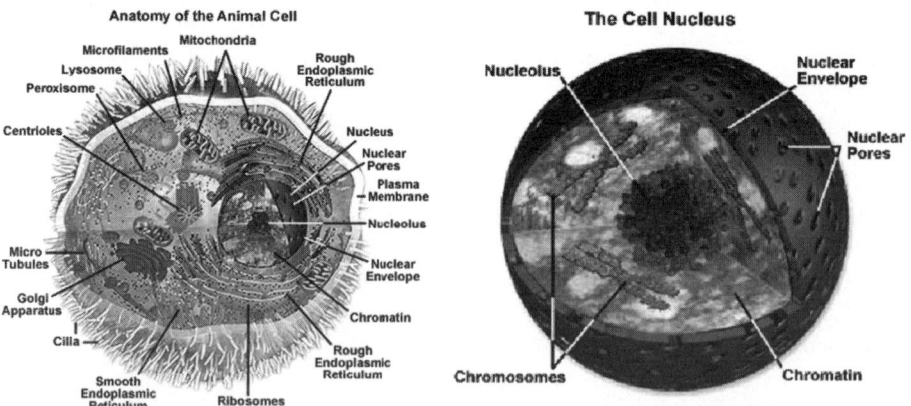

Fig. 2.1 Anatomy of animal cell, cell nucleus

2.2 Biological Background

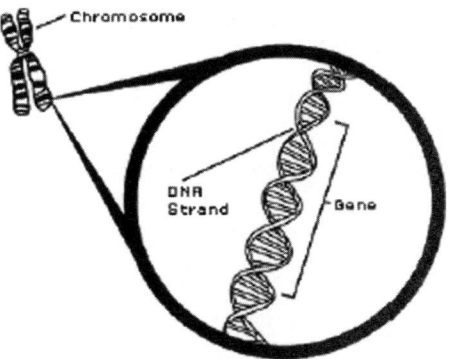

Fig. 2.2 Model of chromosome

the future generations. The size of the gene pool helps in determining the diversity of the individuals in the population. The set of all the genes of a specific species is called genome. Each and every gene has an unique position on the genome called locus. In fact, most living organisms store their genome on several chromosomes, but in the Genetic Algorithms (GAs), all the genes are usually stored on the same chromosomes. Thus chromosomes and genomes are synonyms with one other in GAs. Figure 2.2 shows a model of chromosome.

2.2.3 Genetics

For a particular individual, the entire combination of genes is called genotype. The phenotype describes the physical aspect of decoding a genotype to produce the phenotype. One interesting point of evolution is that selection is always done on the phenotype whereas the reproduction recombines genotype. Thus morphogenesis plays a key role between selection and reproduction. In higher life forms, chromosomes contain two sets of genes. This is known as diploids. In case of conflicts between two values of the same pair of genes, the dominant one will determine the phenotype whereas the other one, called recessive, will still be present and can be passed on to the offspring. Diploidy allows a wider diversity of alleles. This provides a useful memory mechanism in changing or noisy environment. However, most GA concentrates on haploid chromosomes because they are much simple to construct. In haploid representation, only one set of each gene is stored, thus the process of determining which allele should be dominant and which one should be recessive is avoided. Figure 2.3 shows development of genotype to phenotype.

2.2.4 Reproduction

Reproduction of species via genetic information is carried out by,

- Mitosis
- Meiosis

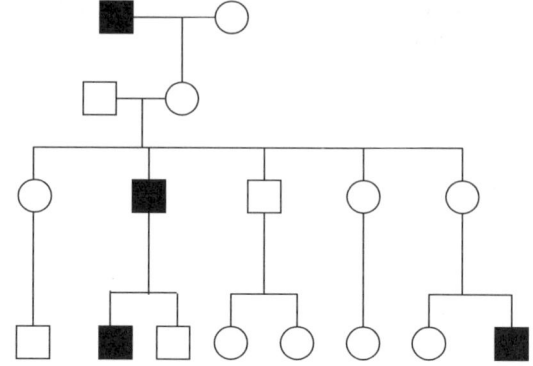

Fig. 2.3 Development of genotype to phenotype

In Mitosis the same genetic information is copied to new offspring. There is no exchange of information. This is a normal way of growing of multi cell structures, like organs. Figure 2.4 shows mitosis form of reproduction.

Meiosis form basis of sexual reproduction. When meiotic division takes place 2 gametes appears in the process. When reproduction occurs, these two gametes conjugate to a zygote which becomes the new individual. Thus in this case, the

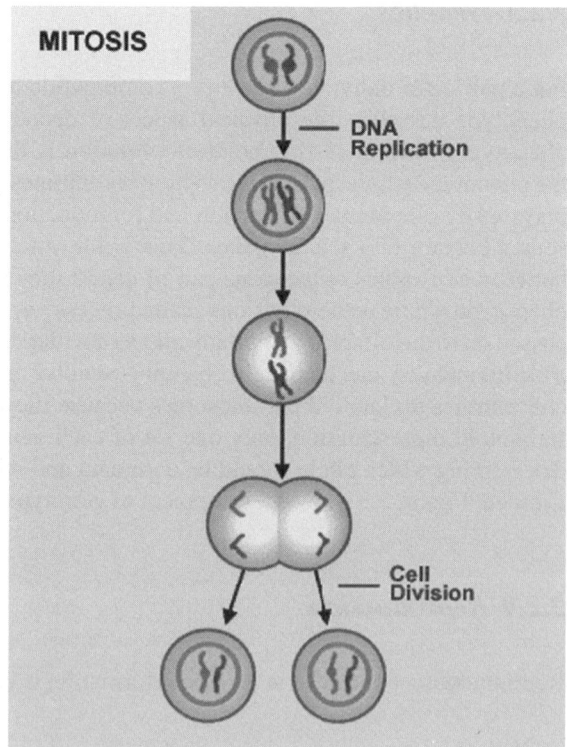

Fig. 2.4 Mitosis form of reproduction

2.2 Biological Background

Fig. 2.5 Meiosis form of reproduction

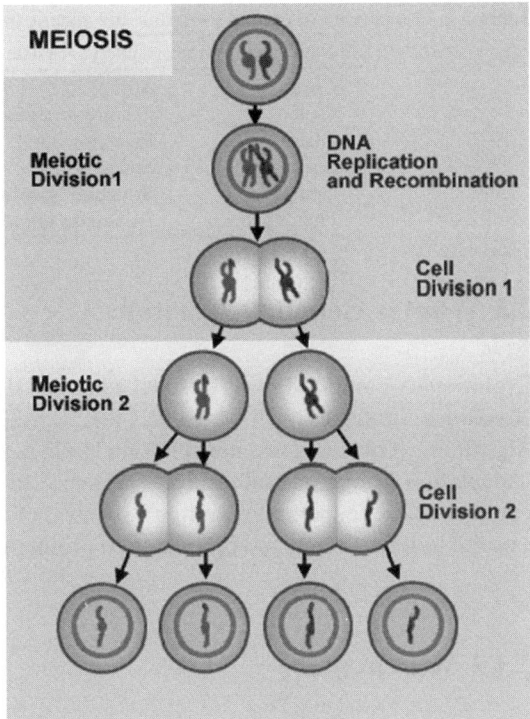

genetic information is shared between the parents in order to create new offspring. Figure 2.5 shows meiosis form of reproduction.

2.2.5 Natural Selection

The origin of species is based on "Preservation of favorable variations and rejection of unfavorable variations". The variation refers to the differences shown by the individual of a species and also by offspring's of the same parents. There are more individuals born than can survive, so there is a continuous struggle for life. Individuals with an advantage have a greater chance for survive i.e., the survival of the fittest. For example, Giraffe with long necks can have food from tall trees as well from grounds, on the other hand goat, deer with small neck have food only from grounds. As a result, natural selection plays a major role in this survival process.

Thus the various biological terminologies to be used in genetic algorithms were discussed in this section.

The following Table 2.1 gives a list of different expressions, which are in common with natural evolution and genetic algorithm.

Table 2.1 Comparison of natural evolution and genetic algorithm terminology

Natural evolution	Genetic algorithm
Chromosome	String
Gene	Feature or character
Allele	Feature value
Locus	String position
Genotype	Structure or coded string
Phenotype	Parameter set, a decoded structure

2.3 What is Genetic Algorithm?

Evolutionary computing was introduced in the 1960s by I. Rechenberg in the work "Evolution strategies". This idea was then developed by other researches. Genetic Algorithms (GAs) was invented by John Holland and developed this idea in his book "Adaptation in natural and artificial systems" in the year 1975. Holland proposed GA as a heuristic method based on "Survival of the fittest". GA was discovered as a useful tool for search and optimization problems.

2.3.1 Search Space

Most often one is looking for the best solution in a specific set of solutions. The space of all feasible solutions (the set of solutions among which the desired solution resides) is called search space (also state space). Each and every point in the search space represents one possible solution. Therefore each possible solution can be "marked" by its fitness value, depending on the problem definition. With Genetic Algorithm one looks for the best solution among a number of possible solutions-represented by one point in the search space i.e.; GAs are used to search the search space for the best solution e.g., minimum. The difficulties in this ease are the local minima and the starting point of the search (see Fig. 2.6).

2.3.2 Genetic Algorithms World

Genetic Algorithm raises a couple of important features. First it is a *stochastic* algorithm; randomness as an essential role in genetic algorithms. Both selection and reproduction needs random procedures. A second very important point is that genetic algorithms always consider a population of solutions. Keeping in memory more than a single solution at each iteration offers a lot of advantages. The algorithm can recombine different solutions to get better ones and so, it can use the benefits of assortment. A population base algorithm is also very amenable for parallelization. The *robustness* of the algorithm should also be mentioned as something essential for the algorithm success. Robustness refers to the ability to perform consistently

2.3 What is Genetic Algorithm?

Fig. 2.6 An example of search space

well on a broad range of problem types. There is no particular requirement on the problem before using GAs, so it can be applied to resolve any problem. All those features make GA a really powerful optimization tool.

With the success of Genetic Algorithms, other algorithms make in use of on the same principle of natural evolution have also emerged. Evolution strategy, Genetic programming are some of those similar of those similar algorithms. The classification is not always clear between those different algorithms, thus to avoid any confusion, they are all gathered in what is called *Evolutionary Algorithms*.

The analogy with nature gives to those algorithms something exciting and enjoyable, Their ability to deal successfully with a wide range of problem area, including those which are difficult for other methods to solve make them quite powerful. But today, GAs are suffering from too much trendiness. GAs are a new field, and parts of the theory have still to be properly established. We can find almost as many opinions on GAs as there are researchers in this field. Things evolve quickly in genetic algorithms, and some comments might not be very accurate in few years.

It is also important to mention in this introduction GA limits. Like most stochastic methods, GAs are not guaranteed to find the global optimum solution to a problem, they are satisfied with finding "acceptably good" solutions to the problem. GAs are an extremely general too, and so specific techniques for solving particular problems are likely to out-perform GAs in both speed and accuracy of the final result.

GAs are something worth trying when everything else as failed or when we know absolutely nothing of the search space. Nevertheless, even when such specialized techniques exist, it often interesting to hybridise them with a GA in order to possibly gain some improvements. It is important always to keep an objective point of view; do not consider that GAs are a panacea for resolving all optimization problems. This warning is for those who might have the temptation to resolve anything with GA. The proverb says "If we have a hammer, all the problems looks like a nails". GAs do work and give excellent results if they are applied properly on appropriate problems.

2.3.3 Evolution and Optimization

We are now 45 millions years ago examining a Basilosaurus :

The Basilosaurus was quite a prototype of a whale (Fig. 2.7). It was about 15 meters long for 5 tons. It still had a quasi-independent head and posterior paws. He moved using undulatory movements and hunted small preys. Its anterior members were reduced to small flippers with an elbow articulation. Movements in such a viscous element (water) are very hard and require big efforts. People concerned must have enough energy to move and control its trajectory. The anterior members of basilosaurus were not really adapted to swimming. To adapt them, a double phenomenon must occur: the shortening of the "arm" with the locking of the elbow articulation and the extension of the fingers which will constitute the base structure of the flipper (refer Fig. 2.8).

The image shows that two fingers of the common dolphin are hypertrophied to the detriment of the rest of the member. The basilosaurus was a hunter, he had to be fast and precise. Through time, subjects appeared with longer fingers and short arms. They could move faster and more precisely than before, and therefore, live longer and have many descendants.

Meanwhile, other improvements occurred concerning the general aerodynamic like the integration of the head to the body, improvement of the profile, strengthening of the caudal fin ... finally producing a subject perfectly adapted to the constraints of an aqueous environment. This process of adaptation, this morphological optimization is so perfect that nowadays, the similarity between a shark, a dolphin or a submarine is striking. But the first is a cartilaginous fish (Chondrichtyen) originating in the Devonian (–400 million years), long before the apparition of the first mammal whose Cetacean descends

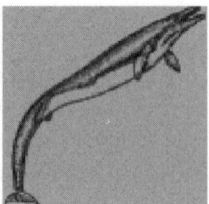

Fig. 2.7 Basilosaurus

Fig. 2.8 Tursiops flipper

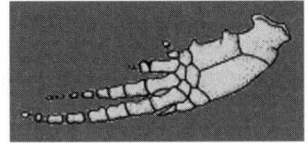

Darwinian mechanism hence generate an optimization process, Hydrodynamic optimization for fishes and others marine animals, aerodynamic for pterodactyls, birds or bats. This observation is the basis of genetic algorithms.

2.3.4 Evolution and Genetic Algorithms

John Holland, from the University of Michigan began his work on genetic algorithms at the beginning of the 60s. A first achievement was the publication of *Adaptation in Natural and Artificial System* in 1975. Holland had a double aim: to improve the understanding of natural adaptation process, and to design artificial systems having properties similar to natural systems.

The basic idea is as follows: the genetic pool of a given population potentially contains the solution, or a better solution, to a given adaptive problem. This solution is not "active" because the genetic combination on which it relies is split between several subjects. Only the association of different genomes can lead to the solution. Simply speaking, we could by example consider that the shortening of the paw and the extension of the fingers of our basilosaurus are controlled by 2 "genes". No subject has such a genome, but during reproduction and crossover, new genetic combination occur and, finally, a subject can inherit a "good gene" from both parents : his paw is now a flipper.

Holland method is especially effective because he not only considered the role of mutation (mutations improve very seldom the algorithms), but he also utilized genetic recombination, (crossover) : these recombination, the crossover of partial solutions greatly improve the capability of the algorithm to approach, and eventually find, the optimum.

Recombination or sexual reproduction is a key operator for natural evolution. Technically, it takes two genotypes and it produces a new genotype by mixing the gene found in the originals. In biology, the most common form of recombination is crossover, two chromosomes are cut at one point and the halves are spliced to create new chromosomes. The effect of recombination is very important because it allows characteristics from two different parents to be assorted. If the father and the mother possess different good qualities, we would expect that all the good qualities will be passed into the child. Thus the offspring, just by combining all the good features from its parents, may surpass its ancestors. Many people believe that this mixing of genetic material via sexual reproduction is one of the most powerful features of Genetic Algorithms. As a quick parenthesis about sexual reproduction, Genetic Algorithms representation usually does not differentiate male and female individuals (without any perversity). As in many livings species (e.g., snails) any individual can be either a male or a female. In fact, for almost all recombination operators, mother and father are interchangeable.

Mutation is the other way to get new genomes. Mutation consists in changing the value of genes. In natural evolution, mutation mostly engenders non-viable genomes. Actually mutation is not a very frequent operator in natural evolution. Nevertheless, is optimization, a few random changes can be a good way of exploring the search space quickly.

Through those low-level notions of genetic, we have seen how living beings store their characteristic information and how this information can be passed into their offspring. It very basic but it is more than enough to understand the Genetic Algorithm Theory.

Darwin was totally unaware of the biochemical basics of genetics. Now we know how the genetic inheritable information is coded in DNA, RNA and proteins and that the coding principles are actually digital much resembling the information storage in computers. Information processing is in many ways totally different, however. The magnificent phenomenon called the evolution of species can also give some insight into information processing methods and optimization in particular. According to Darwinism, inherited variation is characterized by the following properties:

1. Variation must be copying because selection does not create directly anything, but presupposes a large population to work on.
2. Variation must be small-scaled in practice. Species do not appear suddenly.
3. Variation is undirected. This is also known as the blind watchmaker paradigm.

While the natural sciences approach to evolution has for over a century been to analyze and study different aspects of evolution to find the underlying principles, the engineering sciences are happy to apply evolutionary principles, that have been heavily tested over billions of years, to attack the most complex technical problems, including protein folding.

2.4 Conventional Optimization and Search Techniques

The basic principle of optimization is the efficient allocation of scarce resources. Optimization can be applied to any scientific or engineering discipline. The aim of optimization is to find an algorithm, which solves a given class of problems. There exist no specific method, which solves all optimization problems. Consider a function,

$$f(x): \left[x^l, x^u\right] \to [0, 1] : \qquad (2.1)$$

where,

$$f(x) = \begin{cases} 1, & if \ ||x - a|| < \epsilon, \ \epsilon > 0 \\ -1, & elsewhere \end{cases}$$

2.4 Conventional Optimization and Search Techniques

For the above function, f can be maintained by decreasing ϵ or by making the interval of [x^l, x^u] large. Thus a difficult task can be made easier. Therefore, one can solve optimization problems by combining human creativity and the raw processing power of the computers.

The various conventional optimization and search techniques available are discussed as follows:

2.4.1 Gradient-Based Local Optimization Method

When the objective function is smooth and one need efficient local optimization, it is better to use gradient based or Hessian based optimization methods. The performance and reliability of the different gradient methods varies considerably.

To discuss gradient-based local optimization, let us assume a smooth objective function (i.e., continuous first and second derivatives). The objective function is denoted by,

$$f(x): R^n \to R \qquad (2.2)$$

The first derivatives are contained in the gradient vector $\nabla f(x)$

$$\nabla f(x) = \begin{bmatrix} \partial f(x)/\partial x_1 \\ \vdots \\ \partial f(x)/\partial x_n \end{bmatrix} \qquad (2.3)$$

The second derivatives of the objective function are contained in the Hessian matrix H(x).

$$H(x) = \nabla^T \nabla f(x) = \begin{pmatrix} \frac{\partial^2 f(x)}{\partial^2 x_1} & \cdots & \frac{\partial^2 f(x)}{\partial x_1 \partial x_n} \\ \vdots & & \vdots \\ \frac{\partial^2 f(x)}{\partial x_1 \partial x_n} & \cdots & \frac{\partial^2 f(x)}{\partial^2 x_n} \end{pmatrix} \qquad (2.4)$$

Few methods need only the gradient vector, but in the Newton's method we need the Hessian matrix.

The general pseudo code used in gradient methods is as follows:

    Select an initial guess value x^1 and set n=1.
    repeat
        Solve the search direction p^n from (2.5) or (2.6) below.
        Determine the next iteration point using (2.7) below:
$$X^{n+1} = X^n + \lambda_n P^n$$
        Set n=n+1.
    Until $||X^n - X^{n-1}|| < \epsilon$

These gradient methods search for minimum and not maximum. Several different methods are obtained based on the details of the algorithm.

The search direction P^n in conjugate gradient method is found as follows:

$$P^n = -\nabla f(X^n) + \beta_n P^{n-1} \tag{2.5}$$

In secant method,

$$B_n P^n = -\nabla f(x^n) \tag{2.6}$$

is used for finding search direction. The matrix B_n in (2.2) estimates the Hessian. The matrix B_n is updated in each iteration. When B_n is defined as the identity matrix, the steepest descent method occurs. When the matrix B_n is the Hessian $H(x^n)$, we get the Newton's method.

The length λ_n of the search step is computed using:

$$\lambda_n = \arg\min_{\lambda > 0} f(x^n + \lambda P^n) \tag{2.7}$$

The discussed is a one-dimensional optimization problem.

The steepest descent method provides poor performance. As a result, conjugate gradient method can be used. If the second derivatives are easy to compute, then Newton's method may provide best results. The secant methods are faster than conjugate gradient methods, but there occurs memory problems.

Thus these local optimization methods can be combined with other methods to get a good link between performance and reliability.

2.4.2 Random Search

Random search is an extremely basic method. It only explores the search space by randomly selecting solutions and evaluates their fitness. This is quite an unintelligent strategy, and is rarely used by itself. Nevertheless, this method sometimes worth being tested. It doesn't take much effort to implement it, and an important number of evaluations can be done fairly quickly. For new unresolved problems, it can be useful to compare the results of a more advanced algorithm to those obtained just with a random search for the same number of evaluations. Nasty surprises might well appear when comparing for example, genetic algorithms to random search. It's good to remember that the efficiency of GA is extremely dependant on consistent coding and relevant reproduction operators. Building a genetic algorithm, which performs no more than a random search happens more often than we can expect. If the reproduction operators are just producing new random solutions without any concrete links to the ones selected from the last generation, the genetic algorithm is just doing nothing else that a random search.

2.4 Conventional Optimization and Search Techniques

Random search does have a few interesting qualities. However good the obtained solution may be, if it's not optimal one, it can be always improved by continuing the run of the random search algorithm for long enough. A random search never gets stuck in any point such as a local optimum. Furthermore, theoretically, if the search space is finite, random search is guaranteed to reach the optimal solution. Unfortunately, this result is completely useless. For most of problems we are interested in, exploring the whole search space takes far too long an amount of time.

2.4.3 Stochastic Hill Climbing

Efficient methods exist for problems with well-behaved continuous fitness functions. These methods use a kind of gradient to guide the direction of search. *Stochastic Hill Climbing* is the simplest method of these kinds. Each iteration consists in choosing randomly a solution in the neighborhood of the current solution and retains this new solution only if it improves the fitness function. Stochastic Hill Climbing converges towards the optimal solution if the fitness function of the problem is continuous and has only one peak (unimodal function).

On functions with many peaks (multimodal functions), the algorithm is likely to stop on the first peak it finds even if it is not the highest one. Once a peak is reached, hill climbing cannot progress anymore, and that is problematic when this point is a local optimum. Stochastic hill climbing usually starts from a random select point. A simple idea to avoid getting stuck on the first local optimal consists in repeating several hill climbs each time starting from a different randomly chosen points. This method is sometimes known as iterated hill climbing. By discovering different local optimal points, it gives more chance to reach the global optimum. It works well if there is not too many local optima in the search space. But if the fitness function is very "noisy" with many small peaks, stochastic hill climbing is definitely not a good method to use. Nevertheless such methods have the great advantage to be really easy to implement and to give fairly good solutions very quickly.

2.4.4 Simulated Annealing

Simulated Annealing was originally inspired by formation of crystal in solids during cooling i.e., the physical cooling phenomenon. As discovered a long time ago by iron age blacksmiths, the slower the cooling, the more perfect is the crystal formed. By cooling, complex physical systems naturally converge towards a state of minimal energy. The system moves randomly, but the probability to stay in a particular configuration depends directly on the energy of the system and on its temperature. Gibbs law gives this probability formally:

$$p = e^{\frac{E}{kT}} \tag{2.8}$$

Where E stands for the energy, k is the Boltzmann constant and T is the temperature. In the mid 70s, Kirlpatrick by analogy of these physical phenomena laid out the first description of simulated annealing.

As in the stochastic hill climbing, the iteration of the simulated annealing consists of randomly choosing a new solution in the neighborhood of the actual solution. If the fitness function of the new solution is better than the fitness function of the current one, the new solution is accepted as the new current solution. If the fitness function is not improved, the new solution is retained with a probability:

$$p = e^{\frac{-(f(y)-f((x))}{kT}} \qquad (2.9)$$

Where $f(y) - f(x)$ is the difference of the fitness function between the new and the old solution.

The simulated annealing behaves like a hill climbing method but with the possibility of going downhill to avoid being trapped at local optima. When the temperature is high, the probability of deteriorate the solution is quite important, and then a lot of large moves are possible to explore the search space. The more the temperature decreases, the more difficult it is to go downhill, the algorithm tries to climb up from the current solution to reach a maximum. When temperature is lower, there is an exploitation of the current solution. If the temperature is too low, number deterioration is accepted, and the algorithm behaves just like a stochastic hill climbing method. Usually, the simulated annealing starts from a high temperature, which decreases exponentially. The slower the cooling, the better it is for finding good solutions. It even has been demonstrated that with an infinitely slow cooling, the algorithm is almost certain to find the global optimum. The only point is that infinitely slow consists in finding the appropriate temperature decrease rate to obtain a good behavior of the algorithm.

Simulated Annealing by mixing exploration features such as the random search and exploitation features like hill climbing usually gives quite good results. Simulated Annealing is a serious competitor to Genetic Algorithms. It is worth trying to compare the results obtained by each. Both are derived from analogy with natural system evolution and both deal with the same kind of optimization problem. GAs differs by two main features, which should make them more efficient. First GAs uses a population-based selection whereas SA only deals with one individual at each iteration. Hence GAs are expected to cover a much larger landscape of the search space at each iteration, but on the other hand SA iterations are much more simple, and so, often much faster. The great advantage of GA is its exceptional ability to be parallelized, whereas SA does not gain much of this. It is mainly due to the population scheme use by GA. Secondly, GAs uses recombination operators, able to mix good characteristics from different solutions. The exploitation made by recombination operators is supposedly considered helpful to find optimal solutions of the problem.

On the other hand, simulated annealing are still very simple to implement and they give good results. They have proved their efficiency over a large spectrum of difficult problems, like the optimal layout of printed circuit board, or the famous traveling salesman problem. Genetic annealing is developing in the recent years, which is an attempt to combine genetic algorithms and simulated annealing.

2.4.5 Symbolic Artificial Intelligence (AI)

Most symbolic AI systems are very static. Most of them can usually only solve one given specific problem, since their architecture was designed for whatever that specific problem was in the first place. Thus, if the given problem were somehow to be changed, these systems could have a hard time adapting to them, since the algorithm that would originally arrive to the solution may be either incorrect or less efficient. Genetic algorithms (or GA) were created to combat these problems. They are basically algorithms based on natural biological evolution. The architecture of systems that implement genetic algorithms (or GA) is more able to adapt to a wide range of problems.

2.5 A Simple Genetic Algorithm

An algorithm is a series of steps for solving a problem. A genetic algorithm is a problem solving method that uses genetics as its model of problem solving. It's a search technique to find approximate solutions to optimization and search problems.

Basically, an optimization problem looks really simple. One knows the form of all possible solutions corresponding to a specific question. The set of all the solutions that meet this form constitute the search space. The problem consists in finding out the solution that fits the best, i.e. the one with the most payoffs, from all the possible solutions. If it's possible to quickly enumerate all the solutions, the problem does not raise much difficulty. But, when the search space becomes large, enumeration is soon no longer feasible simply because it would take far too much time. In this it's needed to use a specific technique to find the optimal solution. Genetic Algorithms provides one of these methods. Practically they all work in a similar way, adapting the simple genetics to algorithmic mechanisms.

GA handles a population of possible solutions. Each solution is represented through a chromosome, which is just an abstract representation. Coding all the possible solutions into a chromosome is the first part, but certainly not the most straightforward one of a Genetic Algorithm. A set of reproduction operators has to be determined, too. Reproduction operators are applied directly on the chromosomes, and are used to perform mutations and recombinations over solutions of the problem. Appropriate representation and reproduction operators are really something determinant, as the behavior of the GA is extremely dependant on it. Frequently, it can be extremely difficult to find a representation, which respects the structure of the search space and reproduction operators, which are coherent and relevant according to the properties of the problems.

Selection is supposed to be able to compare each individual in the population. Selection is done by using a fitness function. Each chromosome has an associated value corresponding to the fitness of the solution it represents. The fitness should correspond to an evaluation of how good the candidate solution is. The optimal solution is the one, which maximizes the fitness function. Genetic Algorithms deal with the problems that maximize the fitness function. But, if the problem consists in minimizing a cost function, the adaptation is quite easy. Either the cost function can be transformed into a fitness function, for example by inverting it; or the selection can be adapted in such way that they consider individuals with low evaluation functions as better.

Once the reproduction and the fitness function have been properly defined, a Genetic Algorithm is evolved according to the same basic structure. It starts by generating an initial population of chromosomes. This first population must offer a wide diversity of genetic materials. The gene pool should be as large as possible so that any solution of the search space can be engendered. Generally, the initial population is generated randomly.

Then, the genetic algorithm loops over an iteration process to make the population evolve. Each iteration consists of the following steps:

- SELECTION: The first step consists in selecting individuals for reproduction. This selection is done randomly with a probability depending on the relative fitness of the individuals so that best ones are often chosen for reproduction than poor ones.
- REPRODUCTION: In the second step, offspring are bred by the selected individuals. For generating new chromosomes, the algorithm can use both recombination and mutation.
- EVALUATION: Then the fitness of the new chromosomes is evaluated.
- REPLACEMENT: During the last step, individuals from the old population are killed and replaced by the new ones.

The algorithm is stopped when the population converges toward the optimal solution. The basic genetic algorithm is as follows:

- [start] Genetic random population of n chromosomes (suitable solutions for the problem)
- [Fitness] Evaluate the fitness f(x) of each chromosome x in the population
- New population] Create a new population by repeating following steps until the New population is complete
    - [selection] select two parent chromosomes from a population according to their fitness (the better fitness, the bigger chance to get selected).
    - [crossover] With a crossover probability, cross over the parents to form new offspring (children). If no crossover was performed, offspring is the exact copy of parents.
    - [Mutation] With a mutation probability, mutate new offspring at each locus (position in chromosome)
    - [Accepting] Place new offspring in the new population.

2.5 A Simple Genetic Algorithm

- [Replace] Use new generated population for a further sum of the algorithm.
- [Test] If the end condition is satisfied, stop, and return the best solution in current population.
- [Loop] Go to step2 for fitness evaluation.

The Genetic algorithm process is discussed through the GA cycle in Fig. 2.9

Reproduction is the process by which the genetic material in two or more parent is combined to obtain one or more offspring. In fitness evaluation step, the individual's quality is assessed. Mutation is performed to one individual to produce a new version of it where some of the original genetic material has been randomly changed. Selection process helps to decide which individuals are to be used for reproduction and mutation in order to produce new search points.

The flowchart showing the process of GA is as shown in Fig. 2.10.

Before implementing GAs it is important to understand few guidelines for designing a general search algorithm i.e. a global optimization algorithm based on the properties of the fitness landscape and the most common optimization method types:

1. determinism: A purely deterministic search may have an extremely high variance in solution quality because it may soon get stuck in worst case situations from which it is incapable to escape because of its determinism. This can be avoided, but it is a well-known fact that the observation of the worst-case situation is not guaranteed to be possible in general.
2. nondeterminism: A stochastic search method usually does not suffer from the above potential worst case "wolf trap" phenomenon. It is therefore likely that a search method should be stochastic, but it may well contain a substantial portion of determinism, however. In principle it is enough to have as much nondeterminism as to be able to avoid the worst-case wolf traps.

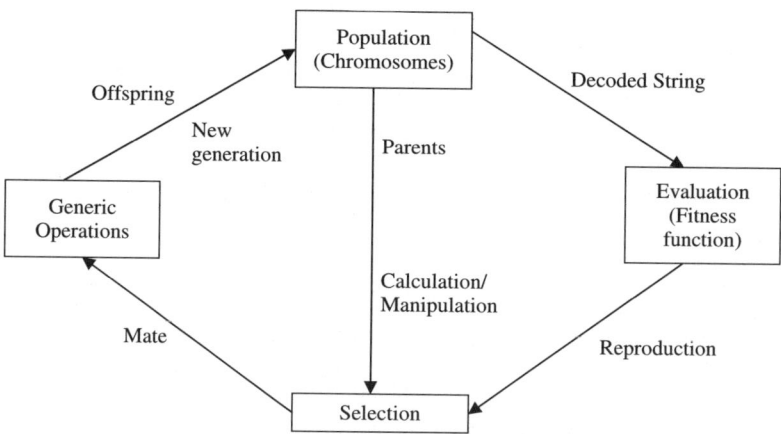

Fig. 2.9 Genetic algorithm cycle

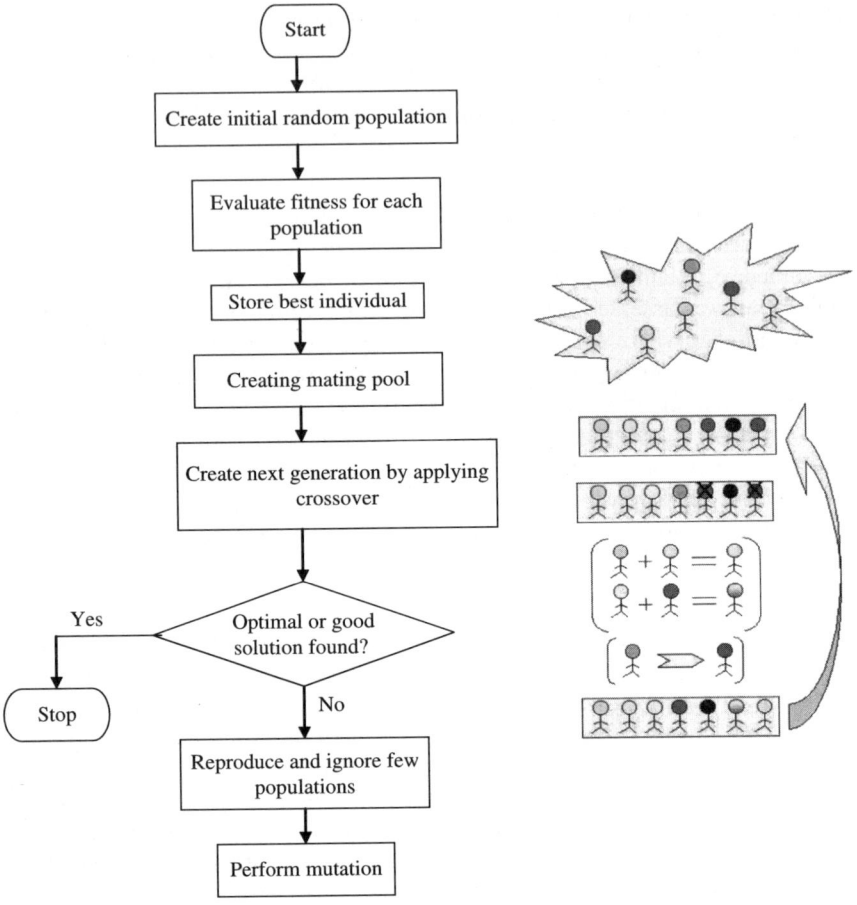

Fig. 2.10 Flowchart of genetic algorithm

3. local determinism: A purely stochastic method is usually quite slow. It is therefore reasonable to do as much as possible efficient deterministic predictions of the most promising directions of (local) proceedings. This is called local hill climbing or greedy search according to the obvious strategies.

Based on the foregoing discussion, the important criteria for GA approach can be formulated as given below:

- Completeness: Any solution should have its encoding
- Non redundancy: Codes and solutions should correspond one to one
- Soundness: Any code (produced by genetic operators) should have its corresponding solution
- Characteristic perseverance: Offspring should inherit useful characteristics from parents.

In short, the basic four steps used in simple Genetic Algorithm to solve a problem are,

1. The representation of the problem
2. The fitness calculation
3. Various variables and parameters involved in controlling the algorithm
4. The representation of result and the way of terminating the algorithm

2.6 Comparison of Genetic Algorithm with Other Optimization Techniques

The principle of GAs is simple: imitate genetics and natural selection by a computer program: The parameters of the problem are coded most naturally as a DNA-like linear data structure, a vector or a string. Sometimes, when the problem is naturally two or three-dimensional also corresponding array structures are used.

A set, called population, of these problem dependent parameter value vectors is processed by GA. To start there is usually a totally random population, the values of different parameters generated by a random number generator. Typical population size is from few dozens to thousands. To do optimization we need a cost function or fitness function as it is usually called when genetic algorithms are used. By a fitness function we can select the best solution candidates from the population and delete the not so good specimens.

The nice thing when comparing GAs to other optimization methods is that the fitness function can be nearly anything that can be evaluated by a computer or even something that cannot! In the latter case it might be a human judgement that cannot be stated as a crisp program, like in the case of eyewitness, where a human being selects among the alternatives generated by GA.

So, there are not any definite mathematical restrictions on the properties of the fitness function. It may be discrete, multimodal etc.

The main criteria used to classify optimization algorithms are as follows: continuous / discrete, constrained / unconstrained and sequential / parallel. There is a clear difference between discrete and continuous problems. Therefore it is instructive to notice that continuous methods are sometimes used to solve inherently discrete problems and vice versa. Parallel algorithms are usually used to speed up processing. There are, however, some cases in which it is more efficient to run several processors in parallel rather than sequentially. These cases include among others such, in which there is high probability of each individual search run to get stuck into a local extreme.

Irrespective of the above classification, optimization methods can be further classified into deterministic and non-deterministic methods. In addition optimization algorithms can be classified as local or global. In terms of energy and entropy local search corresponds to entropy while global optimization depends essentially on the fitness i.e. energy landscape.

Genetic algorithm differs from conventional optimization techniques in following ways:

1. GAs operate with coded versions of the problem parameters rather than parameters themselves i.e., GA works with the coding of solution set and not with the solution itself.
2. Almost all conventional optimization techniques search from a single point but GAs always operate on a whole population of points(strings) i.e., GA uses population of solutions rather than a single solution fro searching. This plays a major role to the robustness of genetic algorithms. It improves the chance of reaching the global optimum and also helps in avoiding local stationary point.
3. GA uses fitness function for evaluation rather than derivatives. As a result, they can be applied to any kind of continuous or discrete optimization problem. The key point to be performed here is to identify and specify a meaningful decoding function.
4. GAs use probabilistic transition operates while conventional methods for continuous optimization apply deterministic transition operates i.e., GAs does not use deterministic rules.

These are the major differences that exist between Genetic Algorithm and conventional optimization techniques.

2.7 Advantages and Limitations of Genetic Algorithm

The advantages of genetic algorithm includes,

1. Parallelism
2. Liability
3. Solution space is wider
4. The fitness landscape is complex
5. Easy to discover global optimum
6. The problem has multi objective function
7. Only uses function evaluations.
8. Easily modified for different problems.
9. Handles noisy functions well.
10. Handles large, poorly understood search spaces easily
11. Good for multi-modal problems Returns a suite of solutions.
12. Very robust to difficulties in the evaluation of the objective function.
13. They require no knowledge or gradient information about the response surface
14. Discontinuities present on the response surface have little effect on overall optimization performance
15. They are resistant to becoming trapped in local optima

2.8 Applications of Genetic Algorithm

16. They perform very well for large-scale optimization problems
17. Can be employed for a wide variety of optimization problems

The limitation of genetic algorithm includes,

1. The problem of identifying fitness function
2. Definition of representation for the problem
3. Premature convergence occurs
4. The problem of choosing the various parameters like the size of the population, mutation rate, cross over rate, the selection method and its strength.
5. Cannot use gradients.
6. Cannot easily incorporate problem specific information
7. Not good at identifying local optima
8. No effective terminator.
9. Not effective for smooth unimodal functions
10. Needs to be coupled with a local search technique.
11. Have trouble finding the exact global optimum
12. Require large number of response (fitness) function evaluations
13. Configuration is not straightforward

2.8 Applications of Genetic Algorithm

Genetic algorithms have been used for difficult problems (such as NP-hard problems), for machine learning and also for evolving simple programs. They have been also used for some art, for evolving pictures and music. A few applications of GA are as follows:

- Nonlinear dynamical systems–predicting, data analysis
- Robot trajectory planning
- Evolving LISP programs (genetic programming)
- Strategy planning
- Finding shape of protein molecules
- TSP and sequence scheduling
- Functions for creating images
- Control–gas pipeline, pole balancing, missile evasion, pursuit
- Design–semiconductor layout, aircraft design, keyboard configuration, communication networks
- Scheduling–manufacturing, facility scheduling, resource allocation
- Machine Learning–Designing neural networks, both architecture and weights, improving classification algorithms, classifier systems
- Signal Processing–filter design
- Combinatorial Optimization–set covering, traveling salesman (TSP), Sequence scheduling, routing, bin packing, graph coloring and partitioning

2.9 Summary

Genetic algorithms are original systems based on the supposed functioning of the Living. The method is very different from classical optimization algorithms.

1. Use of the encoding of the parameters, not the parameters themselves.
2. Work on a population of points, not a unique one.
3. Use the only values of the function to optimize, not their derived function or other auxiliary knowledge.
4. Use probabilistic transition function not determinist ones.

It's important to understand that the functioning of such an algorithm does not guarantee success. The problem is in a stochastic system and a genetic pool may be too far from the solution, or for example, a too fast convergence may halt the process of evolution. These algorithms are nevertheless extremely efficient, and are used in fields as diverse as stock exchange, production scheduling or programming of assembly robots in the automotive industry.

GAs can even be faster in finding global maxima than conventional methods, in particular when derivatives provide misleading information. It should be noted that in most cases where conventional methods can be applied, GAs are much slower because they do not take auxiliary information like derivatives into account. In these optimization problems, there is no need to apply a GA, which gives less accurate solutions after much longer computation time. The enormous potential of GAs lies elsewhere—in optimization of non-differentiable or even discontinuous functions, discrete optimization, and program induction.

It has been claimed that via the operations of selection, crossover, and mutation the GA will converge over successive generations towards the global (or near global) optimum. This simple operation should produces a fast, useful and robust technique largely because of the fact that GAs combine direction and chance in the search in an effective and efficient manner. Since population implicitly contain much more information than simply the individual fitness scores, GAs combine the good information hidden in a solution with good information from another solution to produce new solutions with good information inherited from both parents, inevitably (hopefully) leading towards optimality.

The ability of the algorithm to explore and exploit simultaneously, a growing amount of theoretical justification, and successful application to real-world problems strengthens the conclusion that GAs are a powerful, robust optimization technique.

Review Questions

1. Brief the origin of Genetic Algorithm.
2. Give a suitable example for the Genetic Algorithm principle "Survival of the fittest".

2.9 Summary

3. Discuss in detail about the biological process of natural evolution.
4. Compare the terminologies of natural evolution and Genetic Algorithm.
5. Define: Search space.
6. Describe about various conventional optimization and search techniques.
7. Write short note on simple Genetic Algorithm.
8. Compare and contrast Genetic Algorithm with other optimization techniques.
9. State few advantages and disadvantages of Genetic Algorithm.
10. Mention certain applications of Genetic Algorithm.

Chapter 3
Terminologies and Operators of GA

3.1 Introduction

Genetic Algorithm uses a metaphor where an optimization problem takes the place of an environment and feasible solutions are considered as individuals living in that environment. In genetic algorithms, individuals are binary digits or of some other set of symbols drawn from a finite set. As computer memory is made up of array of bits, anything can be stored in a computer and can also be encoded by a bit string of sufficient length. Each of the encoded individual in the population can be viewed as a representation, according to an appropriate encoding of a particular solution to the problem. For Genetic Algorithms to find a best optimum solution, it is necessary to perform certain operations over these individuals. This chapter discusses the basic terminologies and operators used in Genetic Algorithms to achieve a good enough solution for possible terminating conditions.

3.2 Key Elements

The two distinct elements in the GA are individuals and populations. An individual is a single solution while the population is the set of individuals currently involved in the search process.

3.3 Individuals

An individual is a single solution. Individual groups together two forms of solutions as given below:

1. The chromosome, which is the raw 'genetic' information (genotype) that the GA deals.
2. The phenotype, which is the expressive of the chromosome in the terms of the model.

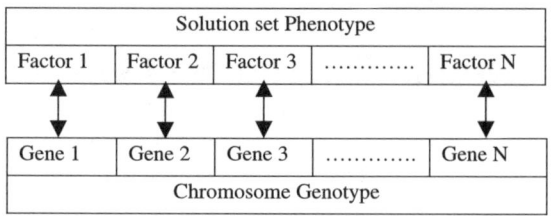

Fig. 3.1 Representation of Genotype and phenotype

Fig. 3.2 Representation of a chromosome

A chromosome is subdivided into genes. A gene is the GA's representation of a single factor for a control factor. Each factor in the solution set corresponds to gene in the chromosome. Figure 3.1 shows the representation of a genotype.

A chromosome should in some way contain information about solution that it represents. The morphogenesis function associates each genotype with its phenotype. It simply means that each chromosome must define one unique solution, but it does not mean that each solution encoded by exactly one chromosome. Indeed, the morphogenesis function is not necessary bijective, and it is even sometimes impossible (especially with binary representation). Nevertheless, the morphogenesis function should at least be subjective. Indeed; all the candidate solutions of the problem must correspond to at least one possible chromosome, to be sure that the whole search space can be explored. When the morphogenesis function that associates each chromosome to one solution is not injective, i.e., different chromosomes can encode the same solution, the representation is said to be degenerated. A slight degeneracy is not so worrying, even if the space where the algorithm is looking for the optimal solution is inevitably enlarged. But a too important degeneracy could be a more serious problem. It can badly affect the behavior of the GA, mostly because if several chromosomes can represent the same phenotype, the meaning of each gene will obviously not correspond to a specific characteristic of the solution. It may add some kind of confusion in the search.

Chromosomes are encoded by bit strings are given below in Fig. 3.2,

3.4 Genes

Genes are the basic "instructions" for building a Generic Algorithms. A chromosome is a sequence of genes. Genes may describe a possible solution to a problem, without actually being the solution. A gene is a bit string of arbitrary lengths. The bit string is a binary representation of number of intervals from a lower bound. A gene is the GA's representation of a single factor value for a control factor, where control factor must have an upper bound and lower bound. This range can be divided

3.6 Populations

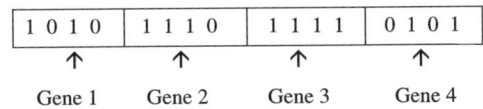

Fig. 3.3 Representation of a gene

into the number of intervals that can be expressed by the gene's bit string. A bit string of length 'n' can represent (2^n-1) intervals. The size of the interval would be (range)/(2^n-1).

The structure of each gene is defined in a record of phenotyping parameters. The phenotype parameters are instructions for mapping between genotype and phenotype. It can also be said as encoding a solution set into a chromosome and decoding a chromosome to a solution set. The mapping between genotype and phenotype is necessary to convert solution sets from the model into a form that the GA can work with, and for converting new individuals from the GA into a form that the model can evaluate. In a chromosome, the genes are represented as in (Fig. 3.3):

3.5 Fitness

The fitness of an individual in a genetic algorithm is the value of an objective function for its phenotype. For calculating fitness, the chromosome has to be first decoded and the objective function has to be evaluated. The fitness not only indicates how good the solution is, but also corresponds to how close the chromosome is to the optimal one.

In the case of multicriterion optimization, the fitness function is definitely more difficult to determine. In multicriterion optimization problems, there is often a dilemma as how to determine if one solution is better than another. What should be done if a solution is better for one criterion but worse for another? But here, the trouble comes more from the definition of a 'better' solution rather than from how to implement a GA to resolve it. If sometimes a fitness function obtained by a simple combination of the different criteria can give good result, it suppose that criterions can be combined in a consistent way. But, for more advanced problems, it may be useful to consider something like Pareto optimally or others ideas from multicriteria optimization theory.

3.6 Populations

A population is a collection of individuals. A population consists of a number of individuals being tested, the phenotype parameters defining the individuals and some information about search space. The two important aspects of population used in Genetic Algorithms are:

1. The initial population generation.
2. The population size.

Fig. 3.4 Population

Population	Chromosome 1	1 1 1 0 0 0 1 0
	Chromosome 2	0 1 1 1 1 0 1 1
	Chromosome 3	1 0 1 0 1 0 1 0
	Chromosome 4	1 1 0 0 1 1 0 0

For each and every problem, the population size will depend on the complexity of the problem. It is often a random initialization of population is carried. In the case of a binary coded chromosome this means, that each bit is initialized to a random zero or one. But there may be instances where the initialization of population is carried out with some known good solutions.

Ideally, the first population should have a gene pool as large as possible in order to be able to explore the whole search space. All the different possible alleles of each should be present in the population. To achieve this, the initial population is, in most of the cases, chosen randomly. Nevertheless, sometimes a kind of heuristic can be used to seed the initial population. Thus, the mean fitness of the population is already high and it may help the genetic algorithm to find good solutions faster. But for doing this one should be sure that the gene pool is still large enough. Otherwise, if the population badly lacks diversity, the algorithm will just explore a small part of the search space and never find global optimal solutions.

The size of the population raises few problems too. The larger the population is, the easier it is to explore the search space. But it has established that the time required by a GA to converge is O (nlogn) function evaluations where n is the population size. We say that the population has converged when all the individuals are very much alike and further improvement may only be possibly by mutation. Goldberg has also shown that GA efficiency to reach global optimum instead of local ones is largely determined by the size of the population. To sum up, a large population is quite useful. But it requires much more computational cost, memory and time. Practically, a population size of around 100 individuals is quite frequent, but anyway this size can be changed according to the time and the memory disposed on the machine compared to the quality of the result to be reached.

Population being combination of various chromosomes is represented as in Fig. 3.4

Thus the above population consists of four chromosomes.

3.7 Data Structures

The main data structures in GA are chromosomes, phenotypes, objective function values and fitness values. This is particularly easy implemented when using MATLAB package as a numerical tool. An entire chromosome population can be stored in a single array given the number of individuals and the length of their genotype representation. Similarly, the design variables, or phenotypes that are obtained by applying some mapping from the chromosome representation into the design space can be stored in a single array. The actual mapping depends upon the decoding scheme used. The objective function values can be scalar or vectorial and are necessarily the same as the fitness values. Fitness values are derived from the object function using scaling or ranking function and can be stored as vectors.

3.8 Search Strategies

The search process consists of initializing the population and then breeding new individuals until the termination condition is met. There can be several goals for the search process, one of which is to find the global optima. This can never be assured with the types of models that GAs work with. There is always a possibility that the next iteration in the search would produce a better solution. In some cases, the search process could run for years and does not produce any better solution than it did in the first little iteration.

Another goal is faster convergence. When the objective function is expensive to run, faster convergence is desirable, however, the chance of converging on local, and possibly quite substandard optima is increased.

Apart from these, yet another goal is to produce a range of diverse, but still good solutions. When the solution space contains several distinct optima, which are similar in fitness, it is useful to be able to select between them, since some combinations of factor values in the model may be more feasible than others. Also, some solutions may be more robust than others.

3.9 Encoding

Encoding is a process of representing individual genes. The process can be performed using bits, numbers, trees, arrays, lists or any other objects. The encoding depends mainly on solving the problem. For example, one can encode directly real or integer numbers.

3.9.1 Binary Encoding

The most common way of encoding is a binary string, which would be represented as in Fig. 3.5

Each chromosome encodes a binary (bit) string. Each bit in the string can represent some characteristics of the solution. Every bit string therefore is a solution but not necessarily the best solution. Another possibility is that the whole string

Fig. 3.5 Binary encoding

Chromosome 1	1 1 0 1 0 0 0 1 1 0 1 0
Chromosome 2	0 1 1 1 1 1 1 1 1 1 0 0

can represent a number. The way bit strings can code differs from problem to problem.

Binary encoding gives many possible chromosomes with a smaller number of alleles. On the other hand this encoding is not natural for many problems and sometimes corrections must be made after genetic operation is completed. Binary coded strings with 1s and 0s are mostly used. The length of the string depends on the accuracy.

In this,

- Integers are represented exactly
- Finite number of real numbers can be represented
- Number of real numbers represented increases with string length

3.9.2 Octal Encoding

This encoding uses string made up of octal numbers (0–7).

Chromosome 1	03467216
Chromosome 2	15723314

Fig. 3.6 Octal encoding

3.9.3 Hexadecimal Encoding

This encoding uses string made up of hexadecimal numbers (0–9, A–F).

Chromosome 1	9CE7
Chromosome 2	3DBA

Fig. 3.7 Hexadecimal encoding

3.9.4 Permutation Encoding (Real Number Coding)

Every chromosome is a string of numbers, which represents the number in sequence. Sometimes corrections have to be done after genetic operation is completed. In permutation encoding, every chromosome is a string of integer/real values, which represents number in a sequence.

Chromosome A	1 5 3 2 6 4 7 9 8
Chromosome B	8 5 6 7 2 3 1 4 9

Fig. 3.8 Permutation encoding

Permutation encoding is only useful for ordering problems. Even for this problems for some types of crossover and mutation corrections must be made to leave the chromosome consistent (i.e., have real sequence in it).

3.9.5 Value Encoding

Every chromosome is a string of values and the values can be anything connected to the problem. This encoding produces best results for some special problems. On the other hand, it is often necessary to develop new genetic operator's specific to the problem. Direct value encoding can be used in problems, where some complicated values, such as real numbers, are used. Use of binary encoding for this type of problems would be very difficult.

In value encoding, every chromosome is a string of some values. Values can be anything connected to problem, form numbers, real numbers or chars to some complicated objects.

Chromosome A	1.2324 5.3243 0.4556 2.3293 2.4545
Chromosome B	ABDJEIFJDHDIERJFDLDFLFEGT
Chromosome C	(back), (back), (right), (forward), (left)

Fig. 3.9 Value encoding

Value encoding is very good for some special problems. On the other hand, for this encoding is often necessary to develop some new crossover and mutation specific for the problem.

3.9.6 Tree Encoding

This encoding is mainly used for evolving program expressions for genetic programming. Every chromosome is a tree of some objects such as functions and commands of a programming language.

3.10 Breeding

The breeding process is the heart of the genetic algorithm. It is in this process, the search process creates new and hopefully fitter individuals.

The breeding cycle consists of three steps:

a. Selecting parents.
b. Crossing the parents to create new individuals (offspring or children).
c. Replacing old individuals in the population with the new ones.

3.10.1 Selection

Selection is the process of choosing two parents from the population for crossing. After deciding on an encoding, the next step is to decide how to perform selection i.e., how to choose individuals in the population that will create offspring for the next generation and how many offspring each will create. The purpose of selection is to emphasize fitter individuals in the population in hopes that their off springs have higher fitness. Chromosomes are selected from the initial population to be parents for reproduction. The problem is how to select these chromosomes. According to Darwin's theory of evolution the best ones survive to create new offspring.

The Fig. 3.10 shows the basic selection process.

Selection is a method that randomly picks chromosomes out of the population according to their evaluation function. The higher the fitness function, the more chance an individual has to be selected. The selection pressure is defined as the degree to which the better individuals are favored. The higher the selection pressured, the more the better individuals are favored. This selection pressure drives the GA to improve the population fitness over the successive generations.

The convergence rate of GA is largely determined by the magnitude of the selection pressure, with higher selection pressures resulting in higher convergence rates.

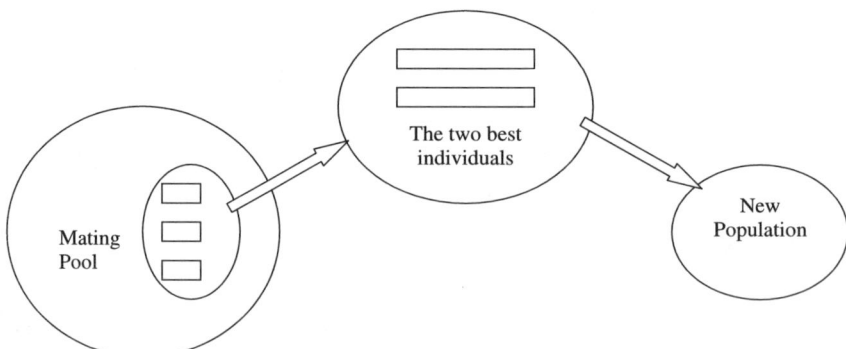

Fig. 3.10 Selection

3.10 Breeding

Genetic Algorithms should be able to identify optimal or nearly optimal solutions under a wise range of selection scheme pressure. However, if the selection pressure is too low, the convergence rate will be slow, and the GA will take unnecessarily longer time to find the optimal solution. If the selection pressure is too high, there is an increased change of the GA prematurely converging to an incorrect (sub-optimal) solution. In addition to providing selection pressure, selection schemes should also preserve population diversity, as this helps to avoid premature convergence.

Typically we can distinguish two types of selection scheme, proportionate selection and ordinal-based selection. Proportionate-based selection picks out individuals based upon their fitness values relative to the fitness of the other individuals in the population. Ordinal-based selection schemes selects individuals not upon their raw fitness, but upon their rank within the population. This requires that the selection pressure is independent of the fitness distribution of the population, and is solely based upon the relative ordering (ranking) of the population.

It is also possible to use a scaling function to redistribute the fitness range of the population in order to adapt the selection pressure. For example, if all the solutions have their fitness in the range [999, 1000], the probability of selecting a better individual than any other using a proportionate-based method will not be important. If the fitness in every individual is brought to the range [0, 1] equitably, the probability of selecting good individual instead of bad one will be important.

Selection has to be balanced with variation form crossover and mutation. Too strong selection means sub optimal highly fit individuals will take over the population, reducing the diversity needed for change and progress; too weak selection will result in too slow evolution. The various selection methods are discussed as follows:

3.10.1.1 Roulette Wheel Selection

Roulette selection is one of the traditional GA selection techniques. The commonly used reproduction operator is the proportionate reproductive operator where a string is selected from the mating pool with a probability proportional to the fitness. The principle of roulette selection is a linear search through a roulette wheel with the slots in the wheel weighted in proportion to the individual's fitness values. A target value is set, which is a random proportion of the sum of the fit nesses in the population. The population is stepped through until the target value is reached. This is only a moderately strong selection technique, since fit individuals are not guaranteed to be selected for, but somewhat have a greater chance. A fit individual will contribute more to the target value, but if it does not exceed it, the next chromosome in line has a chance, and it may be weak. It is essential that the population not be sorted by fitness, since this would dramatically bias the selection.

The above described Roulette process can also be explained as follows: The expected value of an individual is that fitness divided by the actual fitness of the population. Each individual is assigned a slice of the roulette wheel, the size of the slice being proportional to the individual's fitness. The wheel is spun N times, where N is the number of individuals in the population. On each spin, the individual under the wheel's marker is selected to be in the pool of parents for the next generation.

This method is implemented as follows:

1. Sum the total expected value of the individuals in the population. Let it be T.
2. Repeat N times:

 i. Choose a random integer 'r' between o and T.
 ii. Loop through the individuals in the population, summing the expected values, until the sum is greater than or equal to 'r'. The individual whose expected value puts the sum over this limit is the one selected.

Roulette wheel selection is easier to implement but is noisy. The rate of evolution depends on the variance of fitness's in the population.

3.10.1.2 Random Selection

This technique randomly selects a parent from the population. In terms of disruption of genetic codes, random selection is a little more disruptive, on average, than roulette wheel selection.

3.10.1.3 Rank Selection

The Roulette wheel will have a problem when the fitness values differ very much. If the best chromosome fitness is 90%, its circumference occupies 90% of Roulette wheel, and then other chromosomes have too few chances to be selected. Rank Selection ranks the population and every chromosome receives fitness from the ranking. The worst has fitness 1 and the best has fitness N. It results in slow convergence but prevents too quick convergence. It also keeps up selection pressure when the fitness variance is low. It preserves diversity and hence leads to a successful search. In effect, potential parents are selected and a tournament is held to decide which of the individuals will be the parent. There are many ways this can be achieved and two suggestions are,

1. Select a pair of individuals at random. Generate a random number, R, between 0 and 1. If $R < r$ use the first individual as a parent. If the $R>=r$ then use the second individual as the parent. This is repeated to select the second parent. The value of r is a parameter to this method.
2. Select two individuals at random. The individual with the highest evaluation becomes the parent. Repeat to find a second parent.

3.10.1.4 Tournament Selection

An ideal selection strategy should be such that it is able to adjust its selective pressure and population diversity so as to fine-tune GA search performance. Unlike, the Roulette wheel selection, the tournament selection strategy provides selective pressure by holding a tournament competition among N_u individuals.

The best individual from the tournament is the one with the highest fitness, which is the winner of N_u. Tournament competitions and the winner are then inserted into the mating pool. The tournament competition is repeated until the mating pool for generating new offspring is filled. The mating pool comprising of the tournament winner has higher average population fitness. The fitness difference provides the selection pressure, which drives GA to improve the fitness of the succeeding genes. This method is more efficient and leads to an optimal solution.

3.10.1.5 Boltzmann Selection

Simulation annealing is a method of function minimization or maximization. This method simulates the process of slow cooling of molten metal to achieve the minimum function value in a minimization problem. Controlling a temperature like parameter introduced with the concept of Boltzmann probability distribution simulates the cooling phenomenon.

In Boltzmann selection a continuously varying temperature controls the rate of selection according to a preset schedule. The temperature starts out high, which means the selection pressure is low. The temperature is gradually lowered, which gradually increases the selection pressure, thereby allowing the GA to narrow in more closely to the best part of the search space while maintaining the appropriate degree of diversity.

A logarithmically decreasing temperature is found useful for convergence without getting stuck to a local minima state. But to cool down the system to the equilibrium state takes time.

Let f_{max} be the fitness of the currently available best string. If the next string has fitness $f(X_i)$ such that $f(X_i) > f_{max}$, then the new string is selected. Otherwise it is selected with Boltz Mann probability,

$$P = \exp[-(f_{max}-f(X_i))/T] \qquad (3.1)$$

Where $T = T_o(1-\alpha)^k$ and $k = (1 + 100*g/G)$; g is the current generation number; G, the maximum value of g. The value of α can be chosen from the range [0, 1] and T_o from the range [5, 100]. The final state is reached when computation approaches zero value of T, i.e., the global solution is achieved at this point.

The probability that the best string is selected and introduced into the mating pool is very high. However, Elitism can be used to eliminate the chance of any undesired loss of information during the mutation stage. Moreover, the execution time is less.

Elitism

The first best chromosome or the few best chromosomes are copied to the new population. The rest is done in a classical way. Such individuals can be lost if they are not selected to reproduce or if crossover or mutation destroys them. This significantly improves the GA's performance.

3.10.1.6 Stochastic Universal Sampling

Stochastic universal sampling provides zero bias and minimum spread. The individuals are mapped to contiguous segments of a line, such that each individual's segment is equal in size to its fitness exactly as in roulette-wheel selection. Here equally spaced pointers are placed over the line, as many as there are individuals to be selected. Consider *NPointer* the number of individuals to be selected, then the distance between the pointers are 1/*NPointer* and the position of the first pointer is given by a randomly generated number in the range [0, 1/*NPointer*].

For 6 individuals to be selected, the distance between the pointers is $1/6 = 0.167$. Figure 3.11 shows the selection for the above example.

Sample of 1 random number in the range [0, 0.167]: 0.1.

After selection the mating population consists of the individuals,

1, 2, 3, 4, 6, 8.

Stochastic universal sampling ensures a selection of offspring, which is closer to what is deserved than roulette wheel selection.

3.10.2 Crossover (Recombination)

Crossover is the process of taking two parent solutions and producing from them a child. After the selection (reproduction) process, the population is enriched with better individuals. Reproduction makes clones of good strings but does not create new ones. Crossover operator is applied to the mating pool with the hope that it creates a better offspring.

Crossover is a recombination operator that proceeds in three steps:

i. The reproduction operator selects at random a pair of two individual strings for the mating.
ii. A cross site is selected at random along the string length.
iii. Finally, the position values are swapped between the two strings following the cross site.

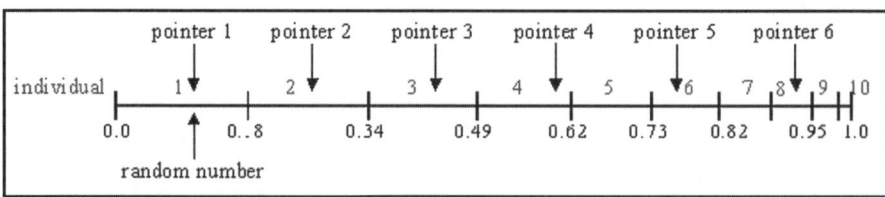

Fig. 3.11 Stochastic universal sampling

3.10 Breeding

That is, the simplest way how to do that is to choose randomly some crossover point and copy everything before this point from the first parent and then copy everything after the crossover point from the other parent. The various crossover techniques are discussed as follows:

3.10.2.1 Single Point Crossover

The traditional genetic algorithm uses single point crossover, where the two mating chromosomes are cut once at corresponding points and the sections after the cuts exchanged. Here, a cross-site or crossover point is selected randomly along the length of the mated strings and bits next to the cross-sites are exchanged. If appropriate site is chosen, better children can be obtained by combining good parents else it severely hampers string quality.

The above Fig. 3.12 illustrates single point crossover and it can be observed that the bits next to the crossover point are exchanged to produce children. The crossover point can be chosen randomly.

3.10.2.2 Two Point Crossover

Apart from single point crossover, many different crossover algorithms have been devised, often involving more than one cut point. It should be noted that adding further crossover points reduces the performance of the GA. The problem with adding additional crossover points is that building blocks are more likely to be disrupted. However, an advantage of having more crossover points is that the problem space may be searched more thoroughly.

In two-point crossover, two crossover points are chosen and the contents between these points are exchanged between two mated parents.

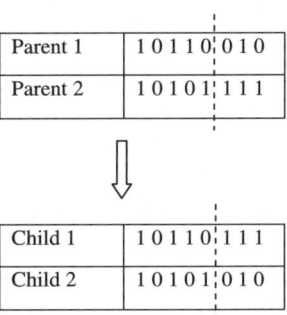

Fig. 3.12 Single point crossover

Fig. 3.13 Two-point Crossover

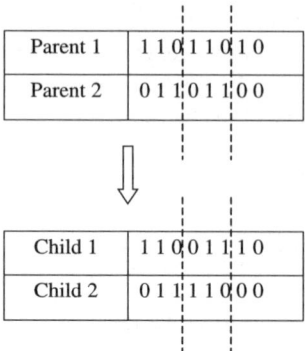

In the above Fig. 3.13 the dotted lines indicate the crossover points. Thus the contents between these points are exchanged between the parents to produce new children for mating in the next generation.

Originally, GAs were using one-point crossover which cuts two chromosomes in one point and splices the two halves to create new ones. But with this one-point crossover, the head and the tail of one chromosome cannot be passed together to the offspring. If both the head and the tail of a chromosome contain good genetic information, none of the offsprings obtained directly with one-point crossover will share the two good features. Using a 2-point crossover avoids this drawback, and then, is generally considered better than 1-point crossover. In fact this problem can be generalized to each gene position in a chromosome. Genes that are close on a chromosome have more chance to be passed together to the offspring obtained through a N-points crossover. It leads to an unwanted correlation between genes next to each other. Consequently, the efficiency of a N-point crossover will depend on the position of the genes within the chromosome. In a genetic representation, genes that encode dependant characteristics of the solution should be close together. To avoid all the problem of genes locus, a good thing is to use a uniform crossover as recombination operator.

3.10.2.3 Multi-Point Crossover (N-Point crossover)

There are two ways in this crossover. One is even number of cross-sites and the other odd number of cross-sites. In the case of even number of cross-sites, cross-sites are selected randomly around a circle and information is exchanged. In the case of odd number of cross-sites, a different cross-point is always assumed at the string beginning.

3.10.2.4 Uniform Crossover

Uniform crossover is quite different from the N-point crossover. Each gene in the offspring is created by copying the corresponding gene from one or the other parent

3.10 Breeding

chosen according to a random generated binary crossover mask of the same length as the chromosomes. Where there is a 1 in the crossover mask, the gene is copied from the first parent, and where there is a 0 in the mask the gene is copied from the second parent. A new crossover mask is randomly generated for each pair of parents. Offsprings, therefore contain a mixture of genes from each parent. The number of effective crossing point is not fixed, but will average L/2 (where L is the chromosome length).

In Fig. 3.14, new children are produced using uniform crossover approach. It can be noticed, that while producing child 1, when there is a 1 in the mask, the gene is copied from the parent 1 else from the parent 2. On producing child 2, when there is a 1 in the mask, the gene is copied from parent 2, when there is a 0 in the mask; the gene is copied from the parent 1.

3.10.2.5 Three Parent Crossover

In this crossover technique, three parents are randomly chosen. Each bit of the first parent is compared with the bit of the second parent. If both are the same, the bit is taken for the offspring otherwise; the bit from the third parent is taken for the offspring. This concept is illustrated in Fig. 3.15.

3.10.2.6 Crossover with Reduced Surrogate

The reduced surrogate operator constrains crossover to always produce new individuals wherever possible. This is implemented by restricting the location of crossover points such that crossover points only occur where gene values differ.

3.10.2.7 Shuffle Crossover

Shuffle crossover is related to uniform crossover. A single crossover position (as in single-point crossover) is selected. But before the variables are exchanged, they are randomly shuffled in both parents. After recombination, the variables in the offspring are unshuffled. This removes positional bias as the variables are randomly reassigned each time crossover is performed.

Parent 1	1 0 1 1 0 0 1 1
Parent 2	0 0 0 1 1 0 1 0
Mask	1 1 0 1 0 1 1 0
Child 1	1 0 0 1 1 0 1 0
Child 2	0 0 1 1 0 0 1 1

Fig. 3.14 Uniform crossover

Fig. 3.15 Three parent crossover

Parent 1	1 1 0 1 0 0 0 1
Parent 2	0 1 1 0 1 0 0 1
Parent 3	0 1 1 0 1 1 0 0
Child	0 1 1 0 1 0 0 1

3.10.2.8 Precedence Preservative Crossover (PPX)

PPX was independently developed for vehicle routing problems by Blanton and Wainwright (1993) and for scheduling problems by Bierwirth et al. (1996). The operator passes on precedence relations of operations given in two parental permutations to one offspring at the same rate, while no new precedence relations are introduced. PPX is illustrated in below, for a problem consisting of six operations A–F.

The operator works as follows:

- A vector of length Sigma, sub i=1to mi, representing the number of operations involved in the problem, is randomly filled with elements of the set {1, 2}.
- This vector defines the order in which the operations are successively drawn from parent 1 and parent 2.
- We can also consider the parent and offspring permutations as lists, for which the operations 'append' and 'delete' are defined.
- First we start by initializing an empty offspring.
- The leftmost operation in one of the two parents is selected in accordance with the order of parents given in the vector.
- After an operation is selected it is deleted in both parents.
- Finally the selected operation is appended to the offspring.
- This step is repeated until both parents are empty and the offspring contains all operations involved.
- Note that PPX does not work in a uniform-crossover manner due to the 'deletion-append' scheme used. Example is shown in Fig. 3.16.

3.10.2.9 Ordered Crossover

Ordered two-point crossover is used when the problem is of order based, for example in U-shaped assembly line balancing etc. Given two parent chromosomes, two random crossover points are selected partitioning them into a left, middle and right portion. The ordered two-point crossover behaves in the following way: child 1 inherits its left and right section from parent 1, and its middle section is determined

Fig. 3.16 Precedence Preservative Crossover (PPX)

Parent permutation 1	A	B	C	D	E	F
Parent permutation 2	C	A	B	F	D	E
Select parent no. (1/2)	1	2	1	1	2	2
Offspring permutation	A	C	B	D	F	E

3.10 Breeding

Fig. 3.17 Ordered crossover
Parent 1 : 4 2 | 1 3 | 6 5   Child 1 : 4 2 | 3 1 | 6 5
Parent 2 : 2 3 | 1 4 | 5 6   Child 2 : 2 3 | 4 1 | 5 6

by the genes in the middle section of parent 1 in the order in which the values appear in parent 2. A similar process is applied to determine child 2. This is shown in Fig. 3.17

3.10.2.10 Partially Matched Crossover (PMX)

PMX can be applied usefully in the TSP. Indeed, TSP chromosomes are simply sequences of integers, where each integer represents a different city and the order represents the time at which a city is visited. Under this representation, known as permutation encoding, we are only interested in labels and not alleles. It may be viewed as a crossover of permutations that guarantees that all positions are found exactly once in each offspring, i.e. both offspring receive a full complement of genes, followed by the corresponding filling in of alleles from their parents.

PMX proceeds as follows:

1. The two chromosomes are aligned.
2. Two crossing sites are selected uniformly at random along the strings, defining a matching section

- The matching section is used to effect a cross through position-by-position exchange operation
- Alleles are moved to their new positions in the offspring
- The following illustrates how PMX works.
- Consider the two strings shown in Fig. 3.18
- Where, the dots mark the selected cross points.
- The matching section defines the position-wise exchanges that must take place in both parents to produce the offspring.
- The exchanges are read from the matching section of one chromosome to that of the other.
- In the example, the numbers that exchange places are 5 and 2, 6 and 3, and 7 and 10.
- The resulting offspring are as shown in Fig. 3.19

PMX is dealt in detail in next chapter.

Fig. 3.18 Strings given
Name 9 8 4 . 5 6 7 . 1 3 2 1 0   Allele 1 0 1 . 0 0 1 . 1 1 0 0
Name 8 7 1 . 2 3 1 0 . 9 5 4 6   Allele 1 1 1 . 0 1 1 . 1 1 0 1

Fig. 3.19 Partially matched crossover

Name 9 8 4 . 2 3 1 0 . 1 6 5 7   **Allele** 1 0 1 . 0 1 0 . 1 0 0 1
Name 8 1 0 1 . 5 6 7 . 9 2 4 3   **Allele** 1 1 1 . 1 1 1 . 1 0 0 1

3.10.2.11 Crossover Probability

The basic parameter in crossover technique is the crossover probability (P_c). Crossover probability is a parameter to describe how often crossover will be performed. If there is no crossover, offspring are exact copies of parents. If there is crossover, offspring are made from parts of both parent's chromosome. If crossover probability is 100%, then all offspring are made by crossover. If it is 0%, whole new generation is made from exact copies of chromosomes from old population (but this does not mean that the new generation is the same!). Crossover is made in hope that new chromosomes will contain good parts of old chromosomes and therefore the new chromosomes will be better. However, it is good to leave some part of old population survive to next generation.

3.10.3 Mutation

After crossover, the strings are subjected to mutation. Mutation prevents the algorithm to be trapped in a local minimum. Mutation plays the role of recovering the lost genetic materials as well as for randomly disturbing genetic information. It is an insurance policy against the irreversible loss of genetic material. Mutation has traditionally considered as a simple search operator. If crossover is supposed to exploit the current solution to find better ones, mutation is supposed to help for the exploration of the whole search space. Mutation is viewed as a background operator to maintain genetic diversity in the population. It introduces new genetic structures in the population by randomly modifying some of its building blocks. Mutation helps escape from local minima's trap and maintains diversity in the population. It also keeps the gene pool well stocked, and thus ensuring ergodicity. A search space is said to be ergodic if there is a non-zero probability of generating any solution from any population state.

There are many different forms of mutation for the different kinds of representation. For binary representation, a simple mutation can consist in inverting the value of each gene with a small probability. The probability is usually taken about 1/L, where L is the length of the chromosome. It is also possible to implement kind of hill-climbing mutation operators that do mutation only if it improves the quality of the solution. Such an operator can accelerate the search. But care should be taken, because it might also reduce the diversity in the population and makes the algorithm converge toward some local optima. Mutation of a bit involves flipping a bit, changing 0 to 1 and vice-versa.

3.10 Breeding

3.10.3.1 Flipping

Flipping of a bit involves changing 0 to 1 and 1 to 0 based on a mutation chromosome generated.

The Fig. 3.20 explains mutation-flipping concept. A parent is considered and a mutation chromosome is randomly generated. For a 1 in mutation chromosome, the corresponding bit in parent chromosome is flipped (0 to 1 and 1 to 0) and child chromosome is produced. In the above case, there occurs 1 at 3 places of mutation chromosome, the corresponding bits in parent chromosome are flipped and child is generated.

3.10.3.2 Interchanging

Two random positions of the string are chosen and the bits corresponding to those positions are interchanged. This is shown in Fig. 3.21.

3.10.3.3 Reversing

A random position is chosen and the bits next to that position are reversed and child chromosome is produced. This is shown in Fig. 3.22.

3.10.3.4 Mutation Probability

The important parameter in the mutation technique is the mutation probability (P_m). The mutation probability decides how often parts of chromosome will be mutated. If there is no mutation, offspring are generated immediately after crossover (or directly copied) without any change. If mutation is performed, one or more parts of a chromosome are changed. If mutation probability is 100%, whole chromosome is changed, if it is 0%, nothing is changed. Mutation generally prevents the GA from falling into local extremes. Mutation should not occur very often, because then GA will in fact change to random search.

3.10.4 Replacement

Replacement is the last stage of any breeding cycle. Two parents are drawn from a fixed size population, they breed two children, but not all four can return to the

Parent	1 0 1 1 0 1 0 1
Mutation chromosome	1 0 0 0 1 0 0 1
Child	0 0 1 1 1 1 0 0

Fig. 3.20 Mutation Flipping

Fig. 3.21 Interchanging

Parent	1 0 1 1 0 1 0 1
Child	1 1 1 1 0 0 0 1

population, so two must be replaced i.e., once off springs are produced, a method must determine which of the current members of the population, if any, should be replaced by the new solutions. The technique used to decide which individual stay in a population and which are replaced in on a par with the selection in influencing convergence. Basically, there are two kinds of methods for maintaining the population; generational updates and steady state updates.

The basic generational update scheme consists in producing N children from a population of size N to form the population at the next time step (generation), and this new population of children completely replaces the parent selection. Clearly this kind of update implies that an individual can only reproduce with individuals from the same generation. Derived forms of generational update are also used like $(\lambda + \mu)$-update and (λ, μ)-update. This time from a parent population of size μ, a little of children is produced of size $\lambda \geq \mu$. Then the μ best individuals from either the offspring population or the combined parent and offspring populations (for (λ, μ)- and $(\lambda + \mu)$-update respectively), form the next generation.

In a steady state update, new individuals are inserted in the population as soon as they are created, as opposed to the generational update where an entire new generation is produced at each time step. The insertion of a new individual usually necessitates the replacement of another population member. The individual to be deleted can be chosen as the worst member of the population. (it leads to a very strong selection pressure), or as the oldest member of the population, but those method are quite radical: Generally steady state updates use an ordinal based method for both the selection and the replacement, usually a tournament method. Tournament replacement is exactly analogous to tournament selection except the less good solutions are picked more often than the good ones. A subtle alternative is to replace the most similar member in the existing population.

3.10.4.1 Random Replacement

The children replace two randomly chosen individuals in the population. The parents are also candidates for selection. This can be useful for continuing the search in small populations, since weak individuals can be introduced into the population.

Fig. 3.22 Reversing

Parent	1 0 1 1 0 1 0 1
Child	1 0 1 1 0 1 1 0

3.10.4.2 Weak Parent Replacement

In weak parent replacement, a weaker parent is replaced by a strong child. With the four individuals only the fittest two, parent or child, return to population. This process improves the overall fitness of the population when paired with a selection technique that selects both fit and weak parents for crossing, but if weak individuals and discriminated against in selection the opportunity will never raise to replace them.

3.10.4.3 Both Parents

Both parents replacement is simple. The child replaces the parent. In this case, each individual only gets to breed once. As a result, the population and genetic material moves around but leads to a problem when combined with a selection technique that strongly favors fit parents: the fit breed and then are disposed of.

3.11 Search Termination (Convergence Criteria)

In short, the various stopping condition are listed as follows:

- **Maximum generations**–The genetic algorithm stops when the specified number of generation's have evolved.
- **Elapsed time**–The genetic process will end when a specified time has elapsed.
  Note: If the maximum number of generation has been reached before the specified time has elapsed, the process will end.
- **No change in fitness**–The genetic process will end if there is no change to the population's best fitness for a specified number of generations.
  Note: If the maximum number of generation has been reached before the specified number of generation with no changes has been reached, the process will end.
- **Stall generations**–The algorithm stops if there is no improvement in the objective function for a sequence of consecutive generations of length **Stall generations**.
- **Stall time limit**–The algorithm stops if there is no improvement in the objective function during an interval of time in seconds equal to **Stall time limit**.

The termination or convergence criterion finally brings the search to a halt. The following are the few methods of termination techniques.

3.11.1 Best Individual

A best individual convergence criterion stops the search once the minimum fitness in the population drops below the convergence value. This brings the search to a faster conclusion guaranteeing at least one good solution.

3.11.2 Worst individual

Worst individual terminates the search when the least fit individuals in the population have fitness less than the convergence criteria. This guarantees the entire population to be of minimum standard, although the best individual may not be significantly better than the worst. In this case, a stringent convergence value may never be met, in which case the search will terminate after the maximum has been exceeded.

3.11.3 Sum of Fitness

In this termination scheme, the search is considered to have satisfaction converged when the sum of the fitness in the entire population is less than or equal to the convergence value in the population record. This guarantees that virtually all individuals in the population will be within a particular fitness range, although it is better to pair this convergence criteria with weakest gene replacement, otherwise a few unfit individuals in the population will blow out the fitness sum. The population size has to be considered while setting the convergence value.

3.11.4 Median Fitness

Here at least half of the individuals will be better than or equal to the convergence value, which should give a good range of solutions to choose from.

3.12 Why do Genetic Algorithms Work?

The search heuristics of GA are based upon Holland's scheme theorem. A schema is defined as templates for describing a subset of chromosomes with similar sections. The schemata consist of bits 0, 1 and meta-character. The template is a suitable way of describing similarities among Patterns in the chromosomes Holland derived an expression that predicts the number of copies of a particular schema would have in the next generation after undergoing exploitation, crossover and mutation. It should be noted that particularly good schemata will propagate in future generations. Thus, schema that are low-order, well defined and have above average fitness are preferred and are termed building blocks. This leads to a building block principle of GA: low order, well-defined, average fitness schemata will combine through crossover to form high order, above average fitness schemata. Since GAs process may schemata in a given generation they are said to have the property of implicit parallelism.

3.12.1 Building Block Hypothesis

Schemata with high fitness values and small defining are called Building Blocks. A genetic algorithm seeks near-optimal performance through the juxtaposition of short, low-order, high-performance schemata, called the building blocks.

The building block hypothesis is one of the most important criteria of "how a genetic algorithm works". The building block hypothesis is said by Goldberg's book as: "A genetic algorithm achieves high performance through the juxtaposition of short, low order, highly fit schemata, or building blocks".

The meaning of "highly fit schemata" is not completely clear. The most obvious interpretation is that a schema is highly fit if its average fitness considerably higher than the average fitness of all strings in the search space. This version of the building block hypothesis might be called the "static building block hypothesis". Under this interpretation, it is easy to give "counterexamples" to the building block hypothesis.

For example, suppose that the string length is 100 and that the defining length and the order of the schema is 10. Then the schema will contain 290 points. First, suppose that every string in the schema except one has relatively low fitness. The single point has very high fitness so that the average schema fitness relative to the search space is high. Then any randomly chosen finite population is highly likely to never see the high fitness point, and so the schema will be very likely to disappear in a few generations. Similarly, one can choose most points to have high fitness, with a few points having sufficiently low fitness that the schema fitness relative to the whole population is low. Then of course, this low-fitness schema will probably grow and may lead to a good solution. It is easy to construct less extreme examples.

Another interpretation is that a schema is highly fit if the average fitness of the schema representatives in the populations of the GA run is higher than the average fitness of all individuals in these populations. This might be called the "relative building block hypothesis".

The meaning of the building block hypothesis can be illustrated by considering the "concatenated trap functions" fitness functions that Goldberg has used as test problems.

For each trap function, the all-zeros string is a global optimum. The schemata that correspond to these strings are the building blocks. For example, suppose that we concatenate 5 trap functions where each trap function has string length 4 (so that the total string length ' is 20). Then the building blocks are the schemata 000**************, ****0000************, etc. If the population size is sufficiently large, then the initial population will contain strings that are in the building block schemata, but it is unlikely for a string to be in very many building block schemata. If the population size is large enough, the GA with one-point crossover will be able to find the global optimum.

If the building block hypothesis is a good explanation of why a GA works on a particular problem, then this suggests that crossover should be designed so that it will not be too disruptive of building blocks, but it needs to be present in order to combine building blocks. Thus, knowledge of the configuration of potential building blocks is important in the design of the appropriate crossover. If the building blocks tend to be contiguous on the string, then one-point crossover is most appropriate. If building blocks are distributed arbitrarily over the string, the GA will not be successful unless the building blocks are identified, either before running the GA or as part of running the GA.

3.12.2 A Macro-Mutation Hypothesis

This is an alternative hypothesis to explain how GAs work. Under this hypothesis, the function of crossover is "macromutation". Macromutation is mutation of many bits rather than just 1 or 2 as is most likely under standard bitwise mutation. The macrosmutation operator that would be similar to one-point or two-point crossover would be to pick a contiguous sequence of positions and then to replace them with a random string. For example, suppose that this kind of macromutation is applied to string x. One choose a contiguous segment of x as shown in the example below. One can choose a random string y of the length of that segment, and replace the segment by the random string y. The result is z.

```
x :   01110101 1101010101 10100100111
y :            0111001110
z :   01110101 0111001110 10100100111
```

In this case it was found that on a limited number of problems without well-defined building blocks, a macormutational hill-climber did as well as the corresponding GA. The macro mutational hill-climber did not need to use a population.

3.12.3 An Adaptive Mutation Hypothesis

In fact, GA has been almost always developed with much regard for the result and with little regard to elegance, proof, or other mathematical niceties. Nevertheless, several hypotheses have been put forward to explain results obtained by GAs.

An adaptive mutation hypothesis is that where crossover in a GA serves as a mutation mechanism that is automatically adapted to the stage of convergence of the population. Crossover produces new individuals in the same region as the current population. Thus, as the population "converges" into a smaller region of the search space, crossover produces new individuals within this region. Thus, crossover can be considered as an adaptive mutation method that reduces the strength of mutation as the run progresses.

Unlike the above hypothesis explanation of how a GA works, this explanation does make use of a population, but not through the building block hypothesis. If this is the more correct explanation of why a GA works on some problem, then this

3.12 Why do Genetic Algorithms Work?

suggests that the GA designer does not need to be so concerned about designing a crossover that will preserve building blocks. Thus, it would seem to suggest the use of a fairly disruptive crossover such as uniform crossover along with a strong selection method, such as a steady-state GA with both selection and worst-element deletion.

There are two GA versions that more or less follow this outline. One is the UMDA, or Uniform Marginal Distribution Algorithm. This algorithm does not do conventional crossover. Instead, it does something called gene pool recombination, which is a form of a multi-parent recombination. Given a population, it first does a selection method on that population. It computes the order-1 schema proportions for the population after selection. Then it selects individuals for the next generation population using only those schema averages. Each bit of each individual for the next generation is selected independently using the schema proportions as probabilities. For example, suppose that the schema proportions for the schema 1********** and 0********** are $7=10$ and $3=10$ respectively. (They must add to 2.) Then when we select the leftmost bit of a new individual, the probability of a one bit is $7=10$, and the probability of a zero bit is $3=10$. Each bit is selected independently of the other bits using the corresponding schema proportions. The UMDA works well on many problems, but it does not work well on the concatenated trap fitness functions. It does not appear that the building block hypothesis is a good explanation for how UMDA works.

Another is CHC, which uses HUX, which is like uniform crossover, except that exactly half of the alleles where the parents differ come from each parent. It also uses truncation selection on the union of the parent and child populations—a very strong selection method. It has an incest-prevention method. When the population "converges" or stagnates, a partial reinitialization is done as follows. The best individual found so far is used as a template for new individuals. Each new individual is created by flipping a fixed proportion (e.g., 35%) of the template's bits. The best individual is also copied into the new population. The CHC algorithm has performed well in practice, and it seems unlikely that the building block hypothesis can be an explanation for the success of CHC.

3.12.4 The Schema Theorem

A schema is a similarity template describing a subset of string displaying similarities at certain string positions. It is formed by the ternary alphabet $\{0.1,*\}$, where $*$ is simply a notation symbol, that allows the description of all possible similarities among strings of a particular length and alphabet. In general, there are 2^l different strings or chromosome of length 1, but schemata display an order of 3^l. A particular string of length 1 inside a population of 'n' individuals into one of the 2^l schemata can be obtained from this string. Thus, in the entire population the number of schemata present in each generation is somewhere between 2^l and $n.2^l$, depending upon the population diversity. J. Holland estimated that in a population of 'n' chromosomes, the Gas process $O(n^3)$ schemata into each generation. This is called as implicit parallel process.

A schema represents an affined variety of the search space: for example the schema 01**11*0 is a sub-space of the space of codes of 8 bits length (* can be 0 or 1).

The GA modeled in schema theory is a canonical GA, which acts on binary strings, and for which the creation of a new generation is based on three operators:

- A proportionate selection, where the fitness function steps in: the probability that a solution of the current population is selected and is proportional to its fitness.
- The genetic operators: single point crossover and bit-flip mutation, randomly applied with probabilities p_c and p_m.

Schemata represent global information about the fitness function. A GA works on a population of N codes, and implicitly uses information on a certain number of schemata. The basic 'schema theorem' presented below is based on the observation that the evaluation of a single code makes it possible to deduce some knowledge about the schemata to which that code belongs.

Theorem :(Schema Theorem (Holland))

The Schema Theorem is called as "The Fundamental Theorem of Genetic Algorithm".

For a given schema H, let:

- $m(H, t)$ be the relative frequency of the schema H in the population of the tth generation.
- $f(H)$ be the mean fitness of the elements of H.
- $O(H)$ be the number of fixed bits in the schema H, called the order of the schema.
- $\delta(H)$ be distance between the first and the last fixed bit of the schema, called the definition length of the schema.
- $\bar{f}$ is the mean fitness of the current population.
- P_c is the crossover probability.
- P_m is the mutation probability.

Then,

$$E[m(H, t+1)] \geq m(H, t)\frac{f(H)}{\bar{f}}\left[1 - P_c\frac{\delta(H)}{1\text{-}1} - O(H)P_m\right] \quad (3.2)$$

Based on qualitative view, the above formula means that the "good" schemata, having a short definition length and a low order, tend to grow very rapidly in the population. These particular schemata are called building blocks.

The application of schema theorem is as follows:

i. It provides some tools to check whether a given representation is well-suited to a GA.

3.12 Why do Genetic Algorithms Work?

ii. The analysis of nature of the "good" schemata gives few ideas on the efficiency of genetic algorithm.

3.12.5 Optimal Allocation of Trials

The Schema Theorem has provided the insight that building blocks receive exponentially increasing trials in future generations. This leads to an important and well-analyzed problem from statistical decision theory—the two-armed bandit problem and its generalization, the k-armed bandit problem.

Consider a gambling machine with two slots for coins and two arms. The gambler can deposit the coin either into the left or the right slot. After pulling the corresponding arm, either a reward is payed or the coin is lost. For mathematical simplicity, working only with outcomes, i.e. the difference between the reward (which can be zero) and the value of the coin. Let us assume that the left arm produces an outcome with mean value μ1 and a variance σ_1^2 while the right arm produces an outcome with mean value μ2 and variance σ_2^2. Without loss of generality, although the gambler does not know this, assume that μ1 ≥ μ2.

The question arises which arm should be played. Since it is not known beforehand which arm is associated with the higher outcome, one is we are faced with an interesting dilemma. Not only one must make a sequence of decisions, which arm to play, he have to collect, at the same time, information about which is the better arm. This trade-off between exploration of knowledge and its exploitation is the key issue in this problem and, as turns out later, in genetic algorithms, too.

A simple approach to this problem is to separate exploration from exploitation. More specifically, it is possible to perform a single experiment at the beginning and thereafter make an irreversible decision that depends on the results of the experiment. Suppose we have N coins. If we first allocate an equal number n (where 2n ≤ N) of trials to both arms, we could allocate the remaining N–2n trials to the observed better arm. Assuming we know all involved parameters, the expected loss is given as,

$$L(N, n) = (\mu_1 - \mu_2) \cdot ((N - n)q(n) + n(1 - q(n))) \tag{3.3}$$

where q(n) is the probability that the worst arm is the observed best arm after the 2n experimental trials. The underlying idea is obvious: In case that we observe that the worse arm is the best, which happens with probability q(n), the total number of trials allocated to the right arm is N–n. The loss is, therefore, (μ1 − μ2) · (N − n). In the reverse case that we actually observe that the best arm is the best, which happens with probability 1-q(n), the loss is only what we get less because we played the worse arm n times, i.e. (μ1 − μ2) · n. Taking the central limit theorem into account, we can approximate q(n) with the tail of a normal distribution:

$$q(n) \approx \frac{1}{\sqrt{2\pi}} \cdot \frac{c^{-e^2/2}}{c}, \quad \text{where } c = \frac{\mu_1 - \mu_2}{\sqrt{\sigma_1^2 + \sigma_2^2}} \cdot \sqrt{n} \tag{3.4}$$

p Now we have to specify a reasonable experiment size n. Obviously, if we choose n = 1, the obtained information is potentially unreliable. If we choose, however, n = N 2 there are no trials left to make use of the information gained through the experimental phase. What we see is again the trade-off between exploitation with almost no exploration (n = 1) and exploration without exploitation (n = N/2). It does not take a Nobel price winner to see that the optimal way is somewhere in the middle. Holland has studied this problem is very detail. He came to the conclusion that the optimal strategy is given by the following equation:

$$n^* \approx b^2 \ln\left(\frac{N^2}{8\pi b^4 \ln N^2}\right), \qquad \text{where } b = \frac{\sigma_1}{\mu_1 - \mu_2}. \qquad (3.5)$$

Making a few transformations, we obtain that

$$N - n^* \approx \sqrt{8\pi b^4 \ln N^2} \cdot e^{<001>}, \qquad (3.6)$$

That is the optimal strategy is to allocate slightly more than an exponentially increasing number of trials to the observed best arm. Although no gambler is able to apply this strategy in practice, because it requires knowledge of the mean values µ1 and µ2, we still have found an important bound of performance a decision strategy should try to approach.

A genetic algorithm, although the direct connection is not yet fully clear, actually comes close to this ideal, giving at least an exponentially increasing number trials to the observed best building blocks. However, one may still wonder how the two-armed bandit problem and GAs are related. Let us consider an arbitrary string position. Then there are two schemata of order one, which have their only specification in this position. According to the Schema Theorem, the GA implicitly decides between these two schemata, where only incomplete data are available (observed average fitness values). In this sense, a GA solves a lot of two-armed problems in parallel.

The Schema Theorem, however, is not restricted to schemata with an order of 2. Looking at competing schemata (different schemata which are specified in the same positions), we observe that a GA is solving an enormous number of k-armed bandit problems in parallel. The k-armed bandit problem, although much more complicated, is solved in an analogous way—the observed better alternatives should receive an exponentially increasing number of trials.

3.12.6 Implicit Parallelism

J. Holland analyzed that in a population of 'n' chromosomes, the Gas process $O(n^3)$ schemata's into each generation. He termed it as "Implicit parallel process" and is as shown in Fig. 3.23.

Even though at each generation one performs a proportional computation to the size of the population n, we obtain useful processing of n^3 schemata's in parallel

3.12 Why do Genetic Algorithms Work?

with memory other than the population itself. At present, the common interpretation is that a GA processes an enormous amount of schemata implicitly. This is accomplished by exploiting the currently available, incomplete information about these schemata continuously, while trying to explore more information about them and other, possibly better schemata.

This remarkable property is commonly called the implicit parallelism of genetic algorithms. A simple GA has only m structures in one time step, without any memory or bookkeeping about the previous generations. We will now try to get a feeling how many schemata a GA actually processes.

Obviously, there are 3^n schemata of length n. A single binary string ful-fills n schema of order 1, $\binom{n}{2}$ schemata of order 2, in general, $\binom{n}{k}$ schemata of order k. Hence, a string fulfills

$$\sum_{k=1}^{n} \binom{n}{k} = 2^n \tag{3.7}$$

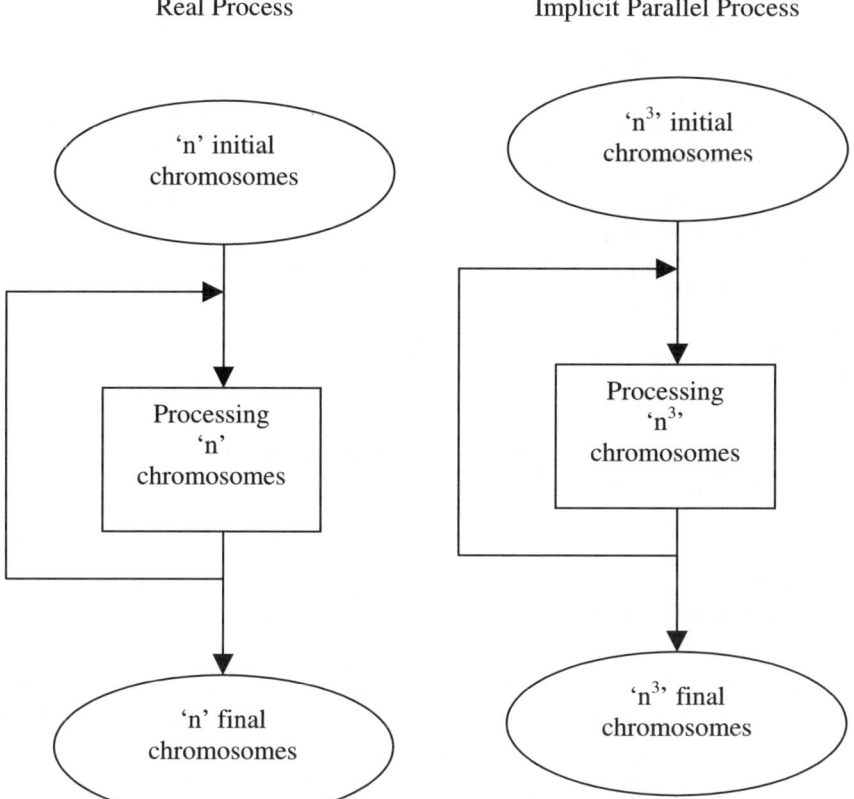

Fig. 3.23 Implicit parallel process

Theorem. Consider a randomly generated start population of a simple GA and let $\varepsilon \in (0, 1)$ be a fixed error bound. Then schemata of length

$$I_s < \varepsilon.(n-1) + 1 \tag{3.8}$$

have a probability of at least $1 - \varepsilon$ to survive one-point crossover (compare with the proof of the Schema Theorem). If the population size is chosen as $m = 2_s^{I}/2$, the number of schemata, which survive for the next generation, is of order $O(m^3)$.

3.12.7 The No Free Lunch Theorem

The No Free Lunch work is a framework that addresses the core aspects of search, focusing on the connection between fitness functions and effective search algorithms. The central importance of this connection is demonstrated by the No Free Lunch theorem, which states that, averaged over all problems, all search algorithms perform equally. This result implies that if we are comparing a genetic algorithm to some other algorithm (e.g., simulated annealing, or even random search) and the genetic algorithm to some other algorithm (e.g., simulated annealing, or even random search) performs better for some class of problems, then the other algorithm necessarily performs better on problems outside the class. Thus it is essential to incorporate knowledge of the problem into the search algorithm. The No Free Lunch framework also does the following:

- it provides a geometric interpretation of what it means for an algorithm to be well matched to a problem;
- it brings insights provided by information theory into the search procedure;
- it investigates time-varying fitness functions;
- it proves that independent of the fitness function, one cannot (without prior domain knowledge) successfully choose between two algorithms based on their previous behavior;
- it provides a number of formal measures of how well an algorithm performs; and
- it addresses the difficulty of optimization problems from a viewpoint outside of traditional computational complexity.

3.13 Solution Evaluation

At the end of the search genetic algorithm displays the final population with their fitnesses, from which it is possible to select a solution and write it back to the system for further generations. In certain systems it is not always practical to declare all the necessary parameters after the search, or perhaps some factors were simply overlooked. Thus once if a solution is obtained, it has to be evaluated for all its various parameters under consideration, which includes fitnesses, median fitness, best individual, maximum fitness and so on.

3.14 Search Refinement

Search parameters like selection, crossover and replacement, which are very effective in the early stages of a search, may not necessarily be the best toward the end of the search. During early search it is desirable to get good spread of points through the solution space in order to find at least the beginning of the various optima. Once the population starts converging on optima it night be better to exercise more stringent selection and replacement to completely cover that region of space.

Alternatively, refinement can also be made in the domain and resolution of the individual genes. A large range and a coarse resolution early in the search will help scatter the points. After certain time period, it may become apparent that few parts of space yield very poor results. Then it would be appropriate to limit the gene ranges and increase the resolution to finely search the better regions. It is possible for the GA to monitor its performance and make alterations to the search parameters when the rate of convergence of fitness values has slowed or after a preset number of generations. A poor looking region in the search space may also contain undiscovered optima.

Sensitivity of solutions is also important in the case where it may not be possible to implement a solution accurately. Two unique solutions may have comparable fit nesses with no undesirable effects; however, one may reside on very steep optima while the other may lie on a broad mound. It may be observed that the solution on the broad mound will be less sensitive to errors in implementation than the one that is steep sided, where even a small deviation results in varying fitness.

3.15 Constraints

If the genetic algorithm that is dealt consists of only objective function and no information about the specifications of variable, then it is called unconstrained optimization problem. Consider, an unconstrained optimization problem of the form,

$$\text{Minimize } f(x) = x^2 \qquad (3.9)$$

and there is no information about 'x' range. Genetic algorithm minimizes this function using its operators in random specifications.

In case of constrained optimization problems, the information's are provided for the variables under consideration. Constraints are classified as,

1. Equality relations.
2. Inequality relations.

A genetic algorithm generates a sequence of parameters to be tested using the system under consideration, objective function (to be maximized or minimized) and the constraints. On running the system, the objective function is evaluated and constraints are checked to see if there are any violations. If there are no violations, the parameter set is assigned the fitness value corresponding to the objective function evaluation. When the constraints are violated, the solution is infeasible and thus has no fitness. Many practical problems are constrained and it is very difficult to find a feasible point that is best. As a result, one should get some information out of infeasible solutions, irrespective of their fitness ranking in relation to the degree of constraint violation. This is performed in penalty method.

Penalty method is that where a constrained optimization problem is transformed to an unconstrained optimization problem by associating a penalty or cost with all constraint violations. This penalty is included in the objective function evaluation.

Consider the original constrained problem in maximization form:

$$\text{Maximize } f(x) \text{ Subject to } g_i(x) \geq 0 \; i = 1, 2, 3, \ldots\ldots\ldots..n \tag{3.10}$$

where x is a k vector.

Transforming this to unconstrained form:

$$\text{Maximize } f(x) + p. \sum_{i=1}^{n} \Phi \left[g_i(x) \right] \tag{3.11}$$

where Φ–penalty function
p–Penalty coefficient

There exist several alternatives for this penalty function. The penalty function can be squared for all violated constraints. In certain situations, the unconstrained solution converges to the constrained solution as the penalty coefficient p tends to infinity.

3.16 Fitness Scaling

Fitness scaling is performed in order to avoid premature convergence and slow finishing. The various types of fitness scaling are:

1. Linear scaling
2. σ–Truncation
3. Power law.

3.16.1 Linear Scaling

Consider,

f–Unscaled raw fitness
f'–Fitness after scaling

$$f' = af + b \tag{3.12}$$

3.16 Fitness Scaling

- In order that the average member gets selection the average of fitness after scaling shall be equal to average of fitness before scaling.

$$\text{fav}' = \text{fav} \tag{3.13}$$

- Inorder not to allow dominance by super individuals the number of copies assigned to them is controlled by taking,

$$f'\max = C * \text{fav}' \tag{3.14}$$

C is the number of copies of highly fit individuals.

Case-1

Initially C is chosen any desired value,

If $fmin > \dfrac{(Cfav - f\max)}{C - 1}$, then

$$a = \frac{fav(C-1)}{f\max - fav}$$

$$b = \frac{f\max - Cfav)fav}{f\max - fav}$$

else

$$a = \frac{fav}{fav - f\min}$$

$$b = \frac{-favf\min}{fav - f\min} \quad \text{and}$$

$$c = \frac{f\max - f\min}{fav - f\min}$$

Case-2

For the entire run, we take $C = 2$

If $fmin > (2fav-fmin)$, then

$$a = \frac{fav}{f\max - fav}$$

$$b = \frac{-favf\min}{f\max - fav}$$

3.16.2 Sigma Truncation

Linear scaling give negative scaling fitness unless special steps are taken as explained above. Negative scaled fitness results at matured runs due to one or two very weak members (low fitness values).

"σ–Truncation" discards such off the average members. Linear scaling is then applied to the remaining members.

$$f" = f - (fav\text{-}C\sigma) \text{ if RHS} > 0 = 0, \text{ otherwise.} \quad (3.15)$$

After this linear scaling is applied without the danger of negative fitness.

$$f' = af" + b \quad (3.16)$$

3.16.3 Power Law Scaling

In power law scaling, the scaled fitness is given by,

$$\text{Scaled fitness } f' = f^k (\text{raw fitness } f) \quad (3.17)$$

K–problem dependent constant. 1.005

Roulette wheel method is adopted after then. The minimum raw fitness (objective function) is subtracted from the raw fitness to obtain the reproductive new fitness roulette wheel method is applied to the new fitness.

3.17 Example Problems

3.17.1 Maximizing a Function

Consider the problem of maximizing the function,

$$f(x) = x^2 \quad (3.18)$$

where x is permitted to vary between 0 to 31. The steps involved in solving this problem are as follows:

Step 1: For using genetic algorithms approach, one must first code the decision variable 'x' into a finite length string. Using a five bit (binary integer) unsigned integer, numbers between 0(00000) and 31(11111) can be obtained.

The objective function here is $f(x) = x^2$ which is to be maximized. A single generation of a genetic algorithm is performed here with encoding, selection, crossover and mutation.

To start with, select initial population at random. Here initial population of size 4 is chosen, but any number of populations can be elected based on the requirement and application. Table 3.1 shows an initial population randomly selected.

Step 2: Obtain the decoded x values for the initial population generated. Consider string 1,

3.17 Example Problems

Table 3.1 Selection

String No.	Initial population (randomly selected)	x value	Fitness value $f(x) = x^2$	$Prob_i$	Percentage probability	Expected count	Actual count
1	01100	12	144	0.1247	12.47%	0.4987	1
2	11001	25	625	0.5411	54.11%	2.1645	2
3	00101	5	25	0.0216	2.16%	0.0866	0
4	10011	19	361	0.3126	31.26%	1.2502	1
Sum			1155	1.0000	100%	4.0000	4
average			288.75	0.2500	25%	1.0000	1
maximum			625	0.5411	54.11%	2.1645	2

$$01100 = 0*2^4 + 1*2^3 + 1*2^2 + 0*2^1 + 0*2^0$$
$$= 0 + 8 + 4 + 0 + 0$$
$$= 12$$

Thus for all the four strings the decoded values are obtained.

Step 3: Calculate the fitness or objective function. This is obtained by simply squaring the 'x' value, since the given function is $f(x) = x^2$.
When, x = 12, the fitness value is,

$$f(x) = x^2 = (12)^2 = 144$$

for x = 25, $\quad f(x) = x^2 = (25)^2 = 625$

and so on, until the entire population is computed

Step 4: Compute the probability of selection,

$$Prob_i = \frac{f(x)_i}{\sum_{i=1}^{n} f(x)_i} \quad (3.19)$$

where n- no of populations
f(x)- fitness value corresponding to a particular individual in the population
$\Sigma f(x)$- Summation of all the fitness value of the entire population.
Considering string 1,

$$\text{Fitness } f(x) = 144$$
$$\Sigma f(x) = 1155$$

The probability that string 1 occurs is given by,

$$P_1 = 144/1155 = 0.1247$$

The percentage probability is obtained as,

$$0.1247 * 100 = 12.47\%$$

The same operation is done for all the strings. It should be noted that, summation of probability select is 1.

Step 5: The next step is to calculate the expected count, which is calculated as,

$$\text{Expected count} = \frac{f(x)_i}{(Avg f(x))_i} \qquad (3.20)$$

where $(Avg\ f(x))_i = \left[\dfrac{\sum_{i=1}^{n} f(x)_i}{n} \right]$

For string 1,

$$\text{Expected count} = \text{Fitness/Average} = 144/288.75 = 0.4987$$

Computing the expected count for the entire population. The expected count gives an idea of which population can be selected for further processing in the mating pool.

Step 6: Now the actual count is to be obtained to select the individuals, which would participate in the crossover cycle using Roulette wheel selection. The Roulette wheel is formed as shown in Fig. 3.24.

Roulette wheel is of 100% and the probability of selection as calculated in step4 for the entire populations are used as indicators to fit into the Roulette wheel. Now the wheel may be spun and the no of occurrences of population is noted to get actual count.

String 1 occupies 12.47%, so there is a chance for it to occur at least once. Hence its actual count may be 1.

With string 2 occupying 54.11% of the Roulette wheel, it has a fair chance of being selected twice. Thus its actual count can be considered as 2.

On the other hand, string 3 has the least probability percentage of 2.16%, so their occurrence for next cycle is very poor. As a result, it actual count is 0.

String 4 with 31.26% has at least one chance for occurring while Roulette wheel is spun, thus its actual count is 1.

The above values of actual count are tabulated as shown is Table 3.1

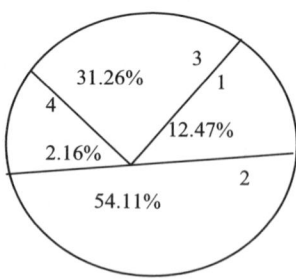

Fig. 3.24 Selection using Roulette wheel

3.17 Example Problems

Table 3.2 Crossover

String No.	Mating pool	Crossover point	Offspring after crossover	x value	Fitness value $f(x) = x^2$
1	0 1 1 0 0	4	0 1 1 0 1	13	169
2	1 1 0 0 1	4	1 1 0 0 0	24	576
2	1 1 0 0 1	2	1 1 0 1 1	27	729
4	1 0 0 1 1	2	1 0 0 0 1	17	289
Sum					1763
average					440.75
maximum					729

Step 7: Now, writing the mating pool based upon the actual count as shown in Table 3.2

The actual count of string no 1 is 1, hence it occurs once in the mating pool. The actual count of string no 2 is 2, hence it occurs twice in the mating pool. Since the actual count of string no 3 is 0, it does not occur in the mating pool. Similarly, the actual count of string no 4 being 1, it occurs once in the mating pool. Based on this, the mating pool is formed.

Step 8: Crossover operation is performed to produce new offspring (children).

The crossover point is specified and based on the crossover point, single point crossover is performed and new offspring is produced. The parents are,

Parent 1	0 1 1 0 0
Parent 2	1 1 0 0 1

The offspring is produced as,

Offspring 1	0 1 1 0 1
Offspring 2	1 1 0 0 0

In a similar manner, crossover is performed for the next strings.

Step 9: After crossover operations, new off springs are produced and 'x' values are decodes and fitness is calculated.

Step 10: In this step, mutation operation is performed to produce new off springs after crossover operation. As discussed in Sect. 3.10.3.1 mutation-flipping operation is performed and new off springs are produced. Table 3.3 shows the new offspring after mutation.

Once the off springs are obtained after mutation, they are decoded to x value and find fitness values are computed.

This completes one generation. The mutation is performed on a bit-bit by basis. The crossover probability and mutation probability was assumed to 1.0 and 0.001 respectively. Once selection, crossover and mutation are performed, the new population is now ready to be tested. This is performed by decoding the new strings created by the simple genetic algorithm after mutation and calculates the fitness

Table 3.3 Mutation

String No.	Offspring after crossover	Mutation chromosomes for flipping	Offspring after Mutation	X value	Fitness value $F(x) = x^2$
1	0 1 1 0 1	1 0 0 0 0	1 1 1 0 1	29	841
2	1 1 0 0 0	0 0 0 0 0	1 1 0 0 0	24	576
2	1 1 0 1 1	0 0 0 0 0	1 1 0 1 1	27	729
4	1 0 0 0 1	0 0 1 0 0	1 0 1 0 0	20	400
Sum					2546
average					636.5
maximum					841

function values from the x values thus decoded. The results for successive cycles of simulation are shown in Tables 3.1–3.3

From the tables, it can be observed how genetic algorithms combine high-performance notions to achieve better performance. In the tables, it can be noted how maximal and average performance has improved in the new population. The population average fitness has improved from 288.75 to 636.5 in one generation. The maximum fitness has increased from 625 to 841 during same period. Although random processes make this best solution, its improvement can also be seen successively. The best string of the initial population (1 1 0 0 1) receives two chances for its existence because of its high, above-average performance. When this combines at random with the next highest string (1 0 0 1 1) and is crossed at crossover point 2 (as shown in Table 3.2), one of the resulting strings (1 1 0 1 1) proves to be a very best solution indeed. Thus after mutation at random, a new offspring (1 1 1 0 1) is produced which is an excellent choice.

This example has shown one generation of a simple genetic algorithm.

3.17.2 Traveling Salesman Problem

The Traveling Salesman Problem is a permutation problem in which the goal is to find the shortest path between N different cities that the salesman takes is called a tour. In other words, the problem deals with finding a route covering all the cities so that the total distance traveled is minimal. The traveling salesman problem finds application in a variety of situations. Suppose we have to route a postal van to pick up mail from mailboxes located at n different cities. A (n + 1) vertex graph can be used to represent the situations. The route taken by the postal van is a tour, and we are interested in finding a tour of minimal length.

3.17.2.1 Encoding

All the cities are sequentially numbered starting from one. The route between the cities is described with an array with each element of the array representing the number of the city. The array represents the sequence in which the cities are traversed to

3.17 Example Problems

Fig. 3.25 Chromosome representing the tour

| 1 | 4 | 2 | 6 | 7 | 3 | 5 |

make up a tour. Each chromosome must contain each and every city exactly once. For instance, for example shown in Fig. 3.25

This chromosome represents the tour starting from city 1 to city 4 and so on and back to city 1.

3.17.2.2 Crossover

To solve the traveling salesman problem, a simple crossover reproduction scheme does not work as it makes the chromosomes inconsistent i.e. some cities may be repeated while others are missed out. The drawback of the simple crossover mechanism is illustrated in Fig. 3.26

As can be seen above, cities 6 and 7 are missing in Child1 while cities 2 and 4 are visited more than once. Child2 too suffers from similar drawbacks. Hence, the need for partially matched crossover.

To avoid this partially matched crossover (PMX) mechanism is used as follows: In this scheme two crossover points are randomly chosen and the region between these is determined. This region is called the crossover region. Crossover is performed in this crossover region to yield transition offspring. This method as applied to the previous example is shown in Fig. 3.27

Consider the crossover points at 3 and 6 and the crossover region between these points is interchanged between the two parents. In the offspring obtained the circled cities are the holes, which are replicated in the crossing region. Cross-referencing with the parent of the alternate chromosome fills these holes. Hence the following two offspring are obtained, which are consistent with our requirements (Fig. 3.28).

3.17.2.3 Mutation

Mutation has a high probability of resulting in a non-viable city order. However, mutation is still applied by accounting for the non-viable city orders in the evaluation function. For this problem, mutation refers to a randomized exchange of cities in the chromosomes. For instance, for example shown in Fig. 3.29.

Here cities 2 and 5 are interchanged because of an inversion operation.

| Parent 1 | 1 2 3 4 5 6 7 |
| Parent 2 | 3 7 6 1 5 2 4 |

⇓

| Offspring 1 | 1 2 3 4 5 2 4 |
| Offspring 2 | 3 7 6 1 5 6 7 |

Fig. 3.26 Crossover

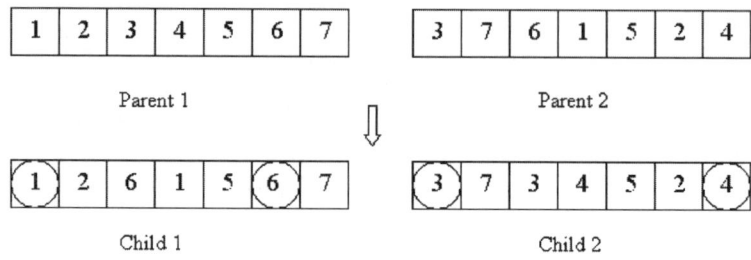

Fig. 3.27 Partially matched crossover

3.17.2.4 Fitness Measure

The fitness function takes a trial solution and returns a fitness value. The shorter the route the higher the fitness value. By using the partially matched crossover and inversion mechanisms non-viable routes are eliminated. Hence the need to punish the low-fitness chromosomes does not arise.

3.17.2.5 Selection Method

Using steady state selection mechanism, two chromosomes from a population are selected for the crossover and mutation operations. The offspring so obtained replace the least fit chromosomes in the existing population. The population size used for this example is 10.

3.17.2.6 Results

Figures 3.30 and 3.31 shows the performance GA applied for 10 and 20 cities case respectively. As can be seen in Table 3.4, the complexity of Genetic Algorithm approach increases nominally with the number of cities.

3.18 Summary

This chapter has laid the basic foundation for understanding genetic algorithms, their terminologies and their operators. The chapter has presented the detailed operation of a simple genetic algorithm. Genetic algorithms operate on populations of strings, with the string coded to represent the underlying parameter set. Selection

Fig. 3.28 Offspring produced after PMX

3.18 Summary

Fig. 3.29 Mutation

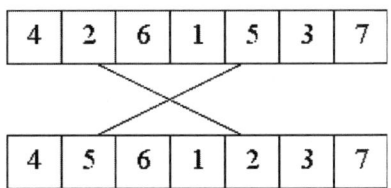

(reproduction), crossover and mutation are applied to string populations to create new string population. A simulation of one generation of the simple genetic algorithm has helped to illustrate the power of the method. Thus the various terminologies and operators used in genetic algorithm are discussed in detail in this chapter.

The working of genetic algorithm has been dealt through the concept of schemata. A schema is a string over an extended alphabet, {0,1,*} where the 0 an the 1 retain the original meaning and the * is a don't care or wild card symbol. The schemata approach simplifies the analysis of the genetic algorithm method because it explicitly recognizes all the possible similarities in a population of strings. The discussion also has been made on building block hypothesis, macro mutation hypothesis, optimal allocation of trials and implicit parallelism.

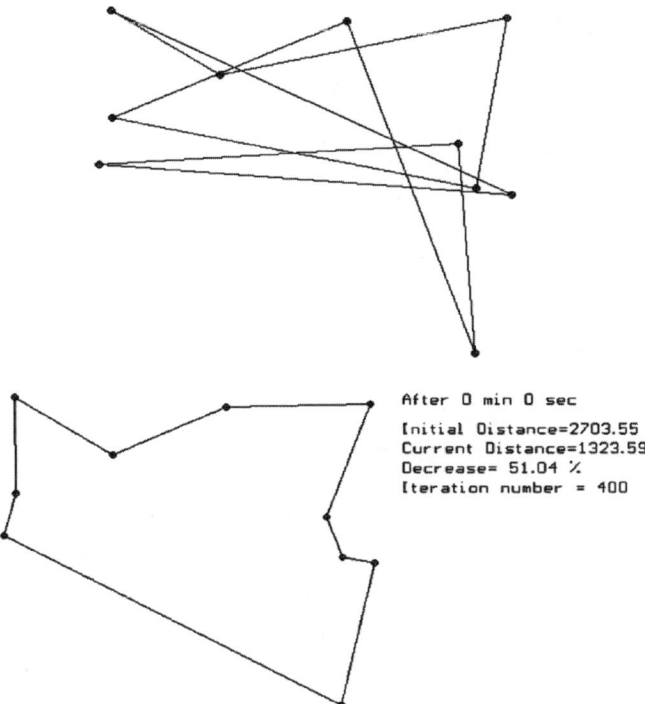

Fig. 3.30 Solution for the traveling salesman problem with 10 cities

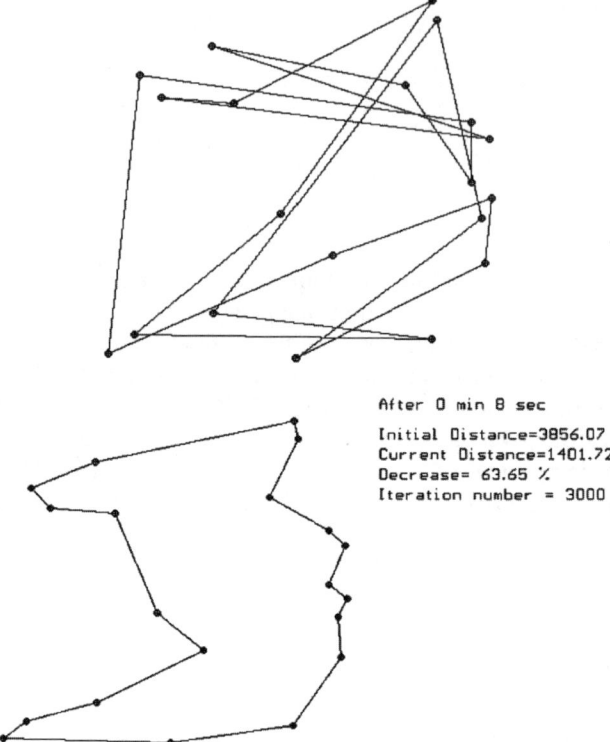

Fig. 3.31 Solution for the traveling salesman problem with 20 cities

Table 3.4 Genetic algorithm approach

Number of cities	Genetic algorithm
5	250
8	500
10	600
12	1100
15	1500
18	2800
20	10000
25	30000
30	100000

Review Questions

1. Mention the key elements of genetic algorithm.
2. Define an individual.
3. Differentiate between phenotype and genotype.
4. What does a gene mean?
5. Define population and fitness.
6. List a few search strategies.

Exercise Problems

7. Write note on the types of encoding techniques.
8. State the importance of breeding cycle.
9. Discuss in detail about the selection process of genetic algorithm.
10. How is crossover operation performed?
11. Give examples to illustrate various crossover techniques.
12. Mention the different types of mutation process.
13. Write short note on replacement cycle of breeding process.
14. How do genetic algorithms work? Explain the building block hypothesis and schema theorem.
15. State the importance of No Free Lunch theorem.
16. Compare and contrast: constrained and unconstrained optimization problem.
17. What is penalty method of transforming constrained optimization problems to unconstrained optimization problems?
18. Define schemata.
19. Differentiate between Roulette wheel selection and tournament selection.
20. List few termination search condition of genetic algorithm.

Exercise Problems

1. Simulate a Genetic Algorithm to minimize a function,

$$F(x) = x^2 + y^2 \qquad (3.21)$$

   where $1 \leq x \leq 15$ and $y \geq 3$ with $x + y = 7$
2. Five strings have the following fitness values: 3,6,9,12,15. Under Roulette wheel selection, compute the expected number of copies of each string in the mating pool if a constant population size, n=5, is maintained.
3. Find the safe light combinations for 8 traffic lights, four of which are vehicle lights having four possible colors (red, yellow/red, yellow and green) and the other four pedestrian lights having only two colors (red and green).
4. Use genetic algorithm to color the nodes of a graph using these colors in such a way that no two nodes connected by an edge are colored using the same color (Fig. 3.32).
5. Consider the strings and schemata of length 11. For the following schemata, calculate the probability of surviving mutation if the probability of mutation is 0.001 at a single bit position: **100****10, 0**********1, 11***00***1, *1111*0000*. Recalculate the survival probabilities for a mutation probability $P_m = 0.1$.

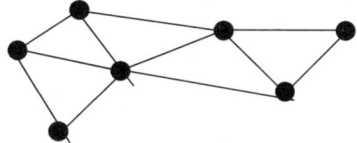

Fig. 3.32 Graph 3 coloring

Chapter 4
Advanced Operators and Techniques in Genetic Algorithm

4.1 Introduction

In the previous chapter we have dealt with simple genetic operators: reproduction, cross over and mutation. In this chapter we consider natural operators and phenomena to improve the robustness of simple genetic algorithms. The low-level operators like dominance, inversion, recording, deletion, segregation and diploidy are discussed here. Also, the higher-level operators like niche and speciation are induced. Multi-objective optimization and knowledge-based techniques are also considered for discussion in this chapter.

4.2 Diploidy, Dominance and Abeyance

Till now, we have considered only the simplest genotype existing in the nature, the haploid or single chromosome. A haploid chromosome contains only one set of genes i.e., one allele to occupy each locus. Nature consists of many haploid organisms, but most of them tend to uncomplicated life form. When nature wants to construct a more complex or animal life to rely upon, a more complex underlying chromosome structure is needed and this is achieved by the diploid or double-stranded chromosomes. In the diploid form, a genotype carries one or more pairs of chromosomes, each containing information for the same function. Consider a diploid chromosome structure where different letters represent different alleles (different gene function values):

$$P\ q\ r\ s\ t$$
$$P\ Q\ R\ S\ t$$

Allele represents the property of a particular gene. Each locus of a letter represents one allele. The uppercase and the lowercase letters mentioned above represent the alternative alleles at that position. Originally, in nature each allele may represent different phenotypic properties. For example Q may represent gray haired gene and q may be black haired gene. This approach even though is not much varied from haploid scheme, one difference is vivid. At present, a pair of genes exists describing each function; there should be some aspect to decide which of the two values to choose because, for example, the phenotype may not have both gray haired and black haired at the same time.

The basic approach for eliminating this conflict of redundancy is through a genetic called dominance. At a locus, it has been noted that one allele (called the dominant allele) will take the precedence over the other alternative allele (called the recessive allele). It can be said, as an allele is dominant if it is expressed when paired with some other allele. Expressing means its occurrence in the phenotype.

In the above example, if it is assumed that all uppercase letters are dominant and all lowercase letters are recessive, the phenotype expressed by the example chromosome is written as,

$$\begin{matrix} P\,q\,r\,S\,t \\ \\ p\,Q\,R\,S\,t \end{matrix} \rightarrow P\,Q\,R\,S\,t$$

The dominant gene is always expressed at each locus and the recessive gene is only expressed when it is present in accordance with another recessive. The dominant gene is expressed when heterozygote (mixed, Pp→P) or homozygote (SS→S) and the recessive genes expressed only when homozygote (tt→ t).

After dominance cross over, the chromosome becomes PQRSt. Thus dominance is genetic operator used to compute phenotype of the allele values possible at a gene position, one is dominant and the others recessive provides a mechanism to remember previously useful genetic material and to protect it from disappearing it is held in abeyance (suspended) has not been found particularly useful in genetic algorithms speculation. It may help if the environment changes over time i.e., the fitness changes or the constraint change.

Dominance and diploidy can be simply implemented in the genetic algorithm. Assume 3 alleles,

* 0: encodes for gene 0
* 1: recessively encodes for gene 1
* 2: dominantly encodes for gene 1.

The above give the following dominance map (Fig. 4.1).

The phenotypic ratio for dominants of 3 to 1 is achieved for allele pairing of 0 with 1, and 0 with 2. Dominance can evolve through allele substitution of 1 for 2 and vice versa.

	0	1	2
0	0	0	1
1	0	1	1
2	1	1	1

Fig. 4.1 Dominance map

Diploid chromosomes lend advantages to individuals where the environment may change over a period of time. Having two genes allows two different solutions to be remembered, and passed on to offspring. One of these will be dominant (i.e., it will be expressed in the phenotype), while the other will be recessive. If environmental conditions change, the dominance can shift, so that the other gene is dominant. This shift can take place much more quickly than would be possible if evolutionary mechanisms had to alter the gene. This mechanism is ideal if the environment regularly switches between two states (e.g., ice-age, non ice-age). The primary advantage of diploidy is that it allows a wider diversity of alleles to be kept in the population, compared with haploidy. Currently harmful, but potentially useful alleles can still be maintained, but in a recessive position. Other genetic mechanisms could achieve the same effect. For example, a chromosome might contain several variants of a gene. Epistasis (in the sense of masking) could be used to ensure that only one of the variants was expressed in any particular individual. A situation like this occurs with haemoglobin production. Different genes code for its production during different stages of development. During the foetal stage, one gene is switched on to produce haemoglobin, whilst later on a different gene is activated. There are a variety of biological metaphors we can use to inspire our development of GAs.

In a GA, diploidy might be useful in an on-line application where the system could switch between different states. Diploidy involves a significant overhead in a GA. As well as carrying twice as much genetic information, the chromosome must also carry dominance information.

4.3 Multiploid

A multiploid genetic algorithm incorporates several candidates for each gene within a single genotype, and uses some form of dominance mechanism to decide which choice of each gene is active in the phenotype. In nature we find that many organisms have poly-ploid genotypes, which consist of multiple sets of chromosomes with some mechanism for determining which gene is expressed i.e., is dominant at each locus. This mechanism seems to confer a number of advantages on a system, mainly by enhancing population diversity; currently unused genes remains in a multiplied genotype, unexpressed, but shielded from extinction until they may become useful later.

A multiploid genotype, shown in Fig. 4.2, contains p chromosomes, each of length L, and a mask which specifies which of the p chromosomes has the dominant gene at a particular position in the chromosome. This information is decoded to yield the phenotype as follows:

Mask	0	0	0	1	1	1	2	2	2
Chromosome [0]:	a	a	a	a	A	a	a	a	a
Chromosome [1]:	b	b	b	b	B	b	b	b	b
Chromosome [2]:	c	c	c	c	C	c	c	c	c
Phenotype:	a	a	a	b	B	b	c	c	c

Fig. 4.2 Multiploid Type 1

Fig. 4.3 Multiploid Type 2

```
Mask             0  1  2
Chromosome [0]:  a  a  a  a  a  a  a  a
Chromosome [1]:  b  b  b  b  b  b  b  b
Chromosome [2]:  c  c  c  c  c  c  c  c
Phenotype:       a  a  a  b  b  b  c  c  c
```

An allele value of *a* at locus *i* in the mask denotes that the *ith* gene in the chromosome with index a becomes the ith gene of the phenotype. The mask length can be shorter than the length of the chromosomes, as in Fig. 4.3.

In Fig. 4.3, if the mask length is m and the chromosome length L, then a gene at locus 'i' in the mask with the value of 'a' indicates that the *i-th* set of L/m consecutives genes in the *a-th* chromosome are dominant.

4.4 Inversion and Reordering

Inversion is a unary, reordering genetic operator. Simple genetic algorithms use stochastic selection, 1-point cross over, and mutation to increase the number of building blocks in the population and to recombine them for even better building blocks. The building blocks being highly fit, low order, short defining length schemes, the encoding scheme chosen must be compatible with this. Can we search for better encoding schemes while searching for building blocks? To answer this question, inversion operator was created.

Inversion operator is a primary natural mechanism to recode a problem. In inversion operator, two points are selected along the length of the chromosome, the chromosome is cut at those points and the end points of the section cut, gets reversed (switched, swapped). To make it clear, consider a chromosome of length 8 where two inverse points are selected random (the points are 2 and 6 denoted by ^ character):

$$1 \quad 1_\wedge 0 \quad 1 \quad 1 \quad 1_\wedge 0 \quad 1$$

On using inversion operator, the string becomes,

$$1\,1\,\mathbf{1}\,\mathbf{1}\,\mathbf{1}\,\mathbf{0}\,0\,1$$

Thus within the specified inversion points, the switching between the chromosomes takes place.

The inversion operator can also be used for extended representation as given by,

$$\begin{array}{cccccccc} 1 & 2 & 3 & 4 & 5 & 6 & 7 & 8 \\ 1 & 1_\wedge 0 & 1 & 1 & 1_\wedge 0 & 1 \end{array}$$

Inversion points are chosen at random (indicated by operator) and the chromosome now becomes,

4.4 Inversion and Reordering

$$\begin{array}{cccccccc} 1 & 2 & 6 & 5 & 4 & 3 & 7 & 8 \\ 1 & 1 & 1 & 1 & 1 & 0 & 0 & 1 \end{array}$$

Original position numbers are retained in the chromosomes because the fitness function is computed using the position numbers, i.e., we are moving the relative position of the genes to examine different potential building blocks but the gene is to be used in the fitness function the same as before inversion and it remains same as after inversion operation also.

The hope is that inversion will reduce the defining length of highly fit schemas so they survive crossover more. Inversion is found to greatly expand the search space. Inversion has not appeared useful in practice, perhaps due to the test cases not being hard enough (conventional GA without inversion worked well).

Inversion was the solution of Holland, intended to bring co-adapted alleles at distant loci closer together on the chromosome. The biological interpretation of the inversion operator is that it maintains linkage disequilibria due to selection in the face of disruption by crossover. Two loci in a population are in linkage equilibrium if the frequency distribution of alleles at one locus is independent of the frequency distribution of alleles at the other locus.

The basic algorithm for inversion can be given as follows:

> Data: l- length of chromosome
> $i_1 \leftarrow$ random integer between 0 and l inclusive;
> $i_2 \leftarrow$ random integer $\neq i_1$ between 0 and ; inclusive;
> if $i_1 > i_2$ then
>   swap i_1 and i_2;
> end
> for I=i_1 to $[(i_1+i_2-1)/2]$ do
>   swap allele and index at locus I with allele and index at locus.
> i_1+i_2-1-I;
> end

Thus an index is needed for each locus to preserve the meaning of the locus independent of its position on the chromosome. Inversion is redundant with operators such as UX (Uniform Crossover), which do not have any positional bias.

There are several other reordering operators, which being variations on inversion. They are:

1. Linear inversion.
2. Liner + end inversion.
3. Continuous inversion.
4. Mass inversion.

Linear inversion is the inversion, which has been discussed earlier. Linear + end inversion is a linear inversion with a specified probability of (0.75). When linear inversion is not performed, end inversion would be done with equal probability

(0.125) at either the left or right end of the string. Linear + end inversion minimizes the property of linear inversion to disrupt the alleles present near the center of the string disproportionately to those of alleles present near the ends. In continuous inversion mode, inversion was applied with a specified inversion probability, P_i, to each new individual as and when it is created. In mass inversion mode no inversion took place until a new population was created, thereafter, one-half of the population is found to undergo identical inversion.

The features of inversion and crossover are combined together to produce a single operator, which lead to the development of other reordering operators. On combining inversion and crossover, the reordering operators formulated are:

1. Partially Matched Crossover (PMX).
2. Order Crossover (OX).
3. Cycle Crossover (CX).

4.4.1 Partially Matched Crossover (PMX)

In Partially Matched Crossover, two strings are aligned, and two crossover points are selected uniformly at random along the length of the strings. The two crossover points give a matching selection, which is used to affect a cross through position-by-position exchange operations.

Consider two strings:

Parent A	4	8	7	3	6	5	1	10	9	2
Parent B	3	1	4	2	7	9	10	8	6	5

Two crossover points were selected at random, and PMX proceeds by position wise exchanges. In-between the crossover points the genes get exchanged i.e., the 3 and the 2, the 6 and the 7, the 5 and the 9 exchange places. This is by mapping parent B to parent A. Now mapping parent A to parent B, the 7 and the 6, the 9 and the 5, the 2 and the 3 exchange places. Thus after PMX, the offspring produced as follows:

Child A	4	8	6	2	7	9	1	10	5	3
Child B	2	1	4	3	6	5	10	8	7	9

where each offspring contains ordering information partially determined by each of its parents. PMX can be applied to problems with permutation representation.

4.4.2 Order Crossover (OX)

The order crossover begins in a manner similar to PMX. But instead of using point-by-point exchanges as PMX does, order crossover applies sliding motion to fill the left out holes by transferring the mapped positions.

4.5 Niche and Speciation

Consider the parent chromosomes,

```
Parent A    4  8  7  3  6  5  1   10  9  2
Parent B    3  1  4  2  7  9  10   8  6  5
```

On mapping parent B with parent A, the places 3,6 and 5 are left with holes.

```
Child B    H  1  4  2  7  9  10  8  H  H
```

These holes are now filled with a sliding motion that starts with the second crossover point.

```
Child B    2  7  9  H  H  H  10  8  1  4
```

The holes are then filled with the matching section taken from the parent A. Thus performing this operation, the offspring produced using order crossover is as given below.

```
Child A    3  6  5 | 2  7  9 | 1   10  4  8
Child B    2  7  9 | 3  6  5 | 10   8  1  4
```

From the examples, it can be noted that PMX tends to respect absolute positions while OX tends to respect relative positions.

4.4.3 Cycle Crossover (CX)

Cycle Crossover is different from PMX and OX. Cycle performs recombination under the constraint that each gene comes from the parent or the other.

4.5 Niche and Speciation

The perennial problem with Genetic Algorithm is that of premature convergence, that is, a non-optimal genotype taking over a population resulting in every individual being either identical or extremely alike, the consequences of which is a population which does not contain sufficient genetic diversity to evolve further.

Simply increasing the population size may not be enough to avoid the problem, while any increase in population size will incur the twofold cost of both extra computation time and more generations to converge on an optimal solution.

Genetic Algorithm then, faces a difficult problem. How can a population be encouraged to converge on a solution while still maintaining diversity? Clearly, those operators, which cause convergence, i.e. crossover and reproduction, must be altered somehow.

Another problem often used as a criticism against Genetic Algorithm is the time involved in deriving a solution. Unlike more deterministic methods such as Neural Networks, hill climbing, rules-based methods etc., Genetic Algorithms contain a large degree of randomness and no guarantee to converge on a solution within a fixed time. It is not unusual for a large proportion of runs not to find an optimal solution.

Fortunately, due to their very nature, Genetic Algorithms are inherently parallel, i.e. individuals can be evaluated in parallel as their performance rarely, if ever, affects that of other individuals. The reproduction phase, however, which commonly involves a sexual free-for-all, during which the individuals of a population dart about in a crazed frenzy vying with one another in attempts to mate as often as possible, represents a serious bottleneck in traditional Genetic Algorithms. The fitness of every individual must be known, and, despite the overwhelming and no doubt impatient ardour present in the population, only one reproduction/crossover may take place at a time. A method that could avoid this sort of bottleneck would lend itself very well to implementation on parallel machines, and hence speed up the whole process.

It is shown in this section how niches are used to solve multimodal and unimodal problems.

Generally speaking the methods adopted to solve the above-discussed problems permit the evolution of individuals, which fill differing environmental niches, with similar individuals congregating together. The correct biological name for such groups in an ecotype, but they tend to be referred to as different species in Genetic Algorithms. The use of the word "species" is not strictly correct, as individuals of each species may freely mate with each other. However, for the sake of consistency, the word species will be used with the meaning usually attributed to it in Genetic Algorithms. In this field, a niche usually refers to that which makes a particular group unique, e.g. having a common fitness rate, genotype etc., while species refers to the individuals in that group.

4.5.1 Niche and Speciation in Multimodal Problems

Suppose a fitness function is multimodal i.e. several peaks a GA will tend to converge to one of the peaks, particularly if one of the peaks is more fit than the others.

Perhaps, one likes to identify the peaks convergence to several peaks simultaneously. A 'niche' can be thought of as one of the peaks and a 'species' is a collection of population members well suited for a particular niche.

We might want a GA to create stable subpopulations (species) that are well suited to the niches. Consider an example of two-armed slot machine, if one person plays a two-armed slot machine, he will try each arm for a while to see which has a bigger payoff, and then play that arm for the rest of time. In this case, it is possible to derive a formula for how many to play on each arm before choosing which arm to play the rest of the time for the optimal overall payoff.

Now suppose that a group of people are all playing the same two-armed slot

4.5 Niche and Speciation

machine and that the group playing arm, one has to share their winnings and likewise for arm too. The players are allowed to switch groups to increase their share. If M people are playing totally,

$$M = M_1 + M_2$$

$$f_i = \text{arm i payoff}$$

and

Then the people forming groups in each arm is given by,

$$f_1/M_1 = f_2/M_2$$

Something similar to this can be implemented in GA during 'crowding' and 'sharing' technique as discussed below.

Genetic Algorithms using panmictic applied to multimodal functions face two main problems, the first being that of distributing individuals evenly across all peaks in the solution as in Fig. 4.4.

Each peak in the solution landscape can be viewed as a separate environmental niche. A successful application of a genetic algorithm to a problem like this would have to result in several individuals spread across each of the environmental niches. What better way to do this than to permit the evolution of several species within the environment, each specializing in its own particular niche?

Using niches and species in applications such as these can lend a more biological meaning to the word species. A second problem now arises as parents of differing species, i.e. from different environmental niches tend to produce unviable children and so, parents occupying different niches must be discouraged from mating. Figure 4.5 below illustrates the problem.

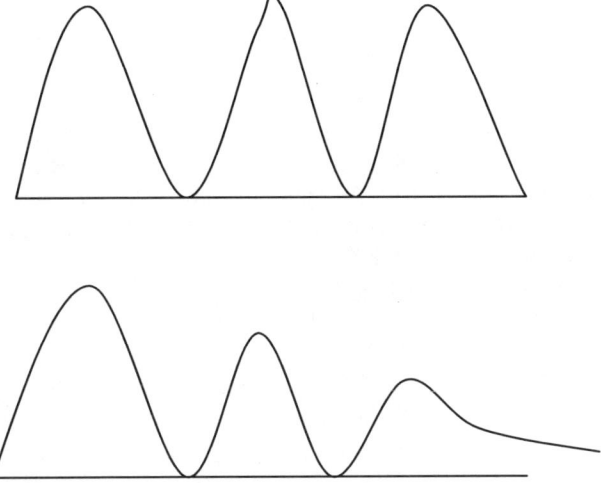

Fig. 4.4 A multimodal function solution landscape

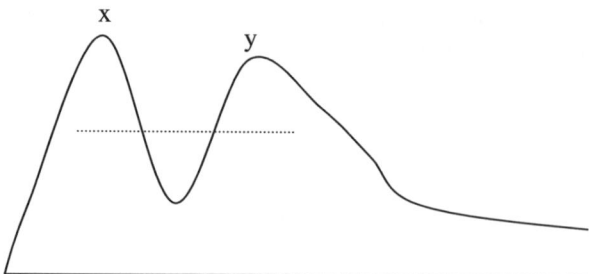

Fig. 4.5 A possible result of mating parents from different peaks is this case, where the resulting offspring appears in a trough between its parents

4.5.1.1 Crowding

The first problem was addressed by [DeJong 75] to prevent a single genotype from dominating a population. By ensuring that a newborn individual will replace one that it is genotypic ally similar to it, Crowding tries to maintain a balanced population. The mechanism for this replacement is quite simple: one randomly selects CF individuals and, through calculation of hamming distances, decides on a suitable victim.

An advantage of Crowding is how few individuals must be examined when choosing a victim, for CF is usually 2 or 3. This replacement of similar individuals' acts to prevent one genotype from taking over a population completely, allowing other, possibly less fit niches to form within the main population.

Crowding does not explicitly create niches, nor does it make any concentrated effort to encourage them, rather it allows them to form.

4.5.1.2 Sharing

A somewhat different approach was adopted in the Sharing scheme, in that individuals in a population which uses Sharing face limited resources as they strive for fitness. To make life more difficult for them, individuals of the same environmental niche, in this case genotypic ally similar, are more inclined to search the same places for resources, so have a more difficult time than unique individuals.

In a similar manner to Crowding, domination of the population by a single genotype is discouraged by the punishing of individuals who are too similar to a large portion of the population.

Sharing, however, is not as simple to calculate as Crowding, and is a very problem-specific as one must know in advance how many peaks there are in the solution landscape. Sharing does encourage the formation of niches and, to prevent the unsavoury prospect of individuals from different niches, mating as in Fig. 4.5 above, uses a form of restricted mating.

Defining a sharing function based on similarity,

>Share (similarity)
>Similarity (x,x)=1 chrom,
>Similarity (x,~x) = 0

(or) base similarity on phenotype (coded value) rather than genotype.

>Share (similarity (x,x))=1.0,
>Share (similarity (x,y))=0.0

$\forall$x,y, with similarity (x,y) $\Leftarrow$cut off value.

The more similar two chromosomes are, the more they have to share their fitness value. Each chromosome x has a share factor calculated for it,

>Share factor (x) = Sum over y in population share (similarity (x,y))

The fitness of a chromosome is recalculated by dividing its original fitness by its share factor,

>New fitness (x) = fitness (x)/share factor (x)

In a sense, similar chromosomes are playing the same arm of the k-armed slot machine so we force them to share the payoff and other subpopulations (species) can develop around other peaks (niches).

Although Crowding is far simpler than Sharing, both in its calculation and execution, the latter has been shown to be far more effective in the area of multimodal functions as the rather gentle powers of persuasion used by Crowding cannot prevent most individuals from ending up on only one or two peaks due to the few individuals that are examined each time. Sharing, on the other hand, aggressively encourages the development of new niches and consequently distributes individuals across all peaks in the landscape. The payoff is a simple one; Crowding is cheap and simple, while Sharing is relatively expensive yet successful.

4.5.2 *Niche and Speciation in Unimodal Problems*

The use of niches in multimodal problems is a very simple mapping, with the evolution of a different species for each peak in the solution landscape. The mapping is not quite so easy in unimodal problems which usually contain only one peak, or at the very least contain one peak higher than the others.

All approaches to unimodal problems, involving niching, attempt to maintain a balanced population, either through restricted mating to prevent inappropriate parents mating, or through replacement methods which hinder the taking over of a population by a single genotype.

Several methods, which do not strictly use niches, but which do imitate their operation to some degree, exist. There are replacement methods, which ensure that newly born individuals are sufficiently different from the rest of the population before allowing them entry, e.g. Clone Prevention, Steady State Genetic Algorithms (SSGA). There is also special selection schemes, which operate in a similar manner to the restricted mating, described above, in that prospective parents are allowed to fulfill their conjugal rites only if they fulfill certain criteria. Restricted mating does permit the evolution of different niches, but typically forces parents to come from differing niches, thus allowing each niche to exert some influence on evolution.

Clone prevention and SSGA operate in a similar manner to Crowding in that before an individual is permitted entry to the population, it is compared to others to verify its uniqueness. While Crowding examines only a few individuals each time, these other methods guarantee uniqueness by comparing a new individual to every other individual in the population. Like Crowding, they do not attempt to explicitly create niches or species, but attempt to prevent the domination of the population by a single species.

Unfortunately, both methods incur high overheads, as the comparing of individuals is a costly affair. It is also fair to say that clones, the presence of which can retard evolution, do not always cause disaster, and in face can sometimes even help direct evolution.

4.5.2.1 Incest Prevention

Eshelman took a similar view, when he suggested the use of Incest Prevention, which only discouraged clones, still permitting them to enter the population. Incest prevention attempts to "matchmake" parents with the intention of their offspring taking the best genes from their parents. It is by mating differing parents that diversity is kept in the population and thus further evolution permitted.

As a population evolves, its individuals become more and more similar, thus it becomes more difficult to find suitable parents. To avoid a situation where there are no such parents in a population, there is a difference threshold set, which can be relaxed if there is some difficulty in selecting parents. It is assumed that difficulty will arise if there is no change in the parent population, and, as incest prevention is used with elitism, i.e. a list of parents is maintained which individuals can only enter if their fitness is sufficiently high, it is a trivial matter to track any changes.

Again, differing species are not explicitly created, nor are guaranteed to appear, but if they do, Incest prevention encourages inter-species mating, as the fitness landscape in unimodal functions tends to be like that of Fig. 4.4.

```
SET THRESHOLD
REPEAT
FOR EACH INDIVIDUAL DO
TEST INDIVIDUAL
ENTER PARENT POPULATION
```

4.5 Niche and Speciation

```
IF NO-NEW-PARENTS THEN LOWER THRESHOLD
FOR EACH INDIVIDUAL DO
   REPEAT
      SELECT PARENTS
   UNTIL DIFFERENT ( )
UNTIL END-CRITERION REACHED OR THRESHOLD=0
```

As soon as an individual is tested it attempts to enter the parent population, as described above, this step is only successful if the individual is fitter than the least fit member of the parent population. After all the new individuals have been tested, one checks to see if the parent population has been changed. An unaltered population will lead to the difference threshold being reduced.

The DIFFERENT ( ) test simply calculates the hamming distance between parents and ensures that, if they are to breed, the difference will be above the threshold. As can be seen from the last step in the algorithm, it is possible for a run to end for reasons other than the reaching of some end criterion. In this case it is common to "reinitialize" the population by mutation of the best performing individuals encountered so far.

4.5.2.2 The Pygmy Algorithm

Although incest prevention avoids the cost of clone prevention, there is still the cost of finding a satisfactory couple each time mating is to be performed. To reduce the cost as much as possible another method, the Pygmy Algorithm has been suggested, which does not explicitly measure differences between parents, but merely suggests that the parents it selects are different.

The Pygmy Algorithm is typically used on problems with two or more requirements, e.g. the evolution of solutions which need to be both efficient and short. Niches are used by having two separate fitness functions, thus creating two species. Individuals from each species are then looked upon as being of distinct genders, and when parents are being chosen for the creation of a new individual, one is drawn from each species with the intention of each parent exerting pressure from its fitness function.

Typically, there is one main fitness function, say efficiency, and a secondary requirement such as shortness. Highly efficient individuals would then enter the first niche, while individuals who are not suited to this niche undergo a second fitness test, which is simply their original fitness function modified to include the secondary requirement. These individuals then attempt to join the second niche, and failure to accomplish this result in a premature death for the individual concerned.

The use of two niches maintains a balanced population and ensures that individuals who are fit in both requirements are produced. Below is the pseudo code for the Pygmy Algorithm.

```
REPEAT
FOR EACH INDIVIDUAL DO
   TEST INDIVIDUAL WITH MAIN FITNESS FUNCTION
```

ENTER PARENT POPULATION #1
IF UNSUCCESSFUL
TEST INDIVIDUAL WITH SECONDARY FITNESS FUNCTION
ENTER PARENT POPULATION #2
FOR EACH INDIVIDUAL DO
SELECT PARENT FROM POPULATION #1
SELECT PARENT FROM POPULATION #2
CREATE NEW INDIVIDUAL
UNTIL END-CRITERION REACHED

Each niche is implemented as a separate, elitist group, because of the elitist nature of each niche, which maintains individuals on a solution landscape similar to Fig. 4.4, there is much pressure on newly born individuals to appear between its parents, and thus outperform them. It is also possible, of course, that a child may be endowed with the worst characteristics of its parent. A child like this will be cast aside by the Pygmy Algorithm but its parents, because they have the potential to produce good children are maintained, outliving their luckless offspring.

As Genetic Algorithms stem directly from natural methods, it is perhaps unsurprising that there are so many benefits to be derived from copying nature once more. Differing niches and species can be evolved and maintained in a number of ways, ranging from decentralized models as close as possible to nature, to highly controlled methods.

Most importantly, once subpopulations have established their environmental niches, they can be put to many uses. Several solutions can be maintained in the population at a time, a diverse array of individuals and, indeed species can easily be persuaded to coexist with one another, thus easing the pressure toward premature convergence.

4.5.3 Restricted Mating

The purpose of restricted mating is to encourage speciation, and reduce the production of lethals. A lethal is a child of parents from two different niches. Although each parent may be highly fit, the combination of their chromosomes may be highly unfit if it falls in the valley between the two maxima. Nature avoids the formation of lethals by preventing mating between different species, using a variety of techniques. (In fact, this is the primary biological definition of a species–a set of individuals that may breed together to produce viable offspring.)

The general philosophy of restricted mating makes the assumption that if two similar parents (i.e., from the same niche) are mated, then the offspring will be similar. However, this will very much depend on the coding scheme-in particular the existence of building blocks, and low epistasis. Under conventional crossover and mutation operators, two parents with similar genotypes will always produce offspring with similar genotypes. But in a highly epistatic chromosome, there is no guarantee that these offspring will not be of low fitness, i.e. lethals. Similarity of genotype does not guarantee similarity of phenotype. These effects limit the use of restricted mating.

4.6 Few Micro-operators

Several other low level micro operators have been proposed for use in genetic algorithm search. The few micro operators to be discussed in this section are as follows:

1. Segregation.
2. Translocation.
3. Duplication.
4. Deletion.
5. Sexual differentiation.

4.6.1 Segregation and Translocation

Consider a process of gamete formation when there is more than one chromosome pair in the genotype. Crossover takes place as dealt in previous chapter and when it is to form a gamete, we randomly select one of each of the haploid chromosomes. This random selection process is called as segregation that disrupts any linking, which might exist between genes on different chromosomes. It is found that segregation exploits the proper organization of the chromosome and it is important to note that how does the chromosome become organized in an appropriate manner. For this purpose, translocation operator is used. Translocation operator can be considered as an interchromosomal crossover operator. This operator can be implemented by connecting alleles with their gene names, so that one can identify their intended meaning when they get shuffled from chromosome to chromosome by the translocation operator.

4.6.2 Duplication and Deletion

There are also a pair of low-level operators for performing genetic algorithm search. Intrachromosomal duplication performs by duplicating a particular gene and placing it along with its ancestor on the chromosome. Deletion performs by removing a duplicate gene from chromosome. The mutation rate can effectively controlled by these operators. When the mutation rate remains constant and intrachromosomal duplication causes 'k' copies of a particular gene, then the effective mutation probability for this gene is multiplied by 'k'. On the other hand, when deletion occurs, the effective mutation rate gets decreased.

4.6.3 Sexual Determination

Originally in mating schemes, we have permitted any individual to mate with any other, and the resulting genetic products are divided so that they have ensured a viable genotype. The sex determination is handled differently in different species, but the human example is sufficient to understand sexual determination. Sex is determined in human by one of the 23 pairs of human chromosomes. Females have two same X and X chromosomes and males have two different X and Y chromosomes. During gametogenesis process, males form sperm (which carry either X or Y chromosomes) and female possess eggs (which carry only X chromosomes). On fertilization, X-chromosome produced by female combined with either X or Y-chromosome produced by females. Thus the method of sex determination in human is simple. Applying the same strategy for sex determination in GA search. The establishment of sex difference effectively divides a species into two or more groups. This allows males and females to specialize, thereby enclosing the range of behaviors necessary for survival more broadly than would be with a single competing population. The research under sex determination and difference to artificial genetic algorithm search is still ongoing.

4.7 Non-binary Representation

A chromosome is a sequence of symbols, and, traditionally, these symbols have been binary digits, so that each symbol has a cardinality of 2. Higher cardinality alphabets have been used in some research, and some believe them to have advantages. Goldberg argues that theoretically, a binary representation gives the largest number of schemata, and so provides the highest degree of implicit parallelism. But Antonisse interprets schemata differently, and concludes that, on the contrary, high-cardinality alphabets contain more schemata than binary ones.

Goldberg has now developed a theory, which explains why high-cardinality representations can perform well. His theory of virtual alphabets says that each symbol converges within the first few generations, leaving only a small number of possible values. In this way, each symbol effectively has only a low cardinality. Empirical studies of high-cardinality alphabets have typically used chromosomes where each symbol represents an integer, or a floating-point number. As Davis points out, problem parameters are often numeric, so representing them directly as numbers, rather than bit-strings, seems obvious, and may have advantages. One advantage is that we can more easily define meaningful, problem-specific crossover and mutation operators. A variety of real-number operators can easily be envisaged, for example:

1. Combination operators

    - Average–take the arithmetic average of the two parent genes.
    - Geometric mean-take the square root of the product of the two values.

- Extension-take the difference between the two values, and add it to the higher, or subtract it from the lower.

2. Mutation operators

- Random replacement-replace the value with a random one
- Creep-add or subtract a small, randomly generated amount.
- Geometric creep-multiply by a random amount close to one.

For both creep operators, the randomly generated number may have a variety of distributions; uniform within a given range, exponential, Gaussian, binomial, etc. Janikow & Michalewicz made a direct comparison between binary and floating-point representations, and found that the floating-point version gave faster, more consistent, and more accurate results.

4.8 Multi-Objective Optimization

Multi-objective optimization problems have received interest form researches since early 1960s. In a multi-objective optimization problem, multiple objective functions need to be optimized simultaneously. In the case of multiple objectives, there does not necessarily exist a solution that is best with respect to all objectives because of differentiation between objectives. A solution may be best in one objective but worst in another. Therefore, there usually exist a set of solutions for the multiple-objective case, which cannot simply be compared with each other. For such solutions, called Pareto optimal solutions or non-dominated solutions, no improvement is possible in any objective function without sacrificing at least one of the other objective functions.

Thus by using the concept of Pareto-optimality we can find a set of solutions that are all optimal compromises between the conflicting objectives. Pareto-optimality is a concept used economics, game theory, etc. A Pareto-optimal solution is one that is not dominated by any other solution i.e. it is one in which no objective can be improved without a deterioration in one or more of the other objectives.

In the past few years, there has been a wide development in applying genetic algorithms to solve the multi-objective optimization problem, known as evolutionary multi-objective optimization or genetic multi-objective optimization. The basic features of genetic algorithms are the multiple directional and global searches, in which a population of potential solutions is maintained from generation to generation. The population-to-population approach is beneficial in the exploration of Pareto-optimal solutions. The main issue in solving multi-objective optimization problems by use of genetic algorithms is how to determine the fitness value of individuals according to multiple objectives.

4.9 Combinatorial Optimizations

Combinatorial optimizations contain a huge body of problems with different features and properties. Although these problems are quite different from each other, the problems can be characterized as one of the following types:

- To determine a permutation of some items associated with the problem.
- To determine a combination of some items.
- To determine both permutation and combination of some items.
- Any one of the above subject to constraints.

The essence of resource-constrained project scheduling problems and vehicle routing and scheduling problems is to determine a permutation of some items subject to some constraints. The essence of the parallel machine-scheduling problem is to determine both a permutation and a combination of items subject to certain constraints. A common feature of the problems is that if the permutation and/or combination can be determined, a solution can then easily be derived with a problem-specific procedure. So the general approach for applying genetic algorithms to these problems is as follows:

- Use genetic algorithms to evolve an appropriate permutation and/or combination of items under consideration.
- Then use a heuristic approach to construct a solution according to the permutation and combination.

4.10 Knowledge Based Techniques

While most research has gone into GAs using the traditional crossover and mutation operators, some have advocated designing new operators for each task, using domain knowledge. This makes each GA more task specific (less robust), but may improve performance significantly. Where a GA is being designed to tackle a real-world problem, and has to compete with other search and optimization techniques, the incorporation of domain knowledge often makes sense. Few researchers argue that problem-specific knowledge can usefully be incorporated into the crossover operation. Domain knowledge may be used to prevent obviously unfit chromosomes, or those, which would violate problem constraints, from being produced in the first place. This avoids wasting time evaluating such individuals, and avoids introducing poor performers into the population.

For example, a researcher designed analogous crossover for his task in robotic trajectory generation. This used local information in the chromosome (i.e., the values of just a few genes) to decide which crossover sites would be certain to yield unfit offspring. Domain knowledge can also be used to design local improvement operators, which allow more efficient exploration of the search space around good

4.10 Knowledge Based Techniques

points. It can also be used to perform heuristic initialization of the population, so that search begins with some reasonably good points, rather than a random set. Goldberg describes techniques for adding knowledge-directed crossover and mutation. He also discusses the hybridization of GAs with other search techniques. Pure genetic algorithms use only the encoding and objective function. This may help to use in problem specific information. The various methods for combining problem specific information with genetic algorithm are as follows:

- Hybrid schemes
- Knowledge directed operators
- Parallel computers.

In hybrid schemes GAs are used to get close to optimum value, then conventional optimization schemes like greedy search, gradient search or stochastic hill climbing may be used to become closer to optimum value. The genetic algorithm may also develop species to each of which conventional optimization can be applied. In gradient like bitwise (G-bit) improvement for one or more highly fit chromosomes, change each bit one at a time to see if the fitness improves, if so, replace the original with the altered chromosome. Also changing pairs or triplets of bits can be tried but in combinatorial explosion. The hybrid scheme can be represented using scheme as shown in Fig. 4.6.

Thus from Fig. 4.6, it can be noted that the genetic algorithm sorts out peak and the local search techniques are used for hill climbing. Considering greedy heuristic crossover for Traveling salesman problem, if chromosomes are permutations of city numbers, then normal crossover may produce infeasible chromosomes. This is done by,

Start at a random city X and go to the closest city to X using the parent's tours; repeat.

Thus the genetic algorithm is now definitely not blind as it was using PMX described easier. The knowledge directed operations were found to use,

- generating Steiner systems
- count preserving mutation and crossover
- goal directed mutation

Using parallel computers in Genetic Algorithms, master/slave operation is performed. Master does selection and mating and slaves evaluated fitness of new chromosomes. Master waits for all the slaves to finish or master can hand out new work as each slave finishes. Thus on a parallel machine the conventional optimization can be done on each species on its own CPU. This is shown in Fig. 4.7.

These knowledge based techniques flourishes with the development of hardware and software.

Fig. 4.6 Genetic Algorithm hybrid scheme

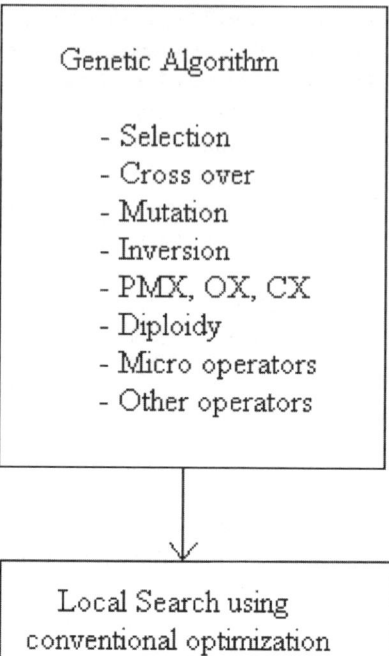

4.11 Summary

In this chapter we have discussed certain advanced operators and techniques available for improving the performance of genetic algorithms. The various genetic operators operating at chromosomal level has been discussed. Diploidy, dominance and abeyance have been dealt as a method involved with long-term population memory. The implementation of different reordering operators has been included in the chapter. Other micro-level operators like segregation and translocation has been discussed in brief. Niche and speciation as applied to multimodal and unimodal problems has been examined through operators acting at population level. In niching, the crowding and sharing techniques has been viewed and the pygmy algorithm for speciation. Multiobjective optimization technique has also been explored in brief. The knowledge-based technique was discussed which is useful for exploiting many search and optimization problems. Thus these advanced operators and techniques leads to the further improvements in the efficiency and competence of genetic algorithms.

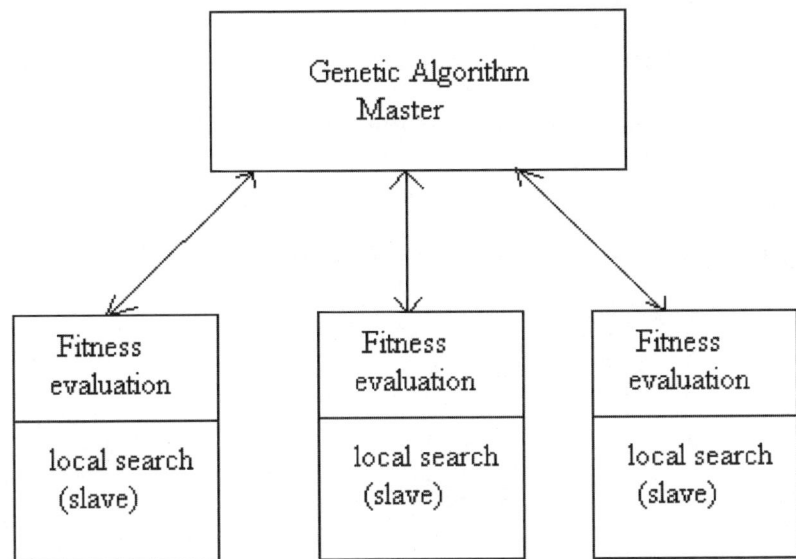

Fig. 4.7 Genetic algorithm hybrid using parallel machine

Review Questions

1. State the importance of advanced operators.
2. Define Diploidy
3. Differentiate between haploid and diploid
4. What is a dominance operator?
5. Under what situation is abeyance operator used?
6. Write a note on multiploid.
7. Discuss in detail about various reordering operators.
8. Explain how niche and speciation are used in multimodal and unimodal problems.
9. Mention the operations of micro operators
10. Compare and contrast: Multiobjective optimization and combinatorial optimization.
11. Give the importance of non-binary representation
12. Describe the various knowledge-based techniques that improve the efficiency of simple genetic algorithm.

Exercise Problems

1. A given chromosome is (1 3 5 2 4 6 8 7 9 0). Perform inversion at points 3 and 5.
2. Consider two parents given by,
   Parent1 2 4 7 1 3 6 8 9 5

Parent2 5 9 8 6 2 4 1 3 7

Choose random points of your own and perform partially matched cross over operation.

3. Implement Traveling Salesman Problem using advanced operators and techniques.
4. Program and implement an inversion operator that treats permutation as a circular string.
5. Implement the order cross over operator for a permutation coding.

Chapter 5
Classification of Genetic Algorithm

5.1 Introduction

Genetic algorithms are search algorithms based on the mechanics of natural selection and natural genetics. Algorithms are nothing but step-by-step procedure to find the solution to the problems. Genetic algorithms also give the step-by-step procedure to solve the problem but they are based on the genetic models. Genetic algorithms are theoretically and empirically proven to provide robust search in complex phases with the above said features. Genetic algorithms are capable of giving rose to efficient and effective search in the problem domain and hence they are now finding more wide spread application in business, scientific and engineering. These algorithms are computationally less complex but more powerful in their search for improvement. These features have enabled the researchers to form different approaches of genetic algorithm. This chapter discusses the various classifications of genetic algorithms like parallel GA, Messy GA, distributed GA and so on.

5.2 Simple Genetic Algorithm (SGA)

Many search techniques required auxiliary information in order to work properly. For e.g. Gradient techniques need derivative in order to chain the current peak and other procedures like greedy technique requires access to most tabular parameters whereas genetic algorithms do not require all these auxiliary information. GA is blind to perform an effective search for better and better structures they only require objective function values associated with the individual strings. This characteristic makes GA a more suitable method than many search schemes. GA uses probabilistic transition rules to guide their search towards regions of search space with likely improvement. Because of these four important characteristics possessed by the GA it is more robust than other commonly used techniques.

The mechanics of simple genetic algorithms (SGA) are surprisingly simple involving nothing more complex than copying strings and swapping partial strings. A simple genetic algorithm that yields good results in many practical problems is composed of three operations. They are

1) Reproduction
2) Cross over and
3) Mutation

The reproduction is a process in which individual string are copied according to their objective function values, f. One can consider the function f as some measure of profit, utility or goodness that we want to maximize. Copying strings according to their fitness will result higher probability of contributing one of more off string in the next generation. This reproduction operator can be implemented in an algorithmic form in a number of ways. The simplest possible way of implementing the reproduction operator is using a biased Roulette Wheel. In this Roulette Wheel each current string in the population has a slot sized in proportional to its fitness. The more-or-less standard procedure for running the simple genetic algorithm is:

> randomly generate population
> select parents (using fitness function)
> > selection methods:
> > > roulette wheel
> > > tournament
> > > demetic
>
> crossover parent chromosomes
> mutate offspring chromosomes
> add offspring back into pool
> > elitism
>
> (select parents)

SGAs are useful and efficient when,

- The search space is large, complex or poorly understood.
- Domain knowledge is scarce or expert knowledge is difficult to encode to narrow the search space.
- No mathematical analysis is available.
- Traditional search methods fail.

The advantage of the SGA approach is the ease with which it can handle arbitrary kinds of constraints and objectives; all such things can be handled as weighted components of the fitness function, making it easy to adapt the SGA scheduler to the particular requirements of a very wide range of possible overall objectives.

5.3 Parallel and Distributed Genetic Algorithm (PGA and DGA)

Parallel execution of various SGAs is called PGA (Parallel Genetic Algorithm). It is used to solve Job shop scheduling problem by making use of various precedence constraints to achieve high optimization. Parallel Genetic Algorithms (PGAs) have been developed to reduce the large execution times that are associated with simple

5.3 Parallel and Distributed Genetic Algorithm (PGA and DGA)

genetic algorithms for finding near-optimal solutions in large search spaces. They have also been used to solve larger problems and to find better solutions. PGAs have considerable gains in terms of performance and scalability. PGAs can easily be implemented on networks of heterogeneous computers or on parallel mainframes. The way in which GAs can be parallelized depends upon the following elements:

- How fitness is evaluated and mutation is applied
- How selection is applied locally or globally
- If single or multiple subpopulations are used
- If multiple populations are used how individuals are exchanged

The simplest way of parallelizing a GA is to execute multiple copies of the same SGA, one on each Processing Element (PE). Each of the PEs starts with a different initial subpopulation, evolves and stops independently. The complete PGA halts when all PE stop. There are no inter-PE communications. The various methods of PGA are:

- Independent PGA
- Migration PGA
- Partition PGA
- Segmentation PGA
- Segmentation–Migration PGA

The advantage of independent PGA approach is that each PE starts with an independent subpopulation. Such subpopulation diversity reduces the chance that all PEs prematurely converge to the same poor quality solution. This approach is equivalent to simply taking the best solution after multiple executions of the SGA on different initial populations.

The second PGA approach is the migration PGA, augments the independent approach with periodic chromosome migrations among the PEs to prevent premature convergence and share high quality solutions. Chromosome migrations occur after certain iterations, with each PE sending a copy of its locally best chromosome to PE P_1 modulo N at the first migration step, then PE P_2 modulo N at the second migration step and so on. The chromosome received replaces the locally worst chromosome unless an identical chromosome already exists in the local population.

Partition PGA is to partition the search space into disjoint subspaces and to force PEs to search in different subspaces. The segmentation PGA starts by segmenting tours into sub tours. Then after sub tour improvements, they are recombined into longer sub tours. The combination of segmentation and migration is the segmentation –migration approach. Recombination occurs at the end of each phase, sub tours are contained by a group of PEs numbered in ascending order.

PGAs are implemented using the standard parallel approach and the decomposition approach. In the first approach, the sequential GA model is implemented on a parallel computer by dividing the task of implementation among the processors. In decomposition approach, the full population exists in distributed form. Either multiple independent or interacting subpopulation exists (coarse grained or distributed GA) or there is only one population with each population member interacting only with limited set of members (fine grained GA). The interactions between the populations or the members of the population, takes place with respect to a spatial structure of a problem. These models maintain more diverse subpopulations mitigating the problem of premature convergence. They also fit in the evolution model, with a large degree of independence in the subpopulation.

Standard parallel approach is also referred as global parallelization or distributed fitness evaluation. This approach uses a single population and the evaluations of the individuals are done in parallel. The selection and mating is done manually with any other. The most common parallelized operation is the evaluation of the fitness function as it requires only the knowledge of the individual being evaluated, hence no communications is needed. It is implemented using the master slave model (Fig. 5.1). The master stores the population and does the selection. The slaves evaluate the fitness and apply the genetic operators like crossover and mutation. Communication occurs only when slaves return the values to the master. It has two modes namely synchronous mode and asynchronous mode. In the synchronous mode, the master waits till it receives the fitness value for the entire population, before preceding to the next generation. On the contrary the master does not stop for any slow processors in the asynchronous. In a distributed memory computer, the master sends the individuals to the slave processors for fitness evaluation gather the results and apply the genetic operators to produce the new generation. The number of individuals assigned to any processor can be static or dynamic.

In decomposition approach, the population is divided into a number of subpopulations called demes. Demes are separated from one another and individuals compete only within a deme. An additional operator called migration is used to move the individuals from one deme to another. If the individuals can migrate to any other deme, the model is called island model. Migration can be controlled by various parameters like migration rate, topology, migration scheme like best/worst/random individuals to migrate and the frequency of migrations.

The other approaches are coarse grained and fine-grained parallel genetic algorithms. Coarse-grained PGA model refers to relatively small number of demes with many individuals. These models are characterized by the relatively long time required for processing a generation within each deme and by their occasional communication for exchanging individual. It is called as distributed GAs as it is usually implemented in distributed memory computers. In case of fine-grained parallel genetic approach, large numbers of processors are required because the population is divided into number of demes. Inter-deme communication is realized either by migration operator or by overlapping demes.

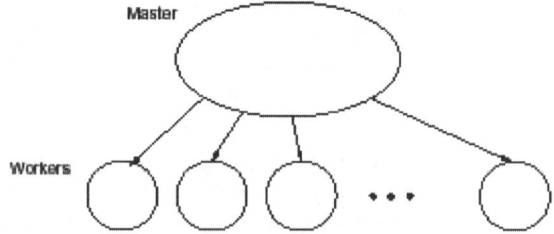

Fig. 5.1 A schematic of a master-slave parallel GA. The master stores the population, executes GA operations, and distributes individuals to the slaves. The slaves only evaluate the fitness of the individuals

5.3 Parallel and Distributed Genetic Algorithm (PGA and DGA)

It is important to emphasize that while the master-slave parallelization method does not affect the behavior of the algorithm, the fine and coarse-grained methods change the way the GA works. For example, in master-slave parallel GAs, selection takes into account all the population, but in the fine and coarse-grained parallel GAs, selection only considers a subset of individuals. Also, in the master-slave any two individuals in the population can mate (i.e., there is random mating), but in fine and coarse-grained methods mating is restricted to a subset of individuals.

5.3.1 Master-Slave Parallelization

This section reviews the master-slave (or global) parallelization method. The algorithm uses a single population and the evaluation of the individuals and/or the application of genetic operators are done in parallel. As in the serial GA, each individual may compete and mate with any other (thus selection and mating are global). Global parallel GAs are usually implemented as master-slave (Fig. 5.1) programs, where the master stores the population and the slaves evaluate the fitness. The most common operation that is parallelized is the evaluation of the individuals, because the fitness of an individual is independent from the rest of the population, and there is no need to communicate during this phase. The evaluation of individuals is parallelized by assigning a fraction of the population to each of the processors available. Communication occurs only as each slave receives its subset of individuals to evaluate and when the slaves return the fitness values. If the algorithm stops and waits to receive the fitness values for all the population before proceeding into the next generation, then the algorithm is synchronous. A synchronous master-slave GA has exactly the same properties as a simple GA, with speed being the only difference. However, it is also possible to implement an asynchronous master-slave GA where the algorithm does not stop to wait for any slow processors, but it does not work exactly like a simple GA. Most global parallel GA implementations are synchronous.

The global parallelization model does not assume anything about the underlying computer architecture, and it can be implemented efficiently on shared-memory and distributed-memory computers. On a shared-memory multiprocessor, the population could be stored in shared memory and each processor can read the individuals assigned to it and write the evaluation results back without any conflicts. On a distributed-memory computer, the population can be stored in one processor. This "master" processor would be responsible for explicitly sending the individuals to the other processors (the "slaves") for evaluation, collecting the results, and applying the genetic operators to produce the next generation. The number of individuals assigned to any processor may be constant, but in some cases (like in a multiuser environment where the utilization of processors is variable) it may be necessary to balance the computational load among the processors by using a dynamic scheduling algorithm (e.g., guided self-scheduling). The following is an informal description of the algorithm:

```
produce an initial population of individuals
for all individuals do in parallel
      evaluate the individual's fitness
end parallel for
while not termination condition do
      select fitter individuals for reproduction
      produce new individuals
      mutate some individuals
      for all individuals do in parallel
            evaluate the individual's fitness
      end parallel for
      generate a new population by inserting some new good individuals
      and by discarding some old bad individuals
end while
```

Master-slave parallel GAs are easy to implement and it can be a very efficient method of parallelization when the evaluation needs considerable computations. Besides, the method has the advantage of not altering the search behavior of the GA, so we can apply directly all the theory available for simple GAs.

5.3.2 Fine Grained Parallel GAs (Cellular GAs)

In the grid or fine-grained model individuals are placed on a large toroidal (the ends wrap around) one or two-dimensional grid, one individual per grid location. The model is also called cellular because of its similarity with cellular automata with stochastic transition rules. Fitness evaluation is done simultaneously for all individuals and selection, reproduction and mating takes place locally within a small neighborhood. In time, semi-isolated niches of genetically homogenous individuals emerge across the grid as a result of slow individual diffusion. This phenomenon is called isolation by distance and is due to the fact that the probability of interaction of two individuals is a fast decaying function of their distance (Fig. 5.2).

The following is the algorithmic description of the process:

```
for each cell j in the grid do in parallel
      generate a random individual j
end parallel for
while not termination condition do
      for each cell j do in parallel
            evaluate individual j
            select a neighboring individual k
            produce offspring from j and k
            assign one of the offspring to j
```

                     mutate j with probability *pmut*
        end parallel for
    end while

In the 1-D case a small number of cells on either side of the central one is taken into account. The selection of the individual in the neighborhood for mating with the central individual can be done in various ways. Tournament selection is commonly used since it matches nicely the spatial nature of the system. The tournament may be probabilistic as well, in which case the probability for an individual to win is generally proportional to its fitness. This makes use of the fully available parallelism and is probably more appropriate if the biological metaphor is to be followed.

5.3.3 Multiple-Deme Parallel GAs (Distributed GAs or Coarse Grained GAs)

Multiple-population (or multiple-deme) GAs are more sophisticated, as they consist on several subpopulations which exchange individuals occasionally (Fig. 5.3). This exchange of individuals is called migration, as discussed above; it is controlled by several parameters. Multiple-deme GAs is very popular, but also are the class of parallel GAs, which is most difficult to understand, because the effects of migration are not fully understood. Multiple-deme parallel GAs introduces fundamental changes in the operation of the GA and has a different behavior than simple GAs.

Multiple-deme parallel GAs is known with different names. Sometimes they are known as "distributed" GAs (DGA), because they are usually implemented on distributed memory MIMD (Multiple input Multiple Data) computers. Since the computation to communication ratio is usually high, they are occasionally called coarse-grained GAs. Finally, multiple-deme GAs resemble the "island model" in Population Genetics which considers relatively isolated demes, so the parallel GAs

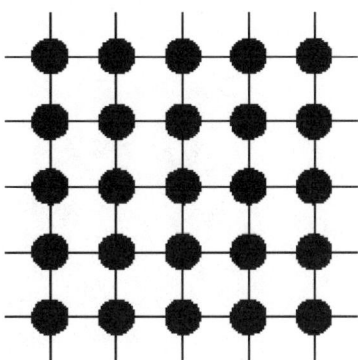

Fig. 5.2 A schematic of a fine-grained parallel GA. This class of parallel GAs has one spatially-distributed population, and it can be implemented very efficiently on massively parallel computers

Fig. 5.3 A schematic of a multiple-population parallel GA. Each process is a simple GA, and there is (infrequent) communication between the populations

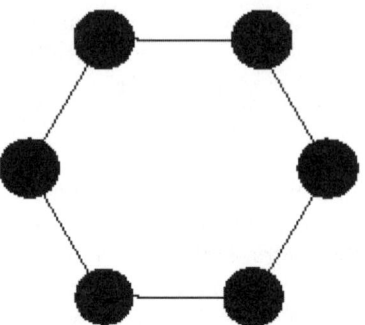

are also known as "island" parallel GAs. Since the size of the demes is smaller than the population used by a serial GA, we would expect that the parallel GA converge faster.

The important characteristics of multiple-deme parallel GAs are the use of a few relatively large subpopulations and migration. The island model features geographically separated subpopulations of relatively large size. Subpopulations may exchange information from time to time by allowing some individuals to migrate from one subpopulation to another according to various patterns. The main reason for this approach is to periodically reinject diversity into otherwise converging subpopulations. It is considered to some extent, different subpopulations will tend to explore different portions of the search space. When the migration takes place between nearest neighbor subpopulations the model is called stepping stone. Within each subpopulation a standard sequential genetic algorithm is executed between migration phases. Several migration topologies have been used: the ring structure, 2-d and 3-d meshes, hyper cubes and random graphs being the most common. The following is a algorithmic description of the process:

```
initialize P subpopulations of size N each
generation number = 1
    while not termination condition do
        for each subpopulation do in parallel
            evaluate and select individuals by fitness
            if generation number mod frequency = 0 then
                send K<N best individuals to
                a neighboring subpopulation
                receive K individuals from a
                neighboring population
                replace K individuals in the subpopulation
            endif
            produce new individuals
            mutate individuals
        end parallel for
        generation number = generation number +1
    end while
```

5.3 Parallel and Distributed Genetic Algorithm (PGA and DGA)

In the above algorithm frequency is the number of generations before an exchange takes place.

5.3.4 Hierarchical Parallel Algorithms

The final method to parallelize GAs combines multiple demes with master-slave or fine-grained GAs. We call this class of algorithms hierarchical parallel GAs, because at a higher level they are multiple-deme algorithms with single-population parallel GAs (either master-slave or fine-grained) at the lower level. A hierarchical parallel GAs combines the benefits of its components, and it promises better performance than any of them alone.

A few researchers have tried to combine two of the methods to parallelize GAs, producing hierarchical parallel GAs. Some of these new hybrid algorithms add a new degree of complexity to the already complicated scene of parallel GAs, but other hybrids manage to keep the same complexity as one of their components. When two methods of parallelizing GAs are combined they form a hierarchy. At the upper level most of the hybrid parallel GAs are multiple-population algorithms. Some hybrids have a fine-grained GA at the lower level (Fig. 5.4).

Another type of hierarchical parallel GA uses a master-slave on each of the demes of a multi-population GA (Fig. 5.5). Migration occurs between demes, and the evaluation of the individuals is handled in parallel. This approach does not introduce new analytic problems, and it can be useful when working with complex applications with objective functions that need a considerable amount of computation time. Bianchini and Brown presented an example of this method of hybridizing parallel GAs, and showed that it can find a solution of the same quality of a master-slave parallel GA or a multi-deme GA in less time.

A third method of hybridizing parallel GAs is to use multi-deme GAs at both the upper and the lower levels (Fig. 5.6). The idea is to force panmictic mixing at the lower level by using a high migration rate and a dense topology, while a low migration rate is used at the high level. The complexity of this hybrid would

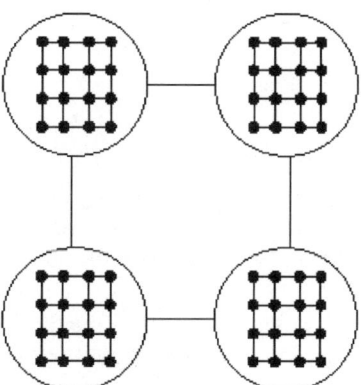

Fig. 5.4 This hierarchical GA combines a multi-deme GA (at the upper level) and a fine-grained GA (at the lower level)

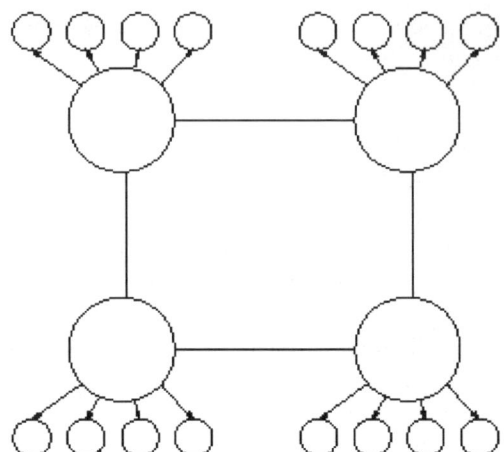

Fig. 5.5 A schematic of a hierarchical parallel GA. At the upper level this hybrid is a multi-deme parallel GA where each node is a master-slave GA

be equivalent to a multiple-population GA if we consider the groups of panmictic subpopulations as a single deme. This method has not been implemented yet. Hierarchical implementations can reduce the execution time more than any of their components alone (Table 5.1).

Thus multiple-deme algorithms dominate the research on parallel GAs. This class of parallel GAs is very complex, and its behavior is affected by many parameters. It seems that the only way to achieve a greater understanding of parallel GAs is to study individual facets independently, and we have seen that some of the most parallel GAs concentrates on migration rates, topology, or deme size. As GAs are applied to larger and more difficult search problems, it becomes necessary to design faster algorithms that retain the capability of finding acceptable solutions. This section presented numerous parallel GAs that are capable of combining speed and efficacy.

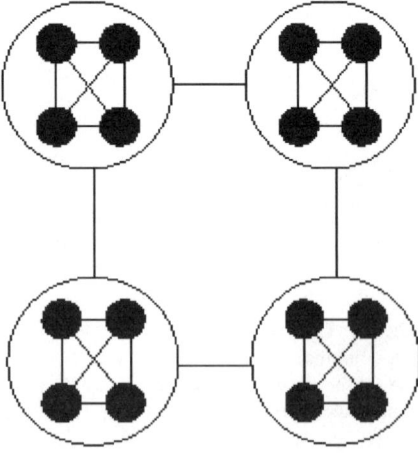

Fig. 5.6 This hybrid uses multi-deme GAs at both the upper and the lower levels. At the lower level the migration rate is faster and the communications topology is much denser than at the upper level

Table 5.1 The main parallel genetic algorithm classes according to their space and time dimensions

	Coarse-grained Population	Fine grained Individual	Fitness
Synchronous	Island GA	Cellular GA Synchronous Stochastic Cellular Automata	Master Slave GA
Asynchronous	Island GA	Asynchronous Stochastic Cellular Automata	Master Slave GA

5.4 Hybrid Genetic Algorithm (HGA)

A Hybrid Genetic Algorithm has been designed by combining a variant of an already existing crossover operator with these heuristics. One of the heuristics is for generating initial population; other two are applied to the offspring either obtained by crossover or by shuffling. The last two heuristics applied to offspring are greedy in nature, hence to prevent getting struck up at local optimum one has to include proper amount of randomness by using the shuffling operator.

The Hybrid Genetic Algorithm in this section is designed to use heuristics for Initialization of population and improvement of offspring produced by crossover for a Traveling Salesman Problem (TSP). The initializationHeuristics algorithm is used to initialize a part of the population; remaining part of the population will be initialized randomly. The offspring is obtained by crossover between two parents selected randomly. The tour improvement heuristics: *RemoveSharp* and *LocalOpt* are used to bring the offspring to a local minimum. If cost of the tour of the offspring thus obtained is less than the cost of the tour of any one of the parents then the parent with higher cost is removed from the population and the offspring is added to the population. If the cost of the tour of the offspring is greater than that of both of its parent then it is discarded. For shuffling, a random number is generated within one and if it is less than the specified probability of the shuffling operator, a tour is randomly selected and is removed from the population. Its sequence is randomized and then added to the population.

The algorithm works as below:

Step 1:

- Initialize a part of population using *InitializationHeuristics* algorithm
- Initialize remaining part of population randomly

Step 2:

- Apply *RemoveSharp* algorithm to all tours in the initial population
- Apply *LocalOpt* algorithm to all tours in the initial population

Step 3:

- Select two parents randomly
- Apply Crossover between parents and generate an offspring
- Apply *RemoveSharp* algorithm to offspring
- Apply *LocalOpt* algorithm to offspring
- If TourCost(offspring) < TourCost(any one of the parents) then replace the weaker parent by the offspring

Step 4:
Shuffle any one randomly selected tour from population
Step 5:
Repeat steps 3 and 4 until end of specified number of iterations.

5.4.1 Crossover

The crossover operator that is used here is a slight variant of the crossover operator devised by Darrell Whitley. The crossover operator uses an "edge map" to construct an offspring which inherits as much information as possible from the parent structures. This edge map stores information about all the connections that lead into and out of a city. Since the distance is same between any two cities, each city will have at least two and atmost four edge associations (two from each parent). The crossover algorithm is as follows:
Step 1:
Choose the initial city from one of the two parent tours. (It can be chosen randomly or according to criteria outlined in step 4). This is the "current city".
Step 2:
Remove all occurrences of the "current city " from the left-hand side of the edge map.
Step 3:
If the "current city" has entries in its *edgelist* go to step 4; otherwise, go to step 5.
Step 4:
Determine which city in the *edgelist* of the "current city", has shortest edge with the "current city". The city with the shortest edge is included in the tour. This city becomes the "current city". Ties are broken randomly. Go to step 2.
Step 5:
If there are no remaining unvisited cities, then STOP. Otherwise, randomly choose an unvisited city and go to step 2.

The difference between the Crossover algorithm of *Darrell Whitley* and this is only in the fourth step of the algorithm. He selected the city with least entries in its *edgelist* as the next city, while we choose the city nearest to the current city. This introduces greedy heuristic in the crossover operator too.

5.4.2 Initialization Heuristics

The *InitializationHeuristics* (IH) algorithm can be applied only to *TSP*. It initializes the population depending upon a greedy algorithm. The greedy algorithm arranges the cities depending on their x and y coordinates. The tours are represented in linked-lists. First an initial list is obtained in the input order (Input List). The linked-list that is obtained after applying the initialization heuristics is the "Output List". During the process of applying the initialization heuristics all the cities in the "Input List" will be moved one by one to the "Output List".

The initialization heuristics algorithm for a TSP is as follows:

Step 1:

Select four cities, first one with largest x-coordinate value, second one with least x-coordinate value, third one with largest y-coordinate and fourth one with least y-coordinate value. Move them from the "Input List" to the "Output List".

Step 2:

From among the possible sequences of the four cities find the sequence of minimum cost and change the sequence of four cities in the "Output List" to the minimum sequence.

Step 3:

Randomize the elements in the "Input List".

Step 4:

Remove the head element of the "Input List" and insert it into the "Output List" at the position where the increase in the cost of the tour is minimum. Suppose M is the cost of the tour before insertion and N be the cost of the tour after insertion. The position of insertion is selected such that $N-M$ is minimum.

Step 5:

Repeat Step 4 until all elements in the "Input List" are moved to the "Output List".

Depending on the sorting criteria in Step 3 of the above algorithm various results will be obtained.

RemoveSharp and *LocalOpt* heuristics are applied to the offspring obtained by this method and added to the initial population.

5.4.3 The RemoveSharp Algorithm

The *RemoveSharp* algorithm removes sharp increase in the tour cost due to a city, which is badly positioned. The algorithm works as below:

Step 1: A list (*NEARLIST*) containing the nearest m cities to a selected city is created.

Step 2: RemoveSharp removes the selected city from the tour and forms a tour with $N - 1$ cities.

Step 3: Now the selected city is reinserted in the tour either before or after any one of the cities in *NEARLIST* and the cost of the new tour length is calculated for each case.

Step 4: The sequence, which produces the least cost, is selected.
Step 5: The above steps are repeated for each city in the tour.

5.4.3.1 Time Complexity of RemoveSharp

As discussed in Step 2 of the algorithm, when a city is removed during *RemoveSharp* there will be a decrease in the tour cost. Suppose the sequence of the cities be

$$- - -P - C - N- - - - - - - - AP - A - AN - - -$$

C is the city to be removed to perform *RemoveSharp*. Let *P* be the city previous to the city *C* and *N* the city next to it. *RemoveSharp* will move the city *C* to a new position, if the increase in the tour length after moving it to the new position is less than the decrease in cost caused due to removing it from the position between *P* and *N*. If city *A* is in the near list then *RemoveSharp* will check possibility of moving to the locations before *A* i.e. *AP* and after *A* i.e. *AN*.

The decrease in tour length will be:

$$DECREASE = Dist(P,C) + Dist(C,N) - Dist(P,N)$$

If *C* is moved to the location previous to *A* i.e. *AP*, increase in tour cost will be:

$$INCREASEP = Dist(AP,C) + Dist(C,A) - Dist(AP,A)$$

If *C* is moved the location next to *A* i.e. *AN* increase in tour cost will be:

$$INCREASEN = Dist(A,C) + Dist(C,AN) - Dist(A,AN)$$

When *RemoveSharp* is applied, *DECREASE* is calculated once, while *INCREASEN* and *INCREASEP* are calculated for every city in the *NEARLIST*. Time complexities for *DECREASE, INCREASEN* and *INCREASEP* are same and let it be x. *INCREASEN* and *INCREASEP* should be compared with *DECREASE* for each city in

the *NEARLIST*. Let *y* be the time taken for one comparison. All these calculations need to be done for every city in the tour.

$$\textit{Time Complexity for RemoveSharp} = n * (x + 2m * x + 2m * y)$$

Here, *m* the size of *NERALIST*, *x* the time taken by for *DECREASE, INCREASEN* and *INCREASEP* and *y* the time taken for comparison are constants. Therefore,

$$\textit{Time Complexity for RemoveSharp} \sim= O(n)$$

5.4.4 The LocalOpt Algorithm

The *LocalOpt* algorithm will select q consecutive cities $(Sp+0, Sp+1, \ldots, Sp+q-1)$ from the tour and it arranges cities $Sp+1, Sp+2, \ldots, Sp+q-2$ in such a way that the distance is minimum between the cities $Sp+0$ and $Sp+q-1$ by searching all possible arrangements. The value of p varies from 0 to $n-q$, where n is the number of cities.

5.4.4.1 Time complexity of LocalOpt

The time complexity of *LocalOpt* varies with value of q, the number of consecutive cities taken for *LocalOpt* at a time. When q cities in a sequence are considered then all possible combinations of $q-2$ cities need to be calculated. There will be $(q-2)!$ combinations, in each case $q-1$ additions need to be done to evaluate the cost of the sequence and one comparison to check whether the sequence is minimum or not. These need to be done for n consecutive sequence of q cities starting from each city in the tour. Therefore,

Time Complexity of LocalOpt = $n * (((q-2)! * (q-1))$ *additions* + $(q-2)!$ *comparisons*)

As n alone is a variable,

Time Complexity $\sim$ = $O(n)$ *(provided q is small ($<=6$))*

Thus the discussed HGA can be extended to various application like VLSI design layout, contour planning and so on. For example in case of VLSI design layout macro cells will be a considering factor for design instead of cities in the above TSP hybrid genetic algorithm.

5.5 Adaptive Genetic Algorithm (AGA)

Adaptive genetic algorithms (AGA) are GAs whose parameters, such as the population size, the crossing over probability, or the mutation probability are varied while the GA is running. A simple variant could be the following: The mutation rate is changed according to changes in the population; the longer the population does not improve, the higher the mutation rate is chosen. Vice versa, it is decreased again as soon as an improvement of the population occurs. The overall Adaptive Genetic Algorithm procedure is as follows:
Step 1: Initial population
    We use the population obtained by random number generation
Step 2: Genetic operators
    Selection: elitist strategy in enlarged sampling space
    Crossover: order-based crossover operator for activity priority
    Mutation: local search-based mutation operator for activity mode

Step 3: Apply the local search using iterative hill-climbing method in GA loop
Step 4: Apply the heuristic for adaptively regulating GA parameters (i.e., the rates of crossover and the mutation operators).
Step 5: Stop condition

If a pre-defined maximum generation number is reached or an optimal solution is located during genetic search process, then stop; otherwise, go to Step 2.
For a multimode function, if it is needed to keep the global search ability it must have balanced search ability. Crossover probability *pc* and mutation probability *pm* are the main factors in affecting balanced search ability (global search ability and local search ability). While we strengthen one ability by increasing or decreasing *pc*, *pm*, we may weaken other abilities. Both *pc* and *pm* in the simple genetic algorithm (SGA) are invariant, so for the complex optimal problem the GA's efficiency is not high. In addition, immature convergence may be caused. Therefore, the goals with adaptive probabilities of crossover and mutation are to maintain the genetic diversity in the population and prevent the genetic algorithms to converge prematurely to local minima. As a result adaptive genetic algorithm was developed, and its basic idea is to adjust *pc* and *pm* according to the individual fitness. This algorithm can better solve the problem of adjusting *pc* and *pm* dynamically and also fits to all kinds of optimal problem. Based on these facts, we adopt the adaptive genetic algorithm to obtain the optimal solution is as follows:

5.5.1 Initialization

Define integer H is the initial random population of chromosomes, and we adopt real code. Every chromosome include n gene bit. Considering the following set:

$$\Omega = \{(x_1, \Lambda, x_x) | u_1 \leq x_1 \leq w_1, \Lambda, u_x \leq x_x \leq w_x\}$$

It produces random number from Ω and tests its feasibility. If it is feasible then it is a member of the initial population, otherwise, we go on produce random number from Ω until obtain feasible solution. After finite sample there are H initial feasible chromosomes $H, V_1 \Lambda V_H$.

5.5.2 Evaluation Function

Through evaluation function *eval(V)* we set probability for every chromosome V, so the selection probability is proportion to their fitness. That is, by roulette wheel selection, chromosomes with a high fitness value have a great chance of being selected to generate children for the next generation, and then the sequence number instead of target value reallocates the chromosome, and chromosome is arrayed from good

5.5 Adaptive Genetic Algorithm (AGA)

to bad. That is say, a chromosome is better sequence number is lower. So we set $\alpha \in (0,1)$ and define evaluating function based on the sequence number,

$$eval(V_f) = \alpha(1-\alpha)^{f-1}, \quad j = 1, 2, \Lambda, H.$$

5.5.3 Selection operator

Step 1: According to the rule that the selection operator chooses individuals with a probability that corresponds to the relative fitness. Chromosomes with a high fitness value have a great chance of being selected to generate children for the next generation., two chosen individuals, called the parents. We define reproduction probability for $v_j(k)$.

$$p_j(k) = \frac{F(v_j(k))}{f(v_1(k)) + F(v_2(k)) + \Lambda + F(v_u(k))}.$$

j=1 means the chromosome is the best chromosome and j=H means the chromosome is the worst chromosome.

Step 2: For $v_j(k)$, $j = 1, 2 \Lambda H$ calculate cumulative probability qj:

$$\begin{cases} q_0 = 0, \\ q_j = \sum_{i=1}^{j} p_i(k), & j = 1, 2, \Lambda, H. \end{cases}$$

Step 3: Produce random number r $(0, q_H)$. If $q_{j-1} < r < q_j$, then we select the chromosome $v_j(k)$, j=1,2,Λ,H.

Step 4: Repeating step 2 and step 3 H times, we can obtain H copied chromosomes, defined by

$$v'(k) = (v_1'(k), v_2'(k), \Lambda, v_H'(k))$$

5.5.4 Crossover operator

Instead of using fixed p_c, we adjust it adaptively based on the following formula:

$$p_c = \begin{cases} p_{c1} = \dfrac{(p_{c1} - p_{c2})(f' - f_{avg})}{f_{max} - f_{avg}} & f' \geq f_{avg}, \\ p_{c1} & f' < f_{avg}, \end{cases}$$

where f max is the highest fitness value in the population; f_{avg} is the average fitness value in every population; f' is higher fitness value between two individuals; in addition we set, P_{c1}=0.9, P_{c2}=0.6.

5.5.5 *Mutation operator*

Instead of using fixed p_m, we adjust it adaptively based on the following formula:

$$p_m = \begin{cases} p_{m1} = \dfrac{(p_{m1} - p_{m2})(f - f_{avg})}{f_{max} - f_{avg}} & f \geq f_{avg} \\ p_{m1} & f < f_{avg} \end{cases}$$

where f max is the highest fitness value in the population; f_{avg} is the average fitness value in every population; f' is higher mutation fitness value; in addition we set, $P_{m1}=0.1$, $P_{m2}=0.001$.

Based on the above operators perform the adaptive genetic algorithm operations until the convergence condition is reached.

In the adaptive GA, low values of p_c and p_m are assigned high fitness solutions, while low fitness solutions have very high values of p_c and p_m. The best solution of every population is 'protected' i.e. it is not subjected to crossover, and receives only a minimal amount of mutation. On the other hand, all solution with a fitness value less than the average fitness value of the population have $p_m= 0.5$. This means that all sub average solutions are completely disrupted and totally new solutions are created. The GA can thus rarely get stuck at a local optimum.

5.6 Fast Messy Genetic Algorithm (FmGA)

The fmGA is a binary, stochastic, variable string length, population based approach to solving optimization problems. The fmGA was developed by Goldberg, Deb and Kargupta and later applied to the PSP problem by Merkle, Gates, Lamont and Pachter. The main difference between the fmGA and other genetic approaches is the ability of the fmGA to explicitly manipulate building blocks (BBs) of genetic material in order to obtain good solutions and potentially the global optimum. The fmGA contains three phases of operation:

- the initialization phase
- the building block filtering (BBF) phase
- the juxtapositional phase, which includes various parameters.

In the initialization phase of the fmGA, a population sizing equation is used to derive a population large enough to overcome the noise present in the BBF process. Once the population size is determined, initial population members are randomly generated and their corresponding fitness values are calculated through the application chosen. These population members are referred to as fully specified since all of the associated genes of the population member contain allelic values. The fully specified population members from the initialization phase are then systematically reduced in length to the user specified BB size through the use of a BBF schedule.

5.6 Fast Messy Genetic Algorithm (FmGA)

The BBF process randomly deletes a certain number of bits from each population member over a number of generations specified in the schedule. This deletion of bits is alternated with tournament selection so that only the best partial strings are kept for processing in the subsequent generations. A CT is used at this stage to evaluate the under specified population members. At the end of the BBF process, the entire population consists of under specified strings of the user specified BB length.

The juxtapositional phase takes the good BBs found from the BBF process and combines them together through a cut-and-splice operator. This operator randomly chooses two strings and based on the probabilities of cut and splice, cuts the strings and splices them together accomplishing the goal of crossing over information between the strings. This process is also alternated with tournament selection so that only the best strings are kept from generation to generation. At the conclusion of this phase, fully specified strings exist in the population and the next BB size is evaluated via an outer loop over these three phases.

5.6.1 Competitive Template (CT) Generation

Competitive templates are an extremely important part of the fmGA. Population members containing very few specified bits (under specified members) with respect to the overall string length, as is the case at the end of the BBF process, are highly dependent on the CT. The reverse holds for strings that have the majority of their bits specified, as they only need to take a few bits from the CT. This illustrates the importance of the CT in the overall execution of the fmGA, especially at the start of the juxtapositional phase in generation fitness values. To evaluate an under specified population member, the CT is copied into a temporary location and the bits that are specified in the population member replace the bits of the CT within this temporary location. Once this is accomplished, the temporary string is evaluated and the resulting fitness is associated with the under specified population member. In the case of an over specified population member, which may occur when the cut-and-splice procedure causes a member to have multiple occurrences of a particular loci, a left-to-right method is employed. In this method, the first allelic value encountered for particular loci is recorded as the value present for evaluation purposes.

A random CT is a natural starting point since the goal of the fmGA work is to generate a robust algorithm that obtains solutions for various optimization problems. In order to increase the effectiveness of the algorithm over this approach, the next step is to incorporate problem domain knowledge into the fmGA or increase the number of CTs utilized. The four CT methods suggested are:

- Randomly generate a CT, and then conduct a localized search on this CT. This memetic approach involves conducting a local search of the competitive template before each template update at the end of the juxtapositional phase.
- The use of a fully specified population member containing specific structures as the CTs. Each seeded CT is hard coded into the fmGA using known alpha helix and beta-sheet dihedral angles. The algorithm is expected to achieve better fitness

values at a faster rate for proteins having either of these secondary structures through this method.
- Utilizing a panmetic CT. The process generates an odd number of CTs and merges these into one template called a panmetic CT in this method. The merge is anticipated to take the best components of each of the CTs and combine them together. The CTs can be all random or a combination of random and those containing a structure.
- Using more than a single CT developed via the aforementioned methods. This approach allows for more exploration since each population member is evaluated using multiple templates and therefore has the potential to find a better solution by searching different areas of the landscape.

5.7 Independent Sampling Genetic Algorithm (ISGA)

One major source of the power of GAs is derived from so called implicit parallelism i.e., the simultaneous allocation of search effort to many regions of the search space. A perfect implementation of implicit parallelism implies that a large number of different short, low order schemata of high fitness are sampled in parallel, thus conferring enough diversity of fundamental building blocks for crossover operators to combine them to form more highly fit, complicated building blocks. However, traditional GAs suffers from premature convergence where considerable fixation occurs at certain schemata of sub optimal regions before attaining more advancement. Among examples of premature convergence, hitchhiking has been identified as a major hindrance, which limits implicit parallelism by reducing the sampling frequency of various beneficial building blocks.

In short, non-relevant alleles hitchhiking on certain schemata could propagate to the next generation and drown out other potentially favorable building blocks, thus preventing independent sampling of building blocks. Consequently, the efficacy of crossover in combining building blocks is restricted by the resulting loss of desired population diversity. As a result researchers considered a so called Idealized Genetic Algorithm (IGA) that allows each individual to evolve completely independently; thus new samples are given independently to each schema region and hitchhiking is suppressed. Then under the assumption that the IGA has the knowledge of the desired schemata in advance, they derived a lower bound for the number of function evaluations that the IGA will need to find the optimum of Royal Road function.. However, the IGA is impractical because it requires the exact knowledge of desired schemata ahead of time.

Partially motivated by the idea of the IGA, a more robust GA is proposed that proceeds in two phases:

- the independent sampling phase
- the breeding phase.

5.7 Independent Sampling Genetic Algorithm (ISGA)

In the independent sampling phase, a core scheme, called Building Block Detecting Strategy (BBDS), to extract relevant building block information of a fitness landscape is designed. In this way, an individual is able to sequentially construct more highly fit partial solutions. For Royal Road Function, the global optimum can be attained easily. For other more complicated fitness landscapes, we allow a number of individuals to adopt the BBDS and independently evolve in parallel so that each schema region can be given samples independently. During this phase, the population is expected to be seeded with promising genetic material. Then follows the breeding phase, in which individuals are paired for breeding based on two mate selection schemes: individuals being assigned mates by natural selection only and individuals being allowed to actively choose their mates. In the latter case, individuals are able to distinguish candidate mates that have the same fitness yet have different string structures, which may lead to quite different performance after crossover. This is not achievable by natural selection alone since it assigns individuals of the same fitness the same probability for being mates, without explicitly taking into account string structures. In short, in the breeding phase individuals manage to construct even more promising schemata through the recombination of highly fit building blocks found in the first phase. Due to the characteristic of independent sampling of building blocks that distinguishes the proposed GAs from conventional GAs, it is called as *independent sampling genetic algorithms* (ISGAs).

5.7.1 Independent Sampling Phase

To implement independent sampling of various building blocks, a number of strings are allowed to evolve in parallel and each individual searches for a possible evolutionary path entirely independent of others.

In this section a new searching strategy called Building Block Detecting Strategy (BBDS) is developed, for each individual to evolve based on the accumulated knowledge for potentially useful building blocks. The idea is to allow each individual to probe valuable information concerning beneficial schemata through testing its fitness increase since each time a fitness increase of a string could come from the presence of useful building blocks on it. In short, by systematically testing each bit to examine whether this bit is associated with the fitness increase during each cycle, a cluster of bits constituting potentially beneficial schemata will be uncovered. Iterating this process guarantees the formation of longer and longer candidate building blocks.

The operation of BBDS on a string can be described as follows.

1. Generate an empty set for collecting genes of candidate schemata and create an initial string with uniform probability for each bit until its fitness exceeds 0. (Record the current fitness as *Fit*.)
2. Except the genes of candidate schemata collected, from left to right, successively °ip all the other bits, one at a time, and evaluate the resulting string. If the

resulting fitness is less than *Fit*, record this bit's position and original value as a gene of candidate schemata.
3. Except the genes recorded, randomly generate all the other bits of the string until the resulting string's fitness exceeds *Fit*. Replace *Fit* by the new fitness.
4. Go to steps 2 and 3 until some end criterion. The idea of this strategy is that the cooperation of certain genes (bits) makes for good fitness.

Once these genes come in sight simultaneously, they contribute a fitness increase to the string containing them; thus any loss of one of these genes leads to the fitness decrease of the string. This is essentially what step 2 does and after this step we should be able to collect a set of genes of candidate schemata. Then at step 3, we keep the collected genes of candidate schemata fixed and randomly generate other bits, awaiting other building blocks to appear and bring forth another fitness in crease.

However, the step 2 in this strategy only emphasizes the fitness drop due to a bitĿip. It ignores the possibility that the same bitĿip leads to a new fitness rise because many loci could interact in an extremely non linear fashion. To take this into account, the second version of BBDS is introduced through the change of step 2 as follows.

Step 2. Except the genes of candidate schemata collected, from left to right, successively Ŀip all the other bits, one at a time, and evaluate the resulting string. If the resulting fitness is less than *Fit*, record this bit's position and original value as a gene of candidate schemata. If the resulting fitness exceeds *Fit*, substitute this bit's new value for the old value, replace *Fit* by this new fitness, record this bit's position and new value as a gene of candidate schemata, and reexecute this step.

Because this version of BBDS takes into consideration the fitness increase resulted from bitĿips, it is expected to take less time for detecting. Other versions of BBDS are of course possible. For ex ample, in step 2, if a bitĿip results in a fitness increase, it can be recorded as a gene of candidate schemata, and the procedure continues to test the residual bits yet without completely traveling back to the first bit to reexamine each bit. However, the empirical results obtained thus far indicate that the performance of this alternative is quite similar to that of the second version. More experimental results are needed to distinguish the difference between them.

The overall implementation of the independent sampling phase of ISGAs is through the proposed BBDS to get autonomous evolution of each string until all individuals in the population have reached some end criterion.

5.7.2 Breeding Phase

After the independent sampling phase, individuals independently build up their own evolutionary avenues by various building blocks. Hence the population is expected to contain diverse beneficial schemata and premature convergence is alleviated to some degree. However, factors such as deception and incompatible schemata (i.e., two schemata have different bit values at common defining positions) still could lead individuals to arrive at sub optimal regions of a fitness landscape. Since building blocks for some strings to leave sub optimal regions may be embedded in other strings, the search for proper mating partners and then exploiting the building blocks on them are critical for overwhelming the difficulty of strings being trapped in undesired regions. The researchers have investigated the importance of mate selection and the results showed that the GAs is able to improve their performance when the individuals are allowed to select mates to a larger degree.

In this section, we adopt two mate selection schemes to breed the population: individuals being assigned mates by natural selection only and individuals being allowed to actively choose their mates. Since natural selection assigns strings of the same fitness the same probability for being parents, individuals of identical fitness yet distinct string structures are treated equally. This may result in significant loss of performance improvement after crossover.

We adopt the tournament selection scheme as the role of natural selection and the mechanism for choosing mates in the breeding phase is as follows:

During each mating event, a binary tournament selection with probability 1.0 the fitter of the two randomly sampled individuals is chosen is run to pick out the first individual, then choosing the mate according to the following two different schemes:

- Run the binary tournament selection again to choose the partner.
- Run another two times of the binary tournament selection to choose two highly fit candidate partners; then the one more dissimilar to the first individual is selected for mating.

The implementation of the breeding phase is through iterating each breeding cycle that consists of (1) Two parents are obtained based on the mate selection schemes above. (2) Two-point crossover operator (crossover rate 1.0) is applied to these parents. (3) Both parents are replaced with both offspring if any of the two offspring is better than them. Then steps 1, 2, and 3 are repeated until the population size is reached and this is a breeding cycle. Thus the Independent Sampling Genetic Algorithm with its two phases is efficient than the conventional GAs.

5.8 Summary

In this chapter we have discussed on the various types of existing genetic algorithms. Multiple-deme algorithms dominating the research on parallel GAs has been discussed in detail. This class of parallel GAs is very complex, and its behavior is affected by many parameters. It seems that the only way to achieve a greater understanding of parallel GAs is to study individual facets independently, and we have seen that some of the most influential publications in parallel GAs concentrate on only one aspect (migration rates, communication topology, or deme size) either ignoring or making simplifying assumptions on the others. The chapter also dealt on master-slave and fine-grained parallel GAs and realized that the combination of different parallelization strategies can result in faster algorithms. It is particularly important to consider the hybridization of parallel techniques in the light of recent results, which predict the existence of an optimal number of demes.

Also the hybrid GA, Adaptive GA and Messy GA has been included with the necessary information's. In this chapter we presented an exploratory method (BBDS) to show how the searching speed of individuals can be improved. Through explicitly acquiring relevant knowledge of candidate building blocks, BBDS outperformed several representative hill-climbing algorithms on non-deceptive Royal Road functions. Then a new class of GAs based on BBDS, i.e., ISGAs, is proposed. In the first phase of ISGAs, implicit parallelism is nicely realized by allowing each individual to accomplish independent building block sampling to suppress hitchhiking; thus the population is expected to carry diverse promising schemata. Afterwards, with one mate selection scheme that allows individuals to actively choose their mating partners, the efficacy of crossover is enhanced and the ISGAs have been shown to outperform several different GAs on a benchmark test function that is full of deception.

Review Questions

1. List the various classifications of Genetic Algorithm
2. State the algorithm of Simple Genetic Algorithm
3. Define deme.
4. What is the necessity of Parallel Genetic Algorithms?
5. With neat figure, explain the concept involved in Master–Slave Parallelization.
6. Discuss in detail on the coarse grained and fine-grained genetic algorithms.
7. Why coarse-grained genetic algorithm is called as distributed genetic algorithm?
8. Differentiate between parallel genetic algorithm and distributed genetic algorithm.
9. How are hierarchical genetic algorithms formed using the parallel GAs?
10. Write the basic Hybrid Genetic Algorithm.
11. State the various algorithmic procedures involved in Hybrid Genetic Algorithm.
12. Mention the formulas involved for crossover rate and mutation rate in adaptive genetic algorithm.
13. What are the advantages of fast messy genetic algorithm compared to conventional genetic algorithm?
14. Discuss the operations involved in the Fast messy Genetic Algorithm.
15. Explain in brief on the competitive template generation of FmGA.
16. Write short note on Independent Sampling Genetic Algorithm.
17. Compare and contrast Parallel GA and Hybrid GA
18. State few application areas of Parallel GAs.
19. How are Hybrid GAs used in planning of a VLSI Design Layout.
20. List some application areas of independent sampling genetic algorithms.

Exercise Problems

1. Implement a parallel genetic algorithm for traveling salesman problem.
2. Develop a computer program for Hybrid GA applied to for network design and routing problems.
3. Write a MATLAB program for Fast messy Genetic Algorithm to a Protein structure prediction.
4. Implement adaptive GA for a portfolio selection problem.
5. Build a C program to implement simple genetic algorithm for a multi objective optimization problem.

Chapter 6
Genetic Programming

6.1 Introduction

One of the central challenges of computer science is to get a computer to do what needs to be done, without telling it how to do it. Genetic Programming (GP) addresses this challenge by providing a method for automatically creating a working computer program from a high-level problem statement of the problem. Genetic Programming achieves this goal of *automatic programming* (also sometimes called *program synthesis* or *program induction*) by genetically breeding a population of computer programs using the principles of Darwinian natural selection and biologically inspired operations. The operations include reproduction, crossover (sexual recombination), mutation, and architecture altering operations patterned after gene duplication and gene deletion in nature. For example, an element of a population might correspond to an arbitrary placement of eight queens on a chessboard, and the fitness function might count the number of queens that are not attacked by any other queens. Given an appropriate set of genetic operators by which an initial population of queen placements can spawn new collections of queen placements, a suitably designed system could solve the classic eight-queens problem.

GP's uniqueness comes from the fact that it manipulates populations of structured programs—in contrast to much of the work in evolutionary computation in which population elements are represented using flat strings over some alphabet. In this chapter, the basic concepts, working, representations and applications of genetic programming have been dealt in detail.

6.2 Comparison of GP with Other Approaches

Genetic programming (GP) is a domain independent, problem-solving approach in which computer programs are evolved to find solutions to problems. The solution technique is based on the Darwinian principle of "survival of the fittest" and is closely related to the field of genetic algorithms (GA).

However three important differences exist between GAs and GP:

- *Structure*: GP usually evolves tree structures while GA's evolve binary or real number strings.
- *Active Vs Passive*: Because GP usually evolves computer programs, the solutions can be executed without post processing i.e. active structures, while GA's typically operate on coded binary strings i.e. passive structures, which require post-processing.
- *Variable Vs fixed length*: In traditional GAs, the length of the binary string is fixed before the solution procedure begins. However a GP parse tree can vary in length throughout the run. Although it is recognized that in more advanced GA work, variable length strings are used.

The ability to search the solution space and locate regions that potentially contain optimal solutions for a given problem is one of the fundamental components of most artificial intelligence (AI) systems. There are three primary types of search; the blind search, hill climbing and beam search. GP is classified as a beam search because it maintains a population of solutions that is smaller than all of the available solutions. GP is also usually implemented as a weak search algorithm as it contains no problem specific knowledge, although some research has been directed towards "strongly typed genetic programming". However while GP can find regions containing optimal solutions, an additional local search algorithm is normally required to locate the optima. Memetic algorithms can fulfill this role, by combining an evolutionary algorithm with problem specific search algorithm to locate optimal solutions.

Genetic programming also differs from all other approaches to artificial intelligence, machine learning, neural networks, adaptive systems, reinforcement learning, or automated logic in all (or most) of the following seven ways:

(1) **Representation:** Genetic programming overtly conducts it search for a solution to the given problem in program space.
(2) **Role of point-to-point transformations in the search:** Genetic programming does not conduct its search by transforming a single point in the search space into another single point, but instead transforms a set of points into another set of points.
(3) **Role of hill climbing in the search:** Genetic programming does not rely exclusively on greedy hill climbing to conduct its search, but instead allocates a certain number of trials, in a principled way, to choices that are known to be inferior.
(4) **Role of determinism in the search:** Genetic programming conducts its search probabilistically.
(5) **Role of an explicit knowledge base:** None.
(6) **Role of formal logic in the search:** None.
(7) **Underpinnings of the technique:** Biologically inspired.

First, consider the issue of representation. Most techniques of artificial intelligence, machine learning, neural networks, adaptive systems, reinforcement learning, or

6.2 Comparison of GP with Other Approaches

automated logic employ specialized structures in lieu of ordinary computer programs. These surrogate structures include if-then production rules, Horn clauses, decision trees, Bayesian networks, propositional logic, formal grammars, binary decision diagrams, frames, conceptual clusters, concept sets, numerical weight vectors (for neural nets), vectors of numerical coefficients for polynomials or other fixed expressions (for adaptive systems), genetic classifier system rules, fixed tables of values (as in reinforcement learning), or linear chromosome strings (as in the conventional genetic algorithm).

Tellingly, except in unusual situations, the world's several million-computer programmers do not use any of these surrogate structures for writing computer programs. Instead, for five decades, human programmers have persisted in writing computer programs that intermix a multiplicity of types of computations (e.g., arithmetic and logical) operating on a multiplicity of types of variables (e.g., integer, floating-point, and Boolean). Programmers have persisted in using internal memory to store the results of intermediate calculations in order to avoid repeating the calculation on each occasion when the result is needed. Moreover, they have persisted in passing parameters to subroutines so that they can reuse their subroutines with different instantiations of values. And they have persisted in organizing their subroutines into hierarchies. All of the above tools of ordinary computer programming have been in use since the beginning of the era of electronic computers in the 1940s. Significantly, none has fallen into disuse by human programmers.

Yet, in spite of the manifest utility of these everyday tools of computer programming, these tools are largely absent from existing techniques of automated machine learning, neural networks, artificial intelligence, adaptive systems, reinforcement learning, and automated logic. We believe that the search for a solution to the challenge of getting computers to solve problems without explicitly programming them should be conducted in the space of computer programs. Of course, once you realize that the search should be conducted in program space, you are immediately faced with the task of finding the desired program in the enormous space of possible programs. As will be seen, genetic programming performs this task of program discovery. It provides a problem-independent way to productively search the space of possible computer programs to find a program that satisfactorily solves the given problem.

Second, another difference between genetic programming and almost every automated technique concerns the nature of the search conducted in the technique's chosen search space. Almost all of these non-genetic methods employ a point-to-point strategy that transforms a single point in the search space into another single point. Genetic programming is different in that it operates by explicitly cultivating a diverse population of often-inconsistent and often-contradictory approaches to solving the problem. Genetic programming performs a beam search in program space by iteratively transforming one population of candidate computer programs into a new population of programs.

Third, consider the role of hill climbing. When the trajectory through the search space is from one single point to another single point, there is a nearly irresistible temptation to extend the search only by moving to a point that is known to be superior to the current point. Consequently, almost all automated techniques rely exclusively on greedy hill climbing to make the transformation from the current point in the search space to the next point. The temptation to rely on hill climbing is reinforced because many of the toy problems in the literature of the fields of machine learning and artificial intelligence are so simple that hill climbing can in fact, solve them. However, popularity cannot cure the innate tendency of hill climbing to become trapped on a local optimum that is not a global optimum. Interesting and nontrivial problems generally have high-payoff points that are inaccessible to greedy hill climbing. In fact, the existence of points in the search space that are not accessible to hill climbing is a good working definition of non-triviality. The fact that genetic programming does not rely on a point-to-point search strategy helps to liberate it from the myopia of hill climbing. Genetic programming is free to allocate a certain measured number of trials to points that are known to be inferior. This allocation of trials to known-inferior individuals is not motivated by charity, but in the expectation that it will often unearth an unobvious trajectory through the search space leading to points with an ultimately higher payoff. The fact that genetic programming operates from a population enables it to make a small number of adventurous moves while simultaneously pursuing the more immediately gratifying avenues of advance through the search space.

Fourth, another difference between genetic programming and almost every other technique of artificial intelligence and machine learning is that genetic programming conducts a probabilistic search. Again, genetic programming is not unique in this respect. For example, simulated annealing and genetic algorithms are also probabilistic. However, most existing automated techniques are deterministic.

Fifth, consider the role of a knowledge base in the pursuit of the goal of automatically creating computer programs. Many computer scientists unquestioningly assume that formal logic must play a preeminent role in any system for automatically creating computer programs. Similarly, the vast majority of contemporary researchers in artificial intelligence believe that a system for automatically creating computer programs must employ an explicit knowledge base. Indeed, over the past four decades, the field of artificial intelligence has been dominated by the strongly asserted belief that the goal of getting a computer to solve problems automatically can be achieved only by means of formal logic inference methods and knowledge. This approach typically entails the selection of a knowledge representation, the acquisition of the knowledge, the codification of the knowledge into a knowledge base, the depositing of the knowledge base into a computer, and the manipulation of the knowledge in the computer using the inference methods of formal logic. Conspicuously, genetic programming does not rely on an explicit knowledge base to achieve the goal of automatically creating computer programs. While there are numerous optional ways to incorporate domain knowledge into a run of genetic programming, genetic programming does not require (or usually use) an explicit knowledge base to guide its search.

Sixth, consider the role of the inference methods of formal logic. Many computer scientists unquestioningly assume that every problem-solving technique must be logically sound, deterministic, logically consistent, and parsimonious. Accordingly, most conventional methods of artificial intelligence and machine learning possess these characteristics. However, logic does not govern two of the most important types of complex problem-solving processes, namely, the invention process performed by creative humans and the evolutionary process occurring in nature. A new idea that can be logically deduced from facts that are known in a field, using transformations that are known in a field, is not considered to be an invention. There must be what the patent law refers to as an "illogical step" (i.e., an unjustified step) to distinguish a putative invention from that which is readily deducible from that which is already known.

The design of complex entities by the evolutionary process in nature is another important type of problem solving that is not governed by logic. In nature, solutions to design problems are discovered by the probabilistic process of evolution and natural selection. This is not a logical process. Indeed, inconsistent and contradictory alternatives abound. In fact, such genetic diversity is necessary for the evolutionary process to succeed. Significantly, the solutions created by evolution and natural selection almost always differ from those created by conventional methods of artificial intelligence and machine learning in one very important respect. Evolved solutions are not brittle; they are usually able to grapple with the perpetual novelty of real environments.

Similarly, genetic programming is not guided by the inference methods of formal logic in its search for a computer program to solve a given problem. When the goal is the automatic creation of computer programs, all of our experience has led us to conclude that the non-logical approaches used in the invention process and in natural evolution are far more fruitful than the logic-driven and knowledge-based principles of conventional artificial intelligence. In short, "logic considered harmful."

Seventh, the biological metaphor underlying genetic programming is very different from the underpinnings of all other techniques that have previously been tried in pursuit of the goal of automatically creating computer programs. Many computer scientists and mathematicians are baffled by the suggestion that biology might be relevant to their fields. In contrast, we do not view biology as an unlikely well from which to draw a solution to the challenge of getting a computer to solve a problem without explicitly programming it. Genetic programming work confirms Turing's view that there is indeed a "connection" between machine intelligence and evolution by describing our implementation of Turing's third way to achieve machine intelligence.

6.3 Primitives of Genetic Programming

Every solution evolved by GP is assembled from two sets of primitive's nodes; terminals and functions. The terminal set contains nodes that provide an input to the GP system while the function set contains nodes that process values already in the system. Constants can be used in GP by including them in the terminal set. Once the evolutionary process is started, the GP system randomly selects nodes from either set or thus may not utilize all of the available nodes. However increasing the size of each node set enlarges the search space. Therefore only a relatively simple node set is initially provided and nodes are usually added only if required.

6.3.1 Genetic Operators

There are three major evolutionary operators within a GP system:
- *Reproduction:* selects an individual from within the current population to be copied exactly into the next generation. There are several ways of selecting which individual is to be copied including "fitness proportionate" selection, "rank" selection and "tournament" selection.
- *Crossover:* mimics sexual recombination in nature, where two parent solutions are chosen and parts of their subtree are swapped and because each function exhibits the property "closure" (each tree member is able to process all possible argument values), every crossover operation should result in the formation of a legal structure.
- *Mutation:* causes random changes in an individual before it is introduced into the subsequent population. Unlike crossover, mutation is asexual and thus only operates on one individual. During mutation either all functions or terminals are removed beneath an arbitrarily determined node and a new branch is randomly created, or a single node is swapped for another.

6.3.2 Generational Genetic Programming

GP has developed two main approaches to dealing with the issue of its generations; generational and steady state. In generational GP, there exists well-defined and distinct generations, with each generation being represented by a complete population of individuals. Therefore each new population is created from the older population, which it then replaces. Steady-state GP does not maintain these discrete generations but continuously evolves the current generation using any available genetic operators.

6.3.3 Tree Based Genetic Programming

The primitives of GP, the function and terminal nodes, must be assembled into a structure before they may be executed. Three main types of structure exist: tree, linear and graph. Within this work, the input (the structure to be optimized or designed) actually forms a graph network. However by the duplication of joint data i.e. the same "joint node" can exist in the same tree on more than one occasion, this graph network is converted into a tree structure.

6.3.4 Representation of Genetic Programming

The need for a good representation in evolutionary computation, and in artificial intelligence more generally, is called the *representation problem*. Genetic programming has two forms of representation; the variational and the generative. The variational representation is a static description of a program and is subject to evolutionary variation. The main requirement for a variational representation is evolvability: the evolution of programs of increasing fitness on a generational basis when subjected to genetic variation. The generative representation is a product of the variational representation, and describes the dynamic form of a program. Its main requirement is that it can be executed. Yet, despite the different requirements of variational and generative representations, most GP systems do not distinguish between the two.

6.3.4.1 Biological Representations

Biology does distinguish between variational and generative representations. They are called, respectively, the genetic and the phenotypic. The genetic representation, from a reductionist viewpoint, is a linear, spatially distributed, sequence of heritable attributes. Each heritable attribute describes the amino acid sequence of a protein. Development interprets these descriptions and generates proteins; the fundamental components of the phenotypic representation.

A group of proteins working upon a common task is called a biochemical pathway. The tasks carried out by biochemical pathways fall into three broad classes: metabolic, signaling and gene expression. Of these, metabolic pathways are considered the most fundamental for they implement the processing behaviors of the cell, whilst signaling and gene expression pathways take on a configurational role. Biochemical processing amounts to the manipulation of a cell's chemical state through systems of chemical reactions. Metabolic pathways are composed of enzymes, a group of proteins that carry out catalytic behaviors; enabling reactions that would otherwise not be possible in the relatively low cellular temperatures. Enzymes achieve their catalytic behavior by binding to specific chemicals, the enzyme's substrates, and guiding their reaction. Cooperation within metabolic pathways emerges from product-substrate sharing between enzymes, where the product of one enzyme becomes the substrate of another.

6.3.4.2 Biomimetic Representations

Biological representations possess a number of qualities conceivably useful to, but not usually found, in genetic programming representations. These, include: the specialisation of evolutionary and executable forms; evolvable representations, "designed" for evolution; neutrality, increasing genetic diversity and adaptability; less constrained behaviour, giving more freedom to evolution; and positional independence, not limiting gene function to gene position. An umber of GP systems mimic the genetic representation of biology. Many of these have introduced a developmental stage, allowing the genetic representation to be independent of the executable representation. This has been shown to increase genetic diversity and encourage neutrality. A number of these approaches also allow positionally-independent genic units within the genome.

The mimicry of phenotypic representations is less common. However, computational idioms have been used to describe the action of enzymes and biochemical pathways. Analogues of enzyme activity have been used for computational purposes in the artificial domain. Evolutionary models of pathway development have also been attempted

6.3.4.3 Enzyme Genetic Programming Representation

Enzyme genetic programming is a system that mimics biology in both genetic and phenotypic representations. Phenotypic representation is based upon an abstraction of metabolic pathways. The aim of the system is to evolve analogues of metabolic pathways within combinational logic circuits.

Figure 6.1 shows the relationship between the representations of enzyme GP. During evolution, circuits are encoded as linear sequences of "genes"; where each gene describes the input preferences, the specificities, of a particular logic gate or output terminal. A specificity is a floating point value between zero and one which indicates relative preference for inputs (substrates). Each input-consuming activity has a specificity defined for the products of every output activity.

The phenotypic representation generated by a genotype is visualised in the center of Fig. 6.1. Line weights indicate relative specificities. In practice, the network should be fully connected. However, for clarity only the dominant and a few of the recessive specificities are shown. When the circuit is realised, the dominant speci.cities should map to circuit connections. However, combinational circuits must be non-recurrent and consequently it will not always be possible to express a circuit element's most preferred connections. This approach is taken rather than allowing invalid circuits or constraining the genetic representation.

Without any doubt, programs can be considered as strings. There are, however, two important limitations which make it impossible to use the representations and operations from our simple GA:

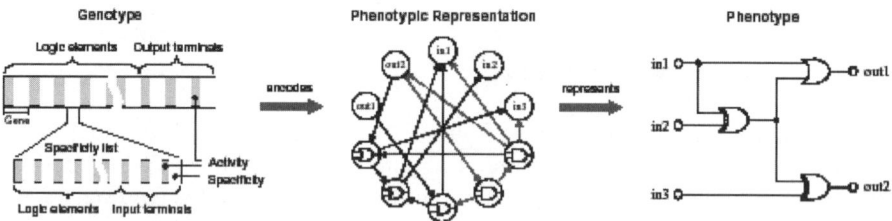

Fig. 6.1 Circuit representations

6.3 Primitives of Genetic Programming

1. It is mostly inappropriate to assume a fixed length of programs.
2. The probability to obtain syntactically correct programs when applying our simple initialization, crossover, and mutation procedures is hopelessly low.

It is, therefore, indispensable to modify the data representation and the operations such that syntactical correctness is easier to guarantee. The common approach to represent programs in genetic programming is to consider programs as trees. By doing so, initialization can be done recursively, crossover can be done by exchanging subtrees, and random replacement of subtrees can serve as mutation operation.

Since the constructs are nested lists, programs in LISP-like languages already have a kind of tree-like structure. Figure 6.2 shows an example how the function $3x + \sin(x + 1)$ can be implemented in a LISP like language and how such a LISP-like function can be split up into a tree. Obviously, the tree representation directly corresponds to the nested lists the program consists of; atomic expressions, like variables and constants, are leaves while functions correspond to non-leave nodes.

There is one important disadvantage of the LISP approach—it is difficult to introduce type checking. In case of a purely numeric function like in the above example, there is no problem at all. However, it can be desirable to process numeric data, strings, and logical expressions simultaneously. This is difficult to handle if we use a tree representation like in Fig. 6.2.

A. Geyer-Schulz has proposed a very general approach, which overcomes this problem allowing maximum flexibility. He suggested representing programs by their syntactical derivation trees with respect to a recursive definition of underlying language in Backus-Naur Form (BNF). This works for any context-free language. It is far beyond the scope of this lecture to go into much detail about formal languages. We will explain the basics with the help of a simple example. Consider the following language, which is suitable for implementing binary logical expressions:

The BNF description consists of so-called syntactical rules. Symbols in angular brackets <> are called non-terminal symbols, i.e. symbols which have to be expanded. Symbols between quotation marks are called terminal symbols, i.e. they cannot be expanded any further. The first rule $S: = <\exp>;$ defines the starting symbol. A BNF rule of the general shape,

$$\langle \text{non-terminal} \rangle : = \langle \text{deriv}_1 \rangle | \langle \text{deriv}_2 \rangle | \ldots | \langle \text{deriv}_n \rangle;$$

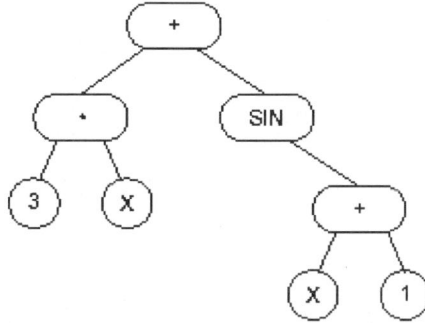

Fig. 6.2 The tree representation of (+ (* 3 X) (SIN (+ X 1)))

defines how a non-terminal symbol may be expanded, where the different variants are separated by vertical bars.

In order to get a feeling how to work with the BNF grammar description, we will now show step by step how the expression (NOT (x OR y)) can be derived from the above language. For simplicity, we omit quotation marks for the terminal symbols:

1. We have to begin with the start symbol: <exp>
2. We replace < exp > with the second possible derivation:

$$< exp > \rightarrow (<neg><exp>)$$

3. The symbol <neg> may only be expanded with the terminal symbol NOT:

$$(<neg><exp>) \rightarrow (\text{NOT} <exp>)$$

4. Next, we replace <exp> with the third possible derivation:

$$(\text{NOT} <exp>) \rightarrow (\text{NOT} (<exp><bin><exp>))$$

5. We expand the second possible derivation for <bin>:

$$(\text{NOT} (<exp><bin><exp>)) \rightarrow (\text{NOT} (<exp> \text{ OR } <exp>))$$

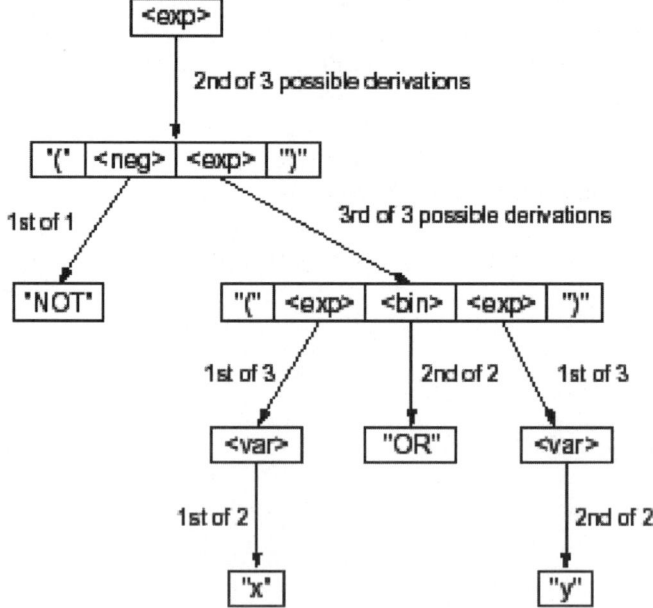

Fig. 6.3 The derivation tree of (NOT (x OR y))

6. The first occurrence of $<\exp>$ is expanded with the first derivation:

$$(\text{NOT } (<\exp> \text{ OR } <\exp>)) \rightarrow (\text{NOT } (<\text{var}> \text{ OR } <\exp>))$$

7. The second occurrence of hexpi is expanded with the first derivation, too:

$$(\text{NOT } (<\text{var}> \text{ OR } <\exp>)) \rightarrow (\text{NOT } (<\text{var}> \text{ OR } <\text{var}>))$$

8. Now we replace the first hvari with the corresponding first alternative:

$$(\text{NOT } (<\text{var}> \text{ OR } <\text{var}>)) \rightarrow (\text{NOT } (x \text{ OR } <\text{var}>))$$

9. Finally, the last non-terminal symbol is expanded with the second alternative:

$$(\text{NOT } (x \text{ OR } <\text{var}>)) \rightarrow (\text{NOT } (x \text{ OR } y))$$

Such a recursive derivation has an inherent tree structure. For the above example, this derivation tree has been visualized in Fig. 6.3.

6.4 Attributes in Genetic Programming

Genetic programming has 16 attributes of what is sometimes called *automatic programming* or *program synthesis* or *program induction*).

One of the central challenges of computer science is to get a computer to solve a problem without explicitly programming it. In particular, it would be desirable to have a problem-independent system whose input is a high-level statement of a problem's requirements and whose output is a working computer program that solves the given problem. When we talk about a computer program, we mean an entity that receives inputs, performs computations, and produces outputs. Computer programs perform basic arithmetic and conditional computations on variables of various types (including integer, floating-point, and Boolean variables), perform iterations and recursions, store intermediate results in memory, organize groups of operations into reusable subroutines, pass information to subroutines in the form of dummy variables (formal parameters), receive information from subroutines in the form of return values, and organize subroutines and a main program into a hierarchy.

We think that a system for automatically creating computer programs should create entities that possess most or all of the above essential features of computer programs (or reasonable equivalents thereof). A non-definitional list of attributes for a system for automatically creating computer programs would include the following 16 items:

- **Attribute No. 1 (Starts with "What needs to be done"):** It starts from a high-level statement specifying the requirements of the problem.

- **Attribute No. 2 (Tells us "How to do it"):** It produces a result in the form of a sequence of steps that can be executed on a computer.
- **Attribute No. 3 (Produces a computer program):** It produces an entity that can run on a computer.
- **Attribute No. 4 (Automatic determination of program size):** It has the ability to automatically determine the exact number of steps that must be performed and thus does not require the user to prespecify the size of the solution.
- **Attribute No. 5 (Code reuse):** It has the ability to automatically organize useful groups of steps so that they can be reused.
- **Attribute No. 6 (Parameterized reuse):** It has the ability to reuse groups of steps with different instantiations of values (formal parameters or dummy variables).
- **Attribute No. 7 (Internal storage):** It has the ability to use internal storage in the form of single variables, vectors, matrices, arrays, stacks, queues, lists, relational memory, and other data structures.
- **Attribute No. 8 (Iterations, loops, and recursions):** It has the ability to implement iterations, loops, and recursions.
- **Attribute No. 9 (Self-organization of hierarchies):** It has the ability to automatically organize groups of steps into a hierarchy.
- **Attribute No. 10 (Automatic determination of program architecture):** It has the ability to automatically determine whether to employ subroutines, iterations, loops, recursions, and internal storage, and the number of arguments possessed by each subroutine, iteration, loop, recursion.
- **Attribute No. 11 (Wide range of programming constructs):** It has the ability to implement analogs of the programming constructs that human computer programmers find useful, including macros, libraries, typing, pointers, conditional operations, logical functions, integer functions, floating-point functions, complex-valued functions, multiple inputs, multiple outputs, and machine code instructions.
- **Attribute No. 12 (Well-defined):** It operates in a well-defined way. It unmistakably distinguishes between what the user must provide and what the system delivers.
- **Attribute No. 13 (Problem-independent):** It is problem-independent in the sense that the user does not have to modify the system's executable steps for each new problem.
- **Attribute No. 14 (Wide applicability):** It produces a satisfactory solution to a wide variety of problems from many different fields.
- **Attribute No. 15 (Scalability):** It scales well to larger versions of the same problem.
- **Attribute No. 16 (Competitive with human-produced results):** It produces results that are competitive with those produced by human programmers, engineers, mathematicians, and designers.

Attribute No. 16 is especially important because it reminds us that the ultimate goal of a system for automatically creating computer programs is to produce useful programs—not merely programs that solve "toy" or "proof of principle" problems.

6.5 Steps of Genetic Programming

The steps of genetic programming are:

- Preparatory steps
- Executional steps

6.5.1 Preparatory Steps of Genetic Programming

The human user communicates the high-level statement of the problem to the genetic programming system by performing certain well-defined preparatory steps.

The five major preparatory steps for the basic version of genetic programming require the human user to specify

(1) the set of terminals (e.g., the independent variables of the problem, zero-argument functions, and random constants) for each branch of the to-be-evolved program,
(2) the set of primitive functions for each branch of the to-be-evolved program,
(3) the fitness measure (for explicitly or implicitly measuring the fitness of individuals in the population),
(4) certain parameters for controlling the run, and
(5) the termination criterion and method for designating the result of the run.

The *preparatory steps* are the problem-specific and domain-specific steps that are performed by the human user prior to launching a run of the problem-solving method.

Figure 6.4 shows the five major preparatory steps for the basic version of genetic programming. The preparatory steps (shown at the top of the figure) are the input to the genetic programming system. A computer program (shown at the bottom) is the output of the genetic programming system. The program that is automatically created by genetic programming may solve, or approximately solve, the user's problem. Genetic programming requires a set of primitive ingredients to get started.

The first two preparatory steps specify the primitive ingredients that are to be used to create the to-be-evolved programs. The universe of allowable compositions of these ingredients defines the search space for a run of genetic programming. The

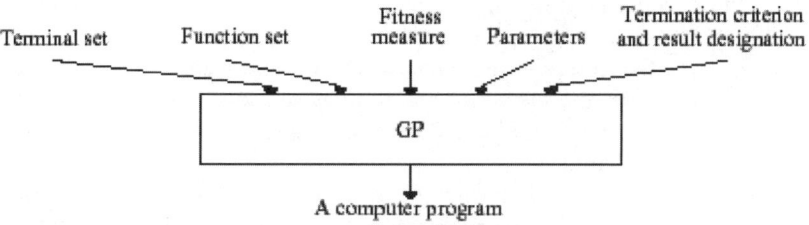

Fig. 6.4 Preparatory steps of genetic programming

identification of the function set and terminal set for a particular problem (or category of problems) is often a mundane and straightforward process that requires only *de minimus* knowledge and platitudinous information about the problem domain. For example, if the goal is to get genetic programming to automatically program a robot to mop the entire floor of an obstacle-laden room, the human user must tell genetic programming that the robot is capable of executing functions such as moving, turning, and swishing the mop.

The human user must supply this information prior to a run because the genetic programming system does not have any built-in knowledge telling it that the robot can perform these particular functions. Of course, the necessity of specifying a problem's primitive ingredients is not a unique requirement of genetic programming. It would be necessary to impart this same basic information to a neural network learning algorithm, a reinforcement-learning algorithm, a decision tree, a classifier system, an automated logic algorithm, or virtually any other automated technique that is likely to be used to solve this problem.

Similarly, if genetic programming is to automatically synthesize an analog electrical circuit, the human user must supply basic information about the ingredients that are appropriate for solving a problem in the domain of analog circuit synthesis. In particular, the human user must inform genetic programming that the components of the to-be-created circuit may include transistors, capacitors, and resistors (as opposed to, say, neon bulbs, relays, and doorbells). Although this information may be second nature to anyone working with electrical circuits, genetic programming does not have any built-in knowledge concerning the fact that transistors, capacitors, and resistors are the workhorse components for nearly all present-day electrical circuits. Once the human user has identified the primitive ingredients, the same function set can be used to automatically synthesize amplifiers, computational circuits, active filters, voltage reference circuits, and any other circuit composed of these basic ingredients.

Likewise, genetic programming does not know that the inputs to a controller include the reference signal and plant output and that controllers are composed of integrators, differentiators, leads, lags, gains, adders, subtractors, and the like. Thus, if genetic programming is to automatically synthesize a controller, the human user must give genetic programming this basic information about the field of control.

The third preparatory step concerns the fitness measure for the problem. The fitness measure specifies what needs to be done. The result that is produced by genetic programming specifies "how to do it." The fitness measure is the primary mechanism for communicating the high-level statement of the problem's requirements to the genetic programming system. If one views the first two preparatory steps as defining the search space for the problem, one can then view the third preparatory step (the fitness measure) as specifying the search's desired direction. The fitness measure is the means of ascertaining that one candidate individual is better than another. That is, the fitness measure is used to establish a partial order among candidate individuals.

The partial order is used during the executional steps of genetic programming to select individuals to participate in the various genetic operations (i.e., crossover, reproduction, mutation, and the architecture-altering operations). The

6.5 Steps of Genetic Programming

fitness measure is derived from the high-level statement of the problem. Indeed, for many problems, the fitness measure may be almost identical to the high level statement of the problem. The fitness measure typically assigns a single numeric value reflecting the extent to which a candidate individual satisfies the problem's high-level requirements. For example:

- If an electrical engineer needs a circuit that amplifies an incoming signal by a factor of 1,000, the fitness measure might assign fitness to a candidate circuit based on how closely the circuit's output comes to a target signal whose amplitude is 1,000 times that of the incoming signal. In comparing two candidate circuits, amplification of 990-to-1 would be considered better than 980-to-1.
- If a control engineer wants to design a controller for the cruise control device in a car, the fitness measure might be based on the time required to bring the car's speed up from 55 to 65 miles per hour. When candidate controllers are compared, a rise time of 10.1 seconds would be considered better than 10.2 seconds.
- If a robot is expected to mop a room, the fitness measure might be based on the percentage of the area of the floor that is cleaned within a reasonable amount of time.
- If a classifier is needed for protein sequences (or any other objects), the fitness measure might be based on the correlation between the category to which the classifier assigns each protein sequence and the correct category.
- If a biochemist wants to find a network of chemical reactions or a metabolic pathway that matches observed data, the fitness measure might assign fitness to a candidate network based on how closely the network's output matches the data.

The fitness measure for a real-world problem is typically multiobjective. That is, there may be more than one element that is considered in ascertaining fitness. For example, the engineer may want an amplifier with 1,000-to-1 gain, but may also want low distortion, low bias, and a low parts count. In practice, the elements of a multiobjective fitness measure usually conflict with one another. Thus, a multiobjective fitness measure must prioritize the different elements so as to reflect the tradeoffs that the engineer is willing to accept. For example, the engineer may be willing to tolerate an additional 1% of distortion in exchange for the elimination of one part from the circuit. One approach is to blend the distinct elements of a fitness measure into a single numerical value (often merely by weighting them and adding them together).

The fourth and fifth preparatory steps are administrative.

The fourth preparatory step entails specifying the control parameters for the run. The major control parameters are the population size and the number of generations to be run. Some analytic methods are available for suggesting optimal population sizes for runs of the genetic algorithm on particular problems. However, the practical reality is that we generally do not use any such analytic method to choose the population size. Instead, we determine the population size such that genetic programming can execute a reasonably large number of generations within the amount of computer time we are willing to devote to the problem. As for other control parameters, we have, broadly speaking, used the same (undoubtedly non-optimal) set of minor control parameters from problem to problem over a period of years.

The fifth preparatory step consists of specifying the termination criterion and the method of designating the result of the run.

6.5.2 Executional Steps of Genetic Programming

Genetic programming typically starts with a population of randomly generated computer programs composed of the available programmatic ingredients. Genetic programming iteratively transforms a population of computer programs into a new generation of the population by applying analogs of naturally occurring genetic operations. These operations are applied to individual(s) selected from the population. The individuals are probabilistically selected to participate in the genetic operations based on their fitness (as measured by the fitness measure provided by the human user in the third preparatory step). The iterative transformation of the population is executed inside the main generational loop of the run of genetic programming.

The executional steps of genetic programming (that is, the flowchart of genetic programming) are as follows:

(1) Randomly create an initial population (generation 0) of individual computer programs composed of the available functions and terminals.
(2) Iteratively perform the following sub-steps (called a *generation*) on the population until the termination criterion is satisfied:

   (a) Execute each program in the population and ascertain its fitness (explicitly or implicitly) using the problem's fitness measure.
   (b) Select one or two individual program(s) from the population with a probability based on fitness (with reselection allowed) to participate in the genetic operations in (c).
   (c) Create new individual programs(s) for the population by applying the following genetic operations with specified probabilities:

      (i) *Reproduction:* Copy the selected individual program to the new population.
      (ii) *Crossover:* Create new offspring program(s) for the new population by recombining randomly chosen parts from two selected programs.
      (iii) *Mutation:* Create one new offspring program for the new population by randomly mutating a randomly chosen part of one selected program.
      (iv) *Architecture-altering operations:* Choose an architecture-altering operation from the available repertoire of such operations and create one new offspring program for the new population by applying the chosen architecture-altering operation to one selected program.

6.5 Steps of Genetic Programming

(3) After the termination criterion is satisfied, the single best program in the population produced during the run (the best-so-far individual) is harvested and designated as the result of the run. If the run is successful, the result may be a solution (or approximate solution) to the problem.

The Fig. 6.5 below is a flowchart showing the executional steps of a run of genetic programming. The flowchart shows the genetic operations of crossover, reproduction, and mutation as well as the architecture-altering operations. This flowchart shows a two-offspring version of the crossover operation.

6.5.2.1 Creation of Initial Population of Computer Programs

Genetic programming starts with a primordial ooze of thousands of randomly-generated computer programs. The set of functions that may appear at the internal

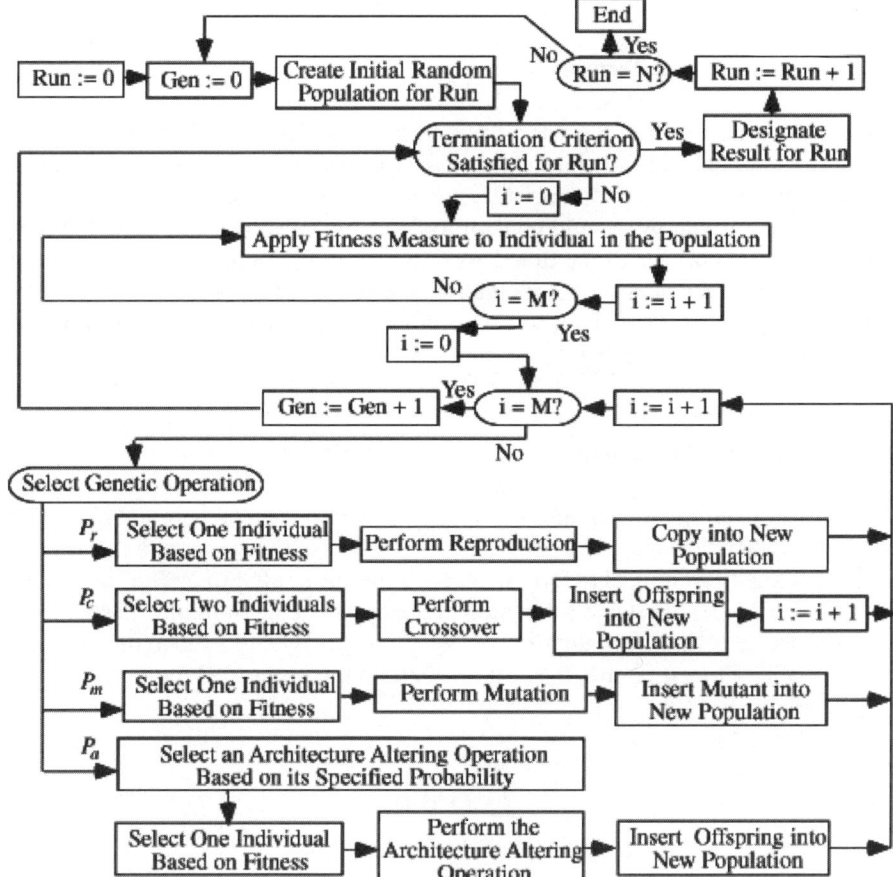

Fig. 6.5 Executional step (flowchart) of genetic programming

points of a program tree may include ordinary arithmetic functions and conditional operators. The set of terminals appearing at the external points typically include the program's external inputs (such as the independent variables X and Y) and random constants (such as 3.2 and 0.4). The randomly created programs typically have different sizes and shapes.

6.5.2.2 Fitness Function

The most difficult and most important concept of genetic programming is the fitness function. The fitness function determines how well a program is able to solve the problem. It varies greatly from one type of program to the next. For example, if one were to create a genetic program to set the time of a clock, the fitness function would simply be the amount of time that the clock is wrong. Unfortunately, few problems have such an easy fitness function; most cases require a slight modification of the problem in order to find the fitness.

6.5.2.3 Functions and Terminals

The terminal and function sets are also important components of genetic programming. The terminal and function sets are the alphabet of the programs to be made. The terminal set consists of the variables and constants of the programs. In the maze example, the terminal set would contain three commands: forward, right and left. The function set consists of the functions of the program. In the maze example the function set would contain: If "dot" then do x else do y. The functions are several mathematical functions, such as addition, subtraction, division, multiplication and other more complex functions.

6.5.2.4 Crossover Operation

Two primary operations exist for modifying structures in genetic programming. The most important one is the crossover operation. In the crossover operation, two solutions are sexually combined to form two new solutions or offspring. The parents are chosen from the population by a function of the fitness of the solutions. Three methods exist for selecting the solutions for the crossover operation. The first method uses probability based on the fitness of the solution. If $f(s_i(t))$ is the fitness of the solution S_i and

$$\sum_{j=1}^{M} f(s_j(t))$$

is the total sum of all the members of the population, then the probability that the solution S_i will be copied to the next generation is:

$$\frac{f(s_i(t))}{\sum_{j=1}^{M} f(s_j(t))}$$

Another method for selecting the solution to be copied is tournament selection. Typically the genetic program chooses two solutions random. The solution with the higher fitness will win. This method simulates biological mating patterns in which, two members of the same sex compete to mate with a third one of a different sex. Finally, the third method is done by rank. In rank selection, selection is based on the rank, (not the numerical value) of the fitness values of the solutions of the population. The creation of the offspring from the crossover operation is accomplished by deleting the crossover fragment of the first parent and then inserting the crossover fragment of the second parent. The second offspring is produced in a symmetric manner. For example consider the two S-expressions in Fig. 6.6, written in a modified scheme programming language and represented in a tree.

An important improvement that genetic programming displays over genetic algorithms is its ability to create two new solutions from the same solution. In the Fig. 6.7 the same parent is used twice to create two new children.

6.5.2.5 Mutation

Mutation is another important feature of genetic programming. Two types of mutations are possible. In the first kind a function can only replace a function or a terminal can only replace a terminal. In the second kind an entire subtree can replace another subtree. Figure 6.8 explains the concept of mutation.

6.6 Characteristics of Genetic Programming

Genetic programming now delivers High-Return Human-Competitive Machine Intelligence. Based on this sentence, it can be noted that the four main characteristics of genetic programming are:

- human-competitive
- high-return
- routine
- machine intelligence.

6.6.1 What We Mean by "Human-Competitive"

In attempting to evaluate an automated problem-solving method, the question arises as to whether there is any real substance to the demonstrative problems that are

Crossover Operation with Different Parents

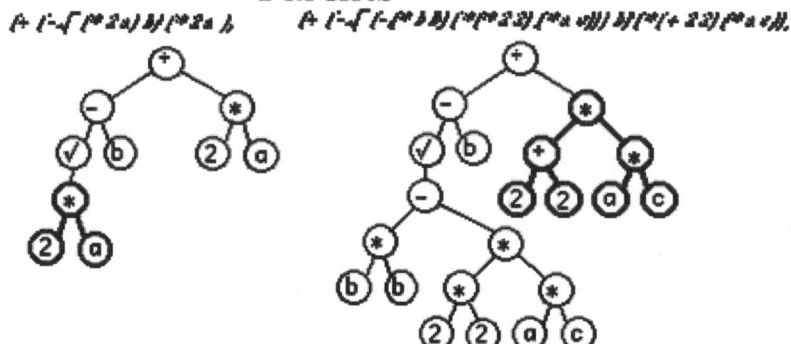

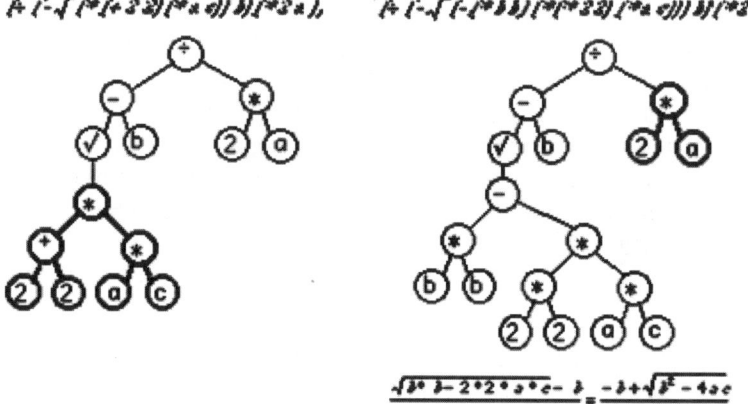

Fig. 6.6 Crossover operation for genetic programming. The bold selections on both parents are swapped to create the offspring or children. (The child on the right is the parse tree representation for the quadratic equation.)

published in connection with the method. Demonstrative problems in the fields of artificial intelligence and machine learning are often contrived toy problems that circulate exclusively inside academic groups that study a particular methodology. These problems typically have little relevance to any issues pursued by any scientist or engineer outside the fields of artificial intelligence and machine learning.

In his 1983 talk entitled "AI: Where It Has Been and Where It Is Going," machine learning pioneer Arthur Samuel said:

> The aim is…to get machines to exhibit behavior, which if done by humans, would be assumed to involve the use of intelligence.

6.6 Characteristics of Genetic Programming

Crossover Operation with Identical Parents

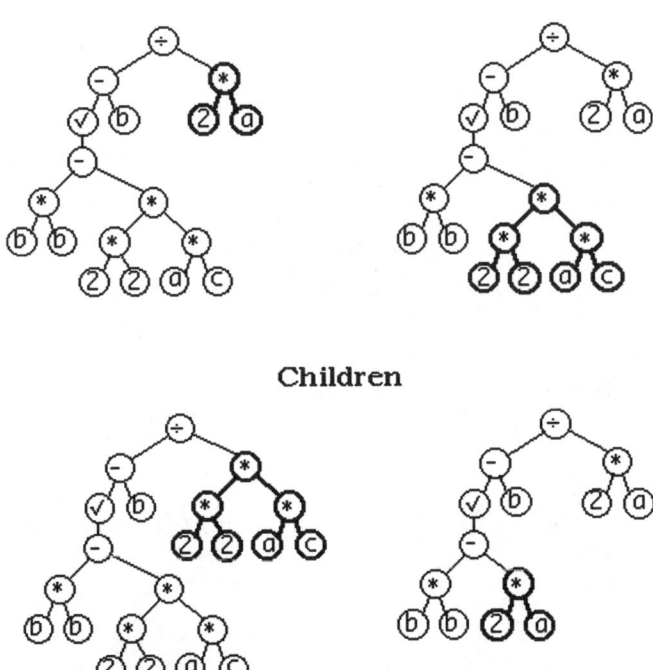

Fig. 6.7 This figure illustrates one of the main advantages of genetic programming over genetic algorithms. In genetic programming identical parents can yield different offspring, while in genetic algorithms identical parents would yield identical offspring. The bold selections indicate the subtrees to be swapped

Samuel's statement reflects the common goal articulated by the pioneers of the 1950s in the fields of artificial intelligence and machine learning. Indeed, getting machines to produce human-like results is *the* reason for the existence of the fields of artificial intelligence and machine learning. To make this goal more concrete, we say that a result is "human-competitive" if it satisfies one or more of the eight criteria in Table 6.1.

The eight criteria in Table 6.1 have the desirable attribute of being at arms-length from the fields of artificial intelligence, machine learning, and genetic programming. That is, a result cannot acquire the rating of "human-competitive" merely because it is endorsed by researchers *inside* the specialized fields that are attempting to create machine intelligence. Instead, a result produced by an automated method must earn the rating of "human-competitive" *independent* of the fact that it was generated by an automated method.

Mutation

Original Individual

{÷{-√{-{*bb}{*{*22}{*ac}}}b}{*2a}}

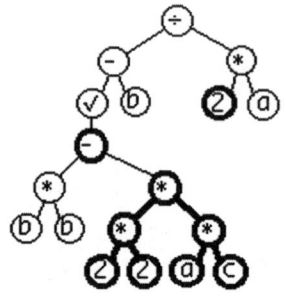

Mutated Individuals

{÷{-√{+{*bb}{*{*22}{*ac}}}b}{*aa}}   {÷{-√{-{*bb}{*2a}}b}{*2a}}

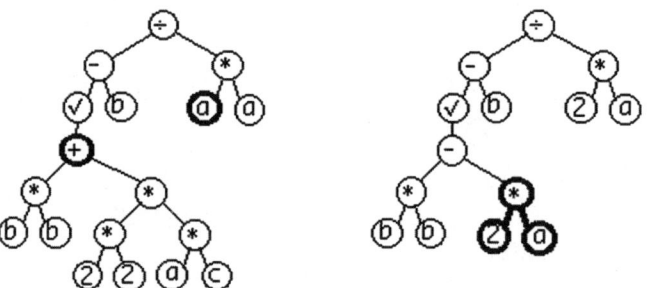

Fig. 6.8 Two different types of mutations. The top parse tree is the original agent. The bottom left parse tree illustrates a mutation of a single terminal (2) for another single terminal (a). It also illustrates a mutation of a single function (−) for another single function (+). The parse tree on the bottom right illustrates a the replacement of a subtree by another subtree

6.6.2 What We Mean by "High-Return"

What is delivered by the actual automated operation of an artificial method in comparison to the amount of knowledge, information, analysis, and intelligence that is pre-supplied by the human employing the method?

We define the *AI ratio* (the "artificial-to-intelligence" ratio) of a problem-solving method as the ratio of that which is delivered by the automated operation of the *artificial* method to the amount of *intelligence* that is supplied by the human applying the method to a particular problem.

The AI ratio is especially pertinent to methods for getting computers to automatically solve problems because it measures the value added by the artificial

6.6 Characteristics of Genetic Programming

Table 6.1 Eight criteria for saying that an automatically created result is human-competitive

	Criterion
A	The result was patented as an invention in the past, is an improvement over a patented invention, or would qualify today as a patentable new invention.
B	The result is equal to or better than a result that was accepted as a new scientific result at the time when it was published in a peer-reviewed scientific journal.
C	The result is equal to or better than a result that was placed into a database or archive of results maintained by an internationally recognized panel of scientific experts.
D	The result is publishable in its own right as a new scientific result—independent of the fact that the result was mechanically created.
E	The result is equal to or better than the most recent human-created solution to a long-standing problem for which there has been a succession of increasingly better human-created solutions.
F	The result is equal to or better than a result that was considered an achievement in its field at the time it was first discovered.
G	The result solves a problem of indisputable difficulty in its field.
H	The result holds its own or wins a regulated competition involving human contestants (in the form of either live human players or human-written computer programs).

problem solving method. Manifestly, the aim of the fields of artificial intelligence and machine learning is to generate human-competitive results with a high AI ratio.

The Chinook checker-playing computer program is an impressive human competitive result. Jonathan Schaeffer recounts the development of Chinook by his eight-member team at the University of Alberta between 1989 and 1996 in his book *One Jump Ahead: Challenging Human Supremacy in Checkers* (Schaeffer, 1997). Schaeffer's team began with analysis. They recognized that the problem could be profitably decomposed into three distinct sub problems. First, an opening book controls the play at the beginning of each game. Second, an evaluation function controls the play during the middle of the game. Finally, when only a small number of pieces are left on the board, an endgame database takes over and dictates the best line of play. Perfecting the opening book entailed an iterative process of identifying "positions where Chinook had problems finding the right move" and looking for "the elusive cooks" (Schaeffer, 1997).

By the time the project ended, the opening book had over 40,000 entries. In a chapter entitled "A Wake-Up Call," Schaeffer refers to the repeated difficulties surrounding the evaluation function by saying "the thought of rewriting the evaluation routine...and tuning it seemed like my worst nightmare come true." Meanwhile, the endgame database was painstakingly extended from five, to six, to seven, and eventually eight pieces using a variety of clever techniques. As Schaeffer observes, "The significant improvements to Chinook came from the knowledge added to the program: endgame databases (computer generated), opening book (human generated but computer refined), and the evaluation function (human generated and tuned). We, too, painfully suffered from the knowledge-acquisition bottleneck of artificial intelligence. Regrettably, our project offered no new insights into this difficult problem, other than to reemphasize how serious a problem it really is."

Chinook defeated world champion Marion Tinsley. However, because of the enormous amount of human "I" invested in the project, Chinook has a low return when measured in terms of the A-to-I ratio.

The aim of the fields of artificial intelligence and machine learning is to get computers to automatically generate human-competitive results with a high AI ratio—not to have humans generate human-competitive results themselves.

6.6.3 What We Mean by "Routine"

Generality is a precondition to what we mean when we say that an automated problem-solving method is "routine." Once the generality of a method is established, "routine ness" means that relatively little human effort is required to get the method to successfully handle new problems within a particular domain and to successfully handle new problems from a different domain. The ease of making the transition to new problems lies at the heart of what we mean by "routine." What fraction of Chinook's highly specialized software, hardware, databases, and evaluation techniques can be brought to bear on different games? For example, can Chinook's three-way decomposition be gainfully applied to a game, such as Go, with a significantly larger number of possible alternative moves at each point in the game? What fraction of these systems can be applied to a game of incomplete information, such as bridge? What more broadly applicable principles are embodied in these two systems? For example, what fraction of these methodologies can be applied to the problem of getting a robot to mop the floor of an obstacle-laden room? Correctly recognizing images or patterns? Devising an algorithm to solve a mathematical problem? Automatically synthesizing a complex structure?

A problem-solving method cannot be considered routine if its executional steps must be substantially augmented, deleted, rearranged, reworked, or customized by the human user for each new problem.

6.6.4 What We Mean by "Machine Intelligence"

We use the term "machine intelligence" to refer to the broad vision articulated in Alan Turing's 1948 paper entitled "Intelligent Machinery" and his 1950 paper entitled "Computing Machinery and Intelligence."

In the 1950s, the terms "machine intelligence," "artificial intelligence," and "machine learning" all referred to the *goal* of getting "machines to exhibit behavior, which if done by humans, would be assumed to involve the use of intelligence" (to again quote Arthur Samuel).

However, in the intervening five decades, the terms "artificial intelligence" and "machine learning" progressively diverged from their original goal-oriented meaning. These terms are now primarily associated with particular *methodologies* for attempting to achieve the goal of getting computers to automatically solve problems.

6.6 Characteristics of Genetic Programming

Thus, the term "artificial intelligence" is today primarily associated with attempts to get computers to solve problems using methods that rely on knowledge, logic, and various analytical and mathematical methods. The term "machine learning" is today primarily associated with attempts to get computers to solve problems that use a particular small and somewhat arbitrarily chosen set of methodologies (many of which are statistical in nature). The narrowing of these terms is in marked contrast to the broad field envisioned by Samuel at the time when he coined the term "machine learning" in the 1950s, the charter of the original founders of the field of artificial intelligence, and the broad vision encompassed by Turing's term "machine intelligence."

Turing's term "machine intelligence" did not undergo this arteriosclerosis because, by accident of history, it was never appropriated or monopolized by any group of academic researchers whose primary dedication is to a particular methodological approach. Thus, Turing's term remains catholic today. We prefer to use Turing's term because it still communicates the broad *goal* of getting computers to automatically solve problems in a human-like way.

In his 1948 paper, Turing identified three broad approaches by which human competitive machine intelligence might be achieved.

The first approach was a logic-driven search. Turing's interest in this approach is not surprising in light of Turing's own pioneering work in the 1930s on the logical foundations of computing.

The second approach for achieving machine intelligence was what he called a "cultural search" in which previously acquired knowledge is accumulated, stored in libraries, and brought to bear in solving a problem—the approach taken by modern knowledge-based expert systems.

Turing's first two approaches have been pursued over the past 50 years by the vast majority of researchers using the methodologies that are today primarily associated with the term "artificial intelligence."

Turing also identified a third approach to machine intelligence in his 1948 paper entitled "Intelligent Machinery", saying:

> "There is the genetical or evolutionary search by which a combination of genes is looked for, the criterion being the survival value."

Thus, Turing correctly perceived in 1948 and 1950 that machine intelligence might be achieved by an evolutionary process in which a description of a computer program (the hereditary material) undergoes progressive modification (mutation) under the guidance of natural selection (i.e., selective pressure in the form of what is now usually called "fitness" by practitioners of genetic and evolutionary computation). Of course, the measurement of fitness in modern genetic and evolutionary computation is usually performed by automated means (as opposed to a human passing judgment on each candidate individual, as suggested by Turing). In addition, modern work generally employs a population (i.e., not just a point-to-point evolutionary progression) and sexual recombination—two key aspects of John Holland's work on genetic algorithms, *Adaptation in Natural and Artificial Systems*.

6.7 Applications of Genetic Programming

The main application areas of Genetic Programming are:

- Computer Science
- Science
- Engineering
- Art and Entertainment.

In Computer Science, the development of algorithms has been a focus of attention. By being able to manipulate symbolic structures, Genetic Programming is one of the few heuristic search methods for algorithms. Sorting algorithms, caching algorithms, random number generators and algorithms for automatic parallelization of code, to name a few, have been studied using Genetic Programming. The spectrum of applications in Computer Science spans from the generation of proofs for predicate calculus to the evolution of machine code for accelerating function evaluation. The general tendency is to try to automate the design process for algorithms of different kinds.

Typical applications in Science are of modeling and pattern recognition type. Modeling of certain processes in Physics and Chemistry with the unconventional help of evolutionary creativity supports research and understanding of the systems under study. Pattern recognition is a key ability in molecular biology and other branches of biology, as well as in Science in general. Here, Genetic Programming has delivered first results that are competitive if not better than human-generated results.

In Engineering, Genetic Programming is used in competition or cooperation with other heuristic methods such as Neural Networks or Fuzzy Systems. The general goal is again to model processes such as production plants, or to classify results of production. Control of man-made apparatus is another area where Genetic Programming has been used successfully. Process control and robot control are primary applications.

In Art and Entertainment has Genetic Programming been used to evolve realistic animation scenes and appealing visual graphics. It also has been used "to extract structural information from musical composition in order to model the process to the extent that automatic composition of music pieces becomes possible."

In this section lets discuss in detail the application of genetic programming to civil engineering.

6.7.1 Applications of Genetic Programming in Civil Engineering

The various application areas of GP in civil engineering are as follows:

- Shear strength prediction of deep RC beams
- Modelling of wastewater treatment plants–Use of genetic programming to model the dynamic performance of municipal activated sludge wastewater treatment plants.

6.7 Applications of Genetic Programming

- Detection of traffic accidents–Detection of accidents on motorways in low flow, high-speed conditions i.e. late at night based on three years of traffic data whilst producing a near zero false alarm rate.
- Flow through a urban basin–Construction of sewage network model in order to calculate the risk posed by rain to the basin and thus provide prior warning of flooding or subsidence.
- Prediction of journey times–Investigation of GP to forecast the motorway journey times.
- Estimation of design intent–Using GP to automatically estimate design intent based on operational and product-specific information monitored throughout the design process.
- Modelling of water supply assets–In order to determine the risk of a pipe burst, a GP is evolved to "data mine" a database containing information about historic pipe bursts.
- Identification of crack profiles–Detection of cracks inside hundreds of heat exchanger tubes in a nuclear power plant's steam generator via analysis of data measured via quantitative non-destructive testing.
- Modelling rainfall runoff–Discovery of rainfall-runoff relationships in two vastly different catchments.
- Prediction of long-term electric power demand
- Evolution of traffic light control laws–Evolution of a new type of adaptive control system for a network of traffic signals depending on variations in traffic flow.
- Identification of crack profiles–Agent generation to detect and track dark regions that could be cracks in grayscale images of textured surfaces.

6.7.1.1 Application of Genetic Programming in Structural Engineering

This section describes how GP can be applied to structural optimization problems by using the tree structure of GP, to represent a building or structure. Earlier an approach (Soh and Yang, 2000) was designed that solved the key issue of how to represent the "node element diagram" of structural analysis as a "point-labeled tree" (as used in GP). However because the structure (the phenotype) is now different from the GP tree (the genotype), an additional decoding step must be included in the solution procedure before any fitness evaluation can occur (Fig. 6.9). This step was not previously required when evolving regression functions, as these solutions could be applied directly to the problem. Although this is a departure from traditional GP, by utilizing this representation Soh and Yang demonstrated a system that produced better results when attempting to simultaneously optimize the geometry, sizing and topology of a truss than work using other evolutionary search techniques. It is also important to note that this tree will not degenerate into a linear search, as its corresponding structure would be a mechanism. A mechanism is not a viable structure in engineering and thus will be heavily penalized by the fitness function and therefore should not be sustained within the population.

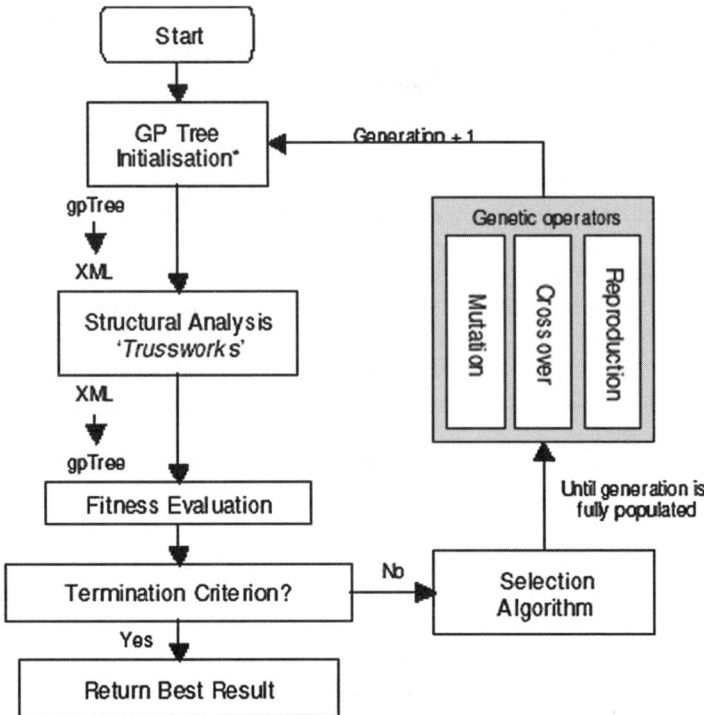

* Each newly created gpTree is checked for uniqueness before it is inserted into the population to ensure the variety of the initial population is 100%.

Fig. 6.9 The solution procedure

Structural Encoding

In this case, the GP tree should compose two types of nodes: inner nodes, which are GP functions representing the cross-sectional areas of the members Ap ($p =$ i, j, k, l, m, n) and outer nodes which are GP terminals representing various node points Ni ($i = 1,2,3,4$) (Fig. 6.10). To create a GP parse tree, one member must be selected from the structure to be the root node. This member then has its' corresponding start and end joints represented by children nodes to its left and right. Then from left to right the members associated with each joint node are removed from the structure and used to replace the relevant joint nodes in the GP tree (Fig. 6.10). This procedure continues until every structural member is represented in the GP tree. However it is important that the left-right relationship of child and parent node is maintained as the GP tree is constructed. Therefore each members' start and end joints are represented by the far left and far right children. For example function Aj connects nodes N1 and N2 (Fig. 6.10). This approach to structural encoding appears very simple when compared to the complex binary string used by a GA to represent member properties. The results for the evolution of a truss were capable of carrying

6.8 Haploid Genetic Programming with Dominance

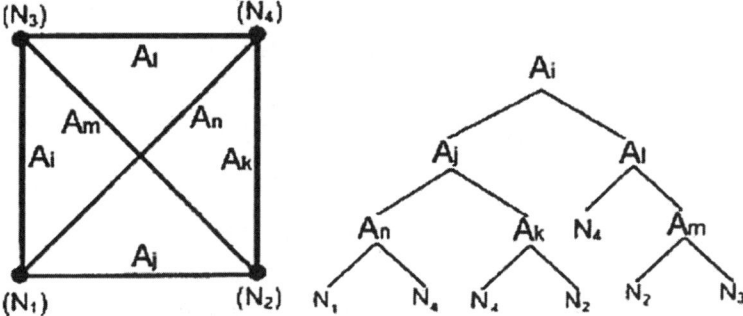

Fig. 6.10 GP tree encoding for structural analysis

six specified loads. To achieve this, a GP tree containing 29 nodes (16 joint nodes and 15 member nodes) was required where as the GA representation required a chromosome of 25,200 bits.

6.8 Haploid Genetic Programming with Dominance

Dominance crossover is similar to the use of dominance in nature. In nature, dominance is used as a genotype to phenotype mapping when an organism carries pairs (or more than one) chromosome, but here we use dominance on a haploid structure. The haploid form contains all the information relevant to the problem, and is the structure that is widely used in evolutionary algorithms. Dominance crossover is used as a way of retaining and promoting successful genes (those which increased the individual's fitness in the current generation) into the next generation. Current crossover operators fail to exploit knowledge acquired in previous generations and rely highly on selection pressures. Dominance crossover in theory allows this exploitation to occur during crossover but we highlight a problem with the application of dominance crossover with genetic programming.

Genetic programming traditionally uses a haploid chromosome: the haploid form contains all the information relevant to the problem and the genes do not have associated dominance values. With dominance crossover the parse tree contains the normal defined function and terminal sets. Each use of a function or terminal will have an associated dominance value. This dominance value will reflect how good each one is with respect to the fitness of the entire program. These dominance values are real numbers in the range [0,1] on initialization but can be increased. During crossover two parent trees are selected and the position for crossover is selected at random. Once the subtrees are chosen, the nodes from each subtree are compared, breadth first. The node with greater dominance is used to create a new subtree. This is a recursive process. In the case where one tree is greater than the other the remaining component in the larger tree is simply copied to the new subtree. This new subtree is then attached to the trees of the parent where crossover occurred. Figure 6.11 shows an example of dominance crossover. In dominance crossover a

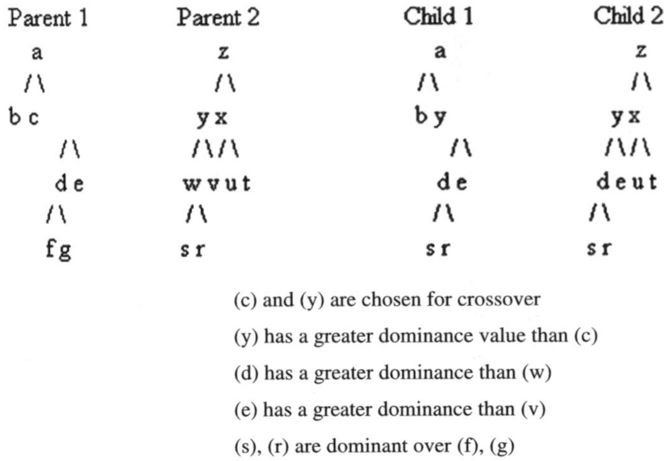

```
Parent 1        Parent 2        Child 1         Child 2
   a               z               a               z
  /\              /\              /\              /\
 b c             y x             b y             y x
    /\          /\ /\              /\           /\ /\
    d e         w v u t             d e         d e u t
    /\          /\                  /\          /\
    f g         s r                 s r         s r
```

(c) and (y) are chosen for crossover

(y) has a greater dominance value than (c)

(d) has a greater dominance than (w)

(e) has a greater dominance than (v)

(s), (r) are dominant over (f), (g)

Fig. 6.11 Crossover using dominance

single subtree is created from the parents and is attached to the original parents at the point chosen for crossover, hence creating two children.

The dominance characteristics of each gene in the tree evolve in each generation. Before "parents" undergo crossover their fitness values are recorded along with those genes that have been changed. After the new children are evaluated their fitness is compared with that of their parents. If the child has a higher fitness than that of at least one of its parents the dominance values of the changed genes are increased. The increase is the difference between the parent fitness and the child fitness. This dominance increase introduces a bias to those genes, which are deemed better for the individual, thus increasing the genes likelihood of being passed to the next generation. Unfortunately dominance crossover was inappropriate for current tree structures in GP. Since functions take different number of arguments they distort the shape of the trees thus the breadth first technique employed by dominance crossover failed to find appropriate points of crossover where the integrity was respected. This led to empirical studies of two alternative methods of crossover "Single-Node Dominance Crossover"(SNDC) and "Sub-Tree Dominance

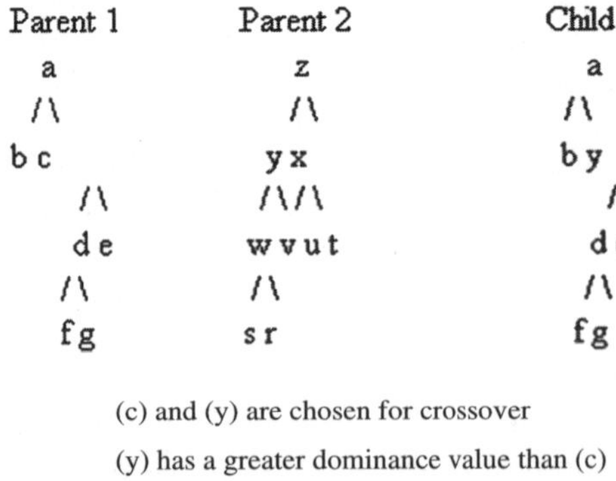

```
              Parent 1        Parent 2        Child 1
                 a               z               a
                /\              /\              /\
               b c             y x             b y
                  /\           /\ /\              /\
                  d e          w v u t            d e
                  /\           /\                 /\
                  f g          s r                f g
```

(c) and (y) are chosen for crossover

(y) has a greater dominance value than (c)

Fig. 6.12 Single-node dominance crossover

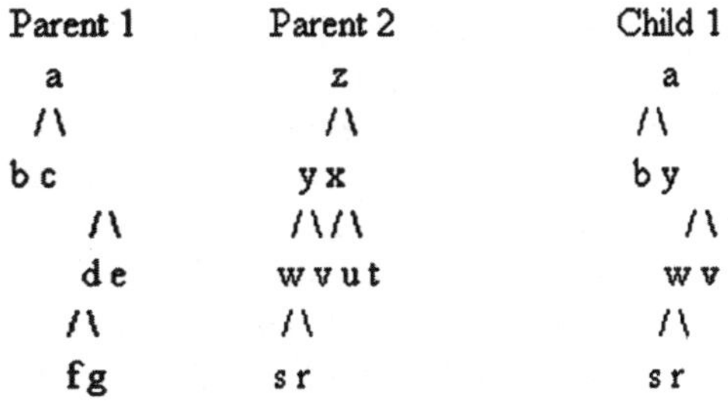

Fig. 6.13 Sub-tree dominance crossover

```
Parent 1      Parent 2       Child 1
   a             z              a
  /\            /\             /\
 b c           y x            b y
    /\        /\/\               /\
   d e       w v u t             w v
   /\         /\                 /\
  f g        s r                 s r
```

(c) and (y) are chosen for crossover

(y) has a greater dominance value than (c)

Crossover"(STDC). Like dominance crossover each use of a function or terminal has associated with it a dominance value.

6.8.1 Single-Node Dominance Crossover

In single-node dominance crossover two parents create a single child. A crossover point is chosen randomly as normal. In the subtrees chosen for crossover the dominance values of the top nodes are compared. The node that has a higher dominance value replaces the other. This is illustrated in Fig. 6.12.

6.8.2 Sub-Tree Dominance Crossover

Sub-tree dominance crossover is very much like SNDC. The difference being that in STDC the top node with the higher dominance value replaces the other tree as depicted in Fig. 6.13.

Dominance crossover is not an appropriate crossover operator to use with current GP structures. Since functions take different number of arguments they distort the shape of the GP trees thus the breadth first technique employed by dominance crossover failed to find appropriate points of crossover where the integrity was respected. Alternative ways of promotion were used in single-node dominance crossover and sub-tree dominance crossover but both operators failed to increase GP performance. The first did not allow tress to grow thus reducing exploration and the second exploited trees rather than genes, thus leading to excessive growth.

6.9 Summary

Genetic programming is much more powerful than genetic algorithms. The output of the genetic algorithm is a quantity, while the output of the genetic programming is another computer program. In essence, this is the beginning of computer programs that program themselves. Genetic programming works best for several types of problems. The first type is where there is no ideal solution, (for example, a program that drives a car). There is no one solution to driving a car. Some solutions drive safely at the expense of time, while others drive fast at a high safety risk. Therefore, driving a car consists of making compromises of speed versus safety, as well as many other variables. In this case genetic programming will find a solution that attempts to compromise and be the most efficient solution from a large list of variables. Furthermore, genetic programming is useful in finding solutions where the variables are constantly changing. In the previous car example, the program will find one solution for a smooth concrete highway, while it will find a totally different solution for a rough unpaved road.

Neutrality, dominance hierarchies and multiplicity are all facets of natural evolution and its products. To this extent enzyme genetic programming succeeds at its aim of biomimicry. The use of an enzyme-like representation for circuit elements, and consequently a pathway-like representation for circuits, illustrates that biological phenotypic representations can be annealed to the artificial domain. Performance-wise, the method is competitive with existing methods, though it has yet to demonstrate a provable performance advantage. Partly this is due to limitations of the results, though it may also be due to deleterious simplifications found within the initial system.

Genetic Programming has been used to model and control a multitude of processes and to govern their behavior according to fitness-based automatically generated algorithms. Most of these applications are characterized by one of the following features:

- Analytical solutions do not exist or cannot be derived
- Relevant variables are poorly understood
- Complexity of solutions is unknown
- Approximate solutions are all that can be expected
- Large amounts of data are available for mining
- Large marginal benefits exist for small improvements in control or modeling and prediction

Theory of Genetic Programming is presently greatly underdeveloped and will need to progress quickly in order to catch up with other evolutionary algorithm paradigms. Most of the obstacles stem from the fact of variable complexity of solutions evolved in Genetic Programming. Implementation of Genetic Programming will benefit in the coming years from new approaches, which include research from developmental biology. Also, it will be necessary to learn to handle the redundancy forming pressures in the evolution of code. Application of Genetic Programming will continue to broaden. Many applications focus on controlling behavior of real or virtual agents. In this role, Genetic Programming may contribute considerably to the growing field of social and behavioral simulations. Genetic Algorithms have already been found beneficial in optimizing strategies of social agents. With its ability to adjust the complexity of a strategy to the environment and to allow competition between agents Genetic Programming is well positioned to play an important role in the study and simulation of societies and their evolution.

Review Questions

1. What is Genetic Programming?
2. Differentiate between Genetic Algorithm and Genetic Programming.
3. Compare and Contrast: Genetic Programming with other traditional approaches.
4. List the genetic operators used in genetic programming.
5. Write short note on Tree based genetic programming.
6. Mention in detail about the various representations used in genetic programming.
7. What are the five preparatory steps involved in genetic programming?
8. With a new flowchart, explain the executional step of genetic programming.
9. Discuss the crossover and mutation operation of GP.
10. Explain with suitable examples, the characteristics of GP.
11. What is dominance crossover in GP?
12. Give few applications of Genetic Programming.

Exercise Problems

1. Implement a Genetic Programming for a Network Design Problem
2. For a traveling Salesman problem, compare the performance of GA and GP.
3. Write a computer program to implement GP for a function optimization problem.
4. Implement Genetic Program approach to obtain solution to a XOR problem.
5. With a computer program, explain the approach of genetic programming to scheduling problems.

Chapter 7
Genetic Algorithm Optimization Problems

7.1 Introduction

Optimization deals with problems of minimizing or maximizing a function with several variables usually subject to equality and/or inequality constraints. It plays a central role in operations research, management science and engineering design. Many industrial engineering design problems are very complex and difficult to solve using conventional optimization techniques. In recent years, genetic algorithms have received considerable attention regarding their potential as a novel optimization technique. Based on their simplicity, ease of operation, minimal requirements and parallel and global perspective, genetic algorithms have been widely applied in a variety of problems. A brief introduction to genetic optimization techniques and their application is described in this section, including major fields of optimization, such as fuzzy, combinatorial and multi objective optimizations.

7.2 Fuzzy Optimization Problems

Fuzzy optimization describes an optimization problem with fuzzy objective function and fuzzy constraints. The results obtained from classical methods of optimization involving deterministic variables exhibit various shortcomings. In particular, the effects of the uncertainty attached to input information is often ignored altogether or only taken into account to a limited degree. The classical deterministic optimization problem according to

$$\text{find } \underline{x}_{OPT} \in \underline{X} \text{ with } z(\underline{x}, \underline{e}) \to \min$$
$$\underline{X} = \{\underline{x} | g_i(\underline{x}, \underline{e}), h_j(\underline{x}, \underline{e})\} \quad g_i(\underline{x}, \underline{e}) \leq r_i \quad i = 1\ldots n$$
$$\text{and} \quad h_j(\underline{x}, \underline{e}) = 0 \quad j = 1\ldots m \qquad (7.1)$$

is considered under the aspect of uncertainty, and extended. For the objective function $z(x, e)$ the optimum solution x_{OPT} from the set of design variables X (design space) is determined under compliance with the equality constraints $h_j(x, e)$ and the inequality constraints $g_i(x, e)$. Input parameters such as geometrical parameters, material parameters, external load parameters, reliability parameters and economic parameters are lumped together in the vector e.

Considering the uncertain parameters to be fuzzy variables, the deterministic optimization problem is extended to a fuzzy optimization problem

$$\text{find } \underline{x}_{OPT} \in \underline{X} \text{ with } \tilde{z}(\underline{x}, \underline{\tilde{e}}) \to \min$$
$$\underline{X} = \left\{\underline{x} | \tilde{g}_i(\underline{x}, \underline{\tilde{e}}), \tilde{h}_j(\underline{x}, \underline{\tilde{e}})\right\} \tilde{g}_i(\underline{x}, \underline{\tilde{e}}) \stackrel{\sim}{\leq} \tilde{r}_i \quad i = 1 \ldots n$$
$$\text{and} \quad \tilde{h}_j(\underline{x}, \underline{\tilde{e}}) = 0 \quad j = 1 \ldots m \quad (7.2)$$

The numerical solution of the fuzzy optimization problem is based on α-level optimization.

7.2.1 Fuzzy Multiobjective Optimization

Suppose we are given a multiobjective mathematical programming problem in which the functional relationship between the decision variables and the objective functions is not completely known. Our knowledge base consists of a block of fuzzy if-then rules, where the antecedent part of the rules contains some linguistic values of the decision variables, and the consequence part consists of a linguistic value of the objective functions. We suggest the use of Tsukamoto's fuzzy reasoning method to determine the crisp functional relationship between the decision variables and objective functions. We model the anding of the objective functions by a t-norm and solve the resulting (usually nonlinear) programming problem to find a fair optimal solution to the original fuzzy multiobjective problem.

Fuzzy multiobjective optimization problems can be stated and solved in many different ways. Consider optimization problems of the form,

$$\max/\min\{G_1(x), \ldots, G_K(x)\}; \text{ subject to } x \in X, \quad (7.3)$$

where G_k, $k = 1, \ldots, K$, or/and X are defined by fuzzy terms. Then they are searching for a crisp x^*, which (in certain) sense maximizes the G_k's under the (fuzzy) constraints X. For example, multiobjective fuzzy linear programming (FMLP) problems can be stated as,

$$\max/\min\{\tilde{c}_1 x, \ldots, \tilde{c}_K x\}; \text{ subject to } \tilde{A}x \leq \tilde{b}, \quad (7.4)$$

where $x \in \mathbb{R}^n$ is the vector of crisp decision variables, $\tilde{A} = (\tilde{a}_{ij})$, $\tilde{b} = (\tilde{b}_i)$ and $\tilde{c}_j = (\tilde{c}_{ij})$ are fuzzy quantities, the inequality relation $\leq$ is given by a certain fuzzy relation and the (implicit) X is a fuzzy set describing the concept "x satisfies $\tilde{A}x \leq \tilde{b}$".

In many important cases (e.g., in strategy formation processes) the values of the objective functions are not known for all $x \in \mathbb{R}^n$, and we are able to describe the

7.2 Fuzzy Optimization Problems

causal link between x and the G_k's linguistically using fuzzy if-then rules. Here we consider a new statement of multiobjective fuzzy optimization problems (FMOP), namely,

$$\max/\min{}_X\{G_1, \ldots, G_K\}; \text{ subject to } \{\Re_1, \ldots, \Re_m\},$$

where $x_1, \ldots, x_n$ are linguistic variables, and

$\Re_i$: if x_1 is A_{i1} and ... and x_n is A_{in} then G_1 is C_{i1} and ... and G_K is C_{iK},

constitutes the only knowledge available about the values of G, and A_{ij} and C_{ik} are fuzzy numbers.

Originally FMLP (Fuzzy Mulitobjective Linear Programming) problems (7.3) are interpreted with fuzzy coefficients and fuzzy inequality relations as multiple fuzzy reasoning schemes, where the antecedents of the scheme correspond to the constraints of the MFLP (Multiobjective Fuzzy Linear Programming) problem and the facts of the scheme are the objectives of the MFLP problem. Generalizing the fuzzy reasoning approach, we determine the crisp value of the Gj's at $y \in X$ by Tsukamoto's fuzzy reasoning method, and obtain an optimal solution to (7.4) by solving the resulting (usually nonlinear) optimization problem,

$$\max/\min \text{ t-norm}(G_1(y), \ldots, G_K(y)); \text{ subject to } y \in X. \qquad (7.5)$$

7.2.1.1 Multiobjective Optimization Under Fuzzy If-then Rules

Consider the FMOP problem (7.4) with continuous A_{ij} representing the linguistic values of x_i, and with strictly monotone and continuous $C_{ik}, i = 1, \ldots, m$ representing the linguistic values of $G_k, k = 1, \ldots, K$. To find a fair solution to the fuzzy optimization problem (7.4) we first determine the crisp value of the k-th objective function G_k at $y \in \Re^n$ from the fuzzy rule-base $\Re$ using Tsukamoto's fuzzy reasoning method as,

$$G_k(y) := \frac{\alpha_1 C_{1k}^{-1}(\alpha_1) + \ldots + \alpha_m C_{mk}^{-1}(\alpha_m)}{\alpha_1 + \ldots + \alpha_m}$$

where

$$\alpha_i = \text{t-norm}(A_{i1}(y_1), \ldots, A_{in}(y_n)) \qquad (7.6)$$

denotes the firing level of the i-th rule, $\Re_i$. To determine the firing level of the rules, we suggest the use of the product t-norm (to have a smooth output function). In this manner the constrained optimization problem (7.4) turns into the crisp (usually nonlinear) mathematical programming problem (7.5). The same principle is applied to constrained maximization problems.

7.2.2 Interactive Fuzzy Optimization Method

An interactive fuzzy optimization method was incorporated into the genetic algorithm to give decision makers a chance to readjust membership functions according to information provided by current genetic search.

> Step 1: Set initial reference membership levels (if it difficult to determine these values, set them to 1.0)
> Step 2: Generate the initial population involving N individuals of double-string type at random.
> Step 3: Calculate the fitness for each individual and apply the reproduction operator based on the fitness.
> Step 4: Insert into the current population the corresponding individual with the optimal solution for each objective function.
> Step 5: Apply a crossover operator according to the probability of crossover P_c.
> Step 6: Apply mutation operator according to the probability of mutation P_m.
> Step 7: Repeat steps 3 to 6 until the termination condition is satisfied. If it is satisfied, regard the individual with maximal fitness as the optimal individual and go to step 8.
> Step 8: If the decision maker is satisfied with the current values of member ship functions and objective functions given by the current optimal individual, stop. Otherwise, ask the decision maker to update reference membership levels by taking account of the current values of membership functions and objective functions and return to step 2.

Thus by the above algorithm, the fuzzy membership functions can be readjusted.

7.2.3 Genetic Fuzzy Systems

In a very broad sense, a Fuzzy System (FS) is any fuzzy logic-based system where fuzzy logic can be used either as the basis for the representation of different forms of system knowledge or to model the interactions and relationships among the system variables. FSs have proven to be an important tool for modeling complex systems in which, due to complexity or imprecision, classical tools are unsuccessful.

Genetic algorithms GAs are search algorithms that use operations found in natural genetics to guide the trek through a search space. GAs are theoretically and empirically proven to provide robust search capabilities in complex spaces, offering a valid approach to problems requiring efficient and effective searching.

Recently, numerous papers and applications combining fuzzy concepts and GAs have appeared, and there is increasing concern about the integration of these two topics. In particular, a great number of publications explore the use of GAs for designing fuzzy systems. These approaches have been given the general name genetic fuzzy systems GFSs.

7.2 Fuzzy Optimization Problems

The automatic design of FSs can be considered in many cases as an optimization or search process on the space of potential solutions FSs. GAs are the best known and most widely used global search technique with an ability to explore and exploit a given operating space using available performance measures. A priori knowledge in the form of linguistic variables, fuzzy membership function parameters, fuzzy rules, number of rules, etc., may be incorporated easily into the genetic design process. The generic code structure and independent performance features of GAs make them suitable candidates for incorporating a priori knowledge. Over the last few years, these advantages have extended the use of GAs in the development of a wide range of approaches for designing fuzzy systems. As in the general case of FSs, the main application area of GFSs is system modeling /control. Regardless of the kind of optimization problem, i.e., given a system to be modeled or controlled, the involved design /tuning /learning process will be based on evolution. Three points are the keys to a genetic process:

- the population of potential solutions
- the pair evolution operators/code, and
- the performance index.

7.2.3.1 The Population of Potential Solutions

The learning search process works on a population of potential solutions to the problem. The individuals of the population are called chromosomes. Different approaches have been considered, but the most widely used is the so-called Pittsburgh approach. In this case, each chromosome represents a complete potential solution, an FS. From this point of view, the learning process will work on a population of FSs. FSs are knowledge-based systems with a processing structure and a knowledge base, and considering that all the systems use an identical processing structure, the individuals in the population will be reduced to rule bases, knowledge bases, etc. In some cases, the process starts off with an initial population obtained from available knowledge, while in other cases the initial population is randomly generated.

7.2.3.2 The Evolution Operators /Code

The second question is the definition of a set of evolution operators that search for new and /or better potential solutions. The search reveals two different aspects: the exploitation of the best solution and the exploration of the search space. The success of evolutionary learning is specifically related to obtaining an adequate balance between exploration and exploitation that finally depends on the selected set of evolution operators. The new potential solutions are obtained by applying the evolution operators to the members of the population; each one of these members is referred to as an individual in the population. Basically, there are three evolution operators that work with a code (chromosome) representing the FS: selection, crossover, and mutation. Since these evolution operators work in a coded representation of the FSs, certain compatibility between the operators and the structure of the chromosomes is required. This compatibility is stated in two different ways: work with chromosomes coded as binary strings (adapting the problem solutions to binary code) using a set of classical (defined for binary-coded chromosomes) genetic operators or adapt the operators to obtain compatible evolution operators using chromosomes with a non-binary code. Consequently, the question of defining a set of evolution operators involves defining a compatible couple of evolution operators and chromosome coding.

7.2.3.3 The Performance Index

Finally, the third question is that of designing an evaluation system capable of generating an appropriate performance index related to each individual in the population and in such a way that a better solution will obtain a higher performance index. This performance index will drive the search /optimization process.

In summary, the points that characterize a specific design/tuning/learning process are: the initial population of solutions obtained randomly or from some initial knowledge, the coding scheme for FSs chromosomes representing the structure according to the design process, as rule bases, membership functions clustering centers for genetic fuzzy clustering, etc., the set of evolution operators, and the evaluation function. In addition to these four points, each evolutionary learning process is characterized by a set of parameters such as the dimension of the population fixed or variable, the parameters regulating the activity of the operators or even their effect, and the parameters or conditions defining the end of the process or the time when a qualitative change in the process occurs.

7.3 Multiobjective Reliability Design Problem

Reliability optimization was generally applied to communication and transportation problems. The system reliability is defined as the probability that the system has operated to its best over a specified interval of time under given conditions. The reliability design problems include reliability analysis, reliability testing, reliability data analysis, reliability growth and so on. During 1993, Gen and Ida first proposed a simple genetic algorithm to handle reliability design problem, after then many researchers developed various reliability design problems based on genetic algorithms. In this section, lets discuss algorithms developed to solve reliability design problems.

7.3.1 Network Reliability Design

Network reliability designs are based on sharing expensive hardware and software resources and provide access to the main server system from distant locations. The

7.3 Multiobjective Reliability Design Problem

important step of network design process is to find the best layout of components to optimize certain performance criteria like cost, reliability, transmission delay or throughput. The network design reliabilities are as follows:

- All terminal network reliability–probability that every node in the network is connected to each other
- Source Sink Network Reliability–probability that the source is connected with the sink, so the source node in the network can communicate with the sink node over a specified mission time.

Genetic algorithm provides solution approaches for the optimal network design considering the above reliabilities into consideration.

7.3.1.1 Problem Description

A computer communication network can be represented by an undirected graph $G = (N, E)$ in which nodes N and edges E represents computer sites and communication cables. A graph G is connected if there is at least one path between every pair of nodes i and j, which minimally requires a spanning tree with (n-1) edges. The number of possible edges is $n(n-1)/2$. The optimal design of all terminal network reliability is defined as follows:

n is the number of nodes
$x_{ij} \in \{0, 1\}$ is a decision variable representing edges between nodes i and j
$x = \{x12, x13, \ldots, xn-1, n\}$ is a topology architecture of network design
x^* is the best solution find so far
p is the edge reliability for all edges
q is edge unreliability for all edges (p+q = 1)
$R(x)$ is the all terminal reliability of network design x
R_{min} is a network reliability requirement
$R_u(x)$ is the upper bound of reliability of the candidate network
c_{ij} is the cost of the edge between nodes i and j
c_{max} is the maximum value of c_{ij}
δ has a value of 1 if $R(x) < R_{min}$, else, has a value of 0
E' is a set of operational edges (E' $\subseteq$ E)
Ω is all operational states.

Assume that the location of each node is given and nodes are perfectly reliable, each c_{ij} and p are fixed and known, each edge is bi-directional, there are no redundant edges in the network, edges are either operational or non-operational, the edge failures are mutually independent and there is no repair.

The optimal design of the network is represented as follows:

$$min \quad Z(x) = \sum_{i=1}^{n-1} \sum_{j=i+1}^{n} c_{ij} x_{ij} \quad with \quad R(x) \geq R_{min} \quad (7.7)$$

At a particular time, only few edges of G might be operational. An operational state of G is a sub-graph G' = (V, E'). The network reliability of state E' ⊆ E is as follows:

$$\sum_{\Omega} \left(\prod_{e \in E'} p_e \right) \left(\prod_{e \in E|E'} q_e \right) \qquad (7.8)$$

7.3.1.2 Genetic Algorithm Approach

This section proposes a genetic algorithm for the design of networks when considering all terminal reliability. The assumption made here is that reliability of all edges in the network are identical, whereas cost depends on which two nodes are connected. It should be noted that only one reliability and cost alternative is available for each pair of nodes. This approach allows edges to be chosen from different components with different costs and reliabilities. The following notations are used to describe the optimal design of the network, allowing edges to be chosen from different edge options:

k is the number of options for the edge connection
t is the option between nodes
x_{ij} is an edge option for the edge between nodes i and j
$p(x_{ij})$ is the reliability design option and
$c(x_{ij})$ is a unit cost of the edge option

Representation: As each network design x is easily formed into an integer vector, it can be used as a chromosome for the genetic algorithm. Each element of the chromosome represents a possible edge in the network design problem, so there are $n \times (n-1)/2$ vector components in each candidate architecture x. The value of each element tells what type of connection the specific edge has with the pair of nodes it connects. The only possible values allowed in each position of the chromosome are $0, 1, \ldots, k-1$. The solution space of possible network architectures is $k^{(n \times (n-1))/2}$.

Fitness: The fitness function is to find the minimum-cost network architecture that meets or exceeds prespecified network reliability, R_{min}. Construction of fitness function is that it may consider infeasible network architectures, because infeasible solutions may contain beneficial information. Also, breeding two infeasible solutions or an infeasible solution with a feasible solution can yield a good feasible solution. The optimal design will lie on the boundary between feasible and infeasible designs, since only one constraint will be active or nearly active for a minimum cost network. Thus the fitness function for this problem is defined as follows:

$$Z_p(x) = Z(x) + Z(x^*)[1 + R_{min} - R(x)]^{r_p + (popsize^* \, gen)/50} \qquad (7.9)$$

7.3 Multiobjective Reliability Design Problem

where,

$Z_p(x)$–penalized cot
$Z(x)$–Unpenalized cost
$Z(x^*)$–cost of the best feasible solution in the population
r_p–penalty rate
popsize–population size
gen–number of generations

Algorithm: The overall algorithm for the minimization of the cost of the network is as follows:

Step 1: Set the parameters, population size (popsize), population percentage mutated (p_{m1}), mutation rate (p_{m2}), penalty rate (r_p), the maximum generation (maxgen) and initialize number of generations gen = 0.

Step 2: Initialize a) Randomly generate the initial population

b) Send the initial population for reliability calculation
c) Send the initial population to the cost calculation function (fitness). If infeasible chromosome exist, they are penalized.
d) Test for the best initial solution. If no chromosome is feasible, the best infeasible chromosome is recorded.

Step 3: Selection a) Insert the best chromosome into the new population

b) Select two distinct candidate chromosomes from the current population by the rank-based selection process.

Step 4: Perform Crossover. Uniform crossover is performed.

Step 5: Perform Mutation. After crossover once a child is created, then mutate it.

Step 6: Check the number of children. If n <popsize-1, goto step 3; else goto step 6, where n represents the number of new children.

Step 7: Form the new population. Replace the parents with children that are created.

Step 8: Evaluate a) Send the new population to the reliability calculation function

b) Calculate fitness function for each chromosome in the new population. If infeasible chromosome exist, they are penalized

Step 9: Check for the best new chromosome. Save the best new chromosome; if no chromosome is feasible, then the best infeasible chromosome is noted.

Step 10: Check the terminating condition. If gen<maxgen, gen=gen+1, and goto step 3 for the next generation. If gen = maxgen, then terminate.

The steps involved for reliability calculation is as follows:

Reliability Calculation: A back tracking algorithm is used to calculate the exact unreliability of the system, 1-R(x), for problems due to their computationally tractable size. The below given back tracking algorithm is used where the probability of all edges within a stack is the product of the failure probabilities of all inoperative edges times the product of 1 minus the failure probabilities of all operative edges.

Step 1: Initialize all edges as free and create a stack S that is empty initialize.

Step 2: Generate a modified cutest

a) Find a set of free edges that together with all inoperative edges will form a network cut.

b) Mark all the edges found in the above step inoperative and add them to the stack.

c) Now the stack represents a modified cut-set; add its probability to a cumulative sum.

Step 3: Backtrack process.

a) If the stack is empty, move to step 4, else, goto step 3- (b) below.

b) Take the edge off the top of the stack.

c) If the edge is inoperative and if when making it operative, a spanning tree of operative edges exists, mark it free and goto step 3-(a).

d) If the edge is inoperative and the condition tested in step 3-(c) does not hold, mark it operative, put it back on the stack, and go to step 2.

e) If the edge is operative, mark it free and go to step 3-(a).

Step 4: Return the network unreliability and end the procedure.

Thus the network reliability design problem can be efficiently solved using Genetic algorithm approach.

7.3.2 Bicriteria Reliability Design

The bicriteria design problem here maximizes the reliability of a series system and simultaneously minimizes the total cost of the system. The problem is a variation of the optimal reliability allocation problem, which is formulated as anon-linear mixed integer programming as follows:

7.3 Multiobjective Reliability Design Problem

$$\max \quad f_1(m, x) = \prod_{i=1}^{n} [1 - (1 - x_i)m_i]$$
$$\min \quad f_2(m, x) = \sum_{i=1}^{n} C(x_i) \left[m_i + exp\left(\frac{m_i}{4}\right) \right]$$
$$such\ that \quad G_1(m) = \sum_{i=1}^{n} w_i m_i exp\left(\frac{m_i}{4}\right) \leq W_s \quad (7.10)$$
$$G2(m) = \sum_{i=1}^{n} v_i (m_i)^2 \leq V_s$$
$$1 \leq m_i \leq 10, \quad 0.5 \leq x_i \leq 1 - 10^{-6} \quad i = 1, \ldots, 4$$

where,

m_i–number of redundant components in subsystem i
x_i–level of component reliability for the ith subsystem
$f_1(m,x)$–reliability of the system with redundant components m and component reliabilities x
$f_2(m,x)$–total cost of the system with component allocation m and component reliability x
v_i–product of weight and volume per element in subsystem i
w_i–weight of each components in subsystem i

and $C(x_i)$–cost of each component with reliability x_i at subsystem i is obtained as follows:

$$C(x_i) = \alpha_i \left(\frac{-O_T}{ln(x_i)} \right)^{\beta} \quad i = 1, \ldots, 4 \quad (7.11)$$

where, α_i and β_i are constants representing the physical characteristics of each component in subsystem i, and O^T is the operating time during which the component must not fail.

7.3.2.1 Genetic Algorithm Approach

The above problem is solved using the GA approach as follows:
Let v_k denote the kth chromosome in a population as follows:

$$v_k = [(m_{k1}, x_{k1})(m_{k2}, x_{k2})(m_{k3}, x_{k3})(m_{k4}, x_{k4})] \quad k = 1, 2, \ldots \text{popsize} \quad (7.12)$$

The initial population is produced such that each gene in a chromosome is generated randomly within its domain. The fitness of chromosomes is calculated by ranking method as follows:

Fitness Evaluation

Step 1: Calculate each objective value for each chromosome
Step 2: Chromosomes are ranked based on their objective function values and obtaining the order $r_i(v_k).r_i(v_k)$ is the rank value of the ith objective value of chromosome v_k and is obtained by setting a value of 1 on the best objective function value and popsize on the worst objective function of the present population.

If objective function is to be maximized, then,
$r_i(v_k)$–set as 1 on the largest objective value
popsize–on smallest objective function value

If objective function is to be minimized, then,
$r_i(v_k)$–set as 1 on the smallest objective value
popsize–on largest objective function value

Step 3: Compute the fitness value using the following equation:

$$eval(v_k) = \sum_{i=1}^{Q} r_i(v_k)$$

where, Q is the number of objective functions.

Calculate the evaluation function $eval(v_k)$, and select chromosomes among the parents and offspring that are superior to the others. The number to be selected is popsize.

Crossover

An arithmetic crossover operator is used, which is a linear combination of two chromosomes.

Mutation

Uniform mutation is performed here. This operator ensures that the GA can search the solution space freely.

Thus, these network design problems that consider a system reliability constraint or objective have many applications in telecommunications, computer networking and in domains like electric, gas and sewer networks.

7.4 Combinatorial Optimization Problem

Combinatorial optimization is a branch of optimization in applied mathematics and computer science, related to operations research, algorithm theory and computational complexity theory that sits at the intersection of several fields, including

7.4 Combinatorial Optimization Problem

artificial intelligence, mathematics and software engineering. Combinatorial optimization algorithms solve instances of problems that are believed to be hard in general, by exploring the usually-large solution space of these instances. Combinatorial optimization algorithms achieve this by reducing the effective size of the space, and by exploring the space efficiently.

Combinatorial optimization problems are concerned with the efficient allocation of limited resources to meet desired objectives when the values of some or all of the variables are restricted to be integral. Constraints on basic resources, such as labor, supplies, or capital restrict the possible alternatives that are considered feasible. Still, in most such problems, there are many possible alternatives to consider and one overall goal determines which of these alternatives is best. For example, most airlines need to determine crew schedules which minimize the total operating cost; automotive manufacturers may want to determine the design of a fleet of cars which will maximize their share of the market; a flexible manufacturing facility needs to schedule the production for a plant without having much advance notice as to what parts will need to be produced that day. In today's changing and competitive industrial environment the difference between using a quickly derived "solution" and using sophisticated mathematical models to find an optimal solution can determine whether or not a company survives.

The versatility of the combinatorial optimization model stems from the fact that in many practical problems, activities and resources, such as machines, airplanes and people, are indivisible. Also, many problems have only a finite number of alternative choices and consequently can appropriately be formulated as combinatorial optimization problems—the word combinatorial referring to the fact that only a finite number of alternative feasible solutions exists. Combinatorial optimization models are often referred to as integer programming models where programming refers to "planning" so that these are models used in planning where some or all of the decisions can take on only a finite number of alternative possibilities.

Combinatorial optimization is the process of finding one or more best (optimal) solutions in a well defined discrete problem space. Such problems occur in almost all fields of management (e.g., finance, marketing, production, scheduling, inventory control, facility location and layout, data-base management), as well as in many engineering disciplines (e.g., optimal design of waterways or bridges, VLSI-circuitry design and testing, the layout of circuits to minimize the area dedicated to wires, design and analysis of data networks, solid-waste management, determination of ground states of spin-glasses, determination of minimum energy states for alloy construction, energy resource-planning models, logistics of electrical power generation and transport, the scheduling of lines in flexible manufacturing facilities, and problems in crystallography).

Combinatorial optimization algorithms are often implemented in an efficient imperative programming language, in an expressive declarative programming language such as Prolog, or some compromise, perhaps a functional programming language such as Haskell, or a multi-paradigm language such as LISP.

A study of computational complexity theory helps to motivate combinatorial optimization. Combinatorial optimization algorithms are typically concerned with problems that are NP-hard. Such problems are not believed to be efficiently solvable in general. However, the various approximations of complexity theory suggest that some instances (e.g., "small" instances) of these problems could be efficiently solved. This is indeed the case, and such instances often have important practical ramifications.

Informal definition: The domain of combinatorial optimization is optimization problems where the set of feasible solutions is discrete or can be reduced to a discrete one, and the goal is to find the best possible solution.

Formal definition: An instance of a combinatorial optimization problem can be described in a formal way as a tuple (X,P,Y,f,extr) where

- X is the solution space (on which f and P are defined)
- P is the feasibility predicate.
- Y is the set of feasible solutions.
- f is the objective function.
- extr is the extreme (usually min or max).

7.4.1 Linear Integer Model

We assume throughout this discussion that both the function to be optimized and the functional form of the constraints restricting the possible solutions are linear functions. Although some research has centered on approaches to problems where some or all of the functions are nonlinear, most of the research to date covers only the linear case. The general linear integer model is

$$\max \sum_{j \in B} c_j x_j + \sum_{j \in I} c_j x_j + \sum_{j \in C} c_j x_j$$

subject to:

$$\sum_{j \in B} a_{ij} x_j + \sum_{j \in I} a_{ij} x_j + \sum_{j \in C} a_{ij} x_j \sim b_i \quad (i = 1, \ldots, m)$$

$$l_j \leq x_j \leq u_j \quad (j \in I \cup c)$$
$$x_i \in \{0, 1\} \quad (j \in B)$$
$$x_j \in integers \quad (j \in I)$$
$$x_j \in reals \quad (j \in C) \tag{7.13}$$

where B is the set of zero-one variables, I is the set of integer variables, C is the set of continuous variables, and the $\sim$ symbol in the first set of constraints denotes the fact that the constraints i = 1, ..., m can be *either approximate, or* =. The data l_j and u_j are the lower and upper bound values, respectively, for variable x_j. As we are discussing the integer case, there must be some variable in B È I. If C = I = f, then

7.4 Combinatorial Optimization Problem

the problem is referred to as a pure 0-1 linear-programming problem; if C = f, the problem is called a pure integer (linear) programming problem. Otherwise, the problem is a mixed integer (linear) programming problem. Here, we call the set of points satisfying all constraints S, and the set of points satisfying all but the integrality restrictions P.

7.4.2 Applications of Combinatorial Optimization

We describe some classical combinatorial optimization models to provide both an overview of the diversity and versatility of this field and to show that the solution of large real-world instances of such problems requires the solution method exploit the specialized mathematical structure of the specific application.

7.4.2.1 Knapsack Problems

Suppose one wants to fill a knapsack that can hold a total weight of W with some combination of items from a list of n possible items each with weight w_i and value v_i so that the value of the items packed into the knapsack is maximized. This problem has a single linear constraint (that the weight of the items in the knapsack not exceed W), a linear objective function which sums the values of the items in the knapsack, and the added restriction that each item either be in the knapsack or not—a fractional amount of the item is not possible.

Although this problem might seem almost too simple to have much applicability, the knapsack problem is important to cryptographers and to those interested in protecting computer files, electronic transfers of funds and electronic mail. These applications use a "key" to allow entry into secure information. Often the keys are designed based on linear combinations of some collection of data items, which must equal a certain value. This problem is also structurally important in that most integer programming problems are generalizations of this problem (i.e., there are many knapsack constraints which together compose the problem). Approaches for the solution of multiple knapsack problems are often based on examining each constraint separately.

An important example of a multiple knapsack problem is the capital budgeting problem. This problem is one of finding a subset of the thousands of capital projects under consideration that yields the greatest return on investment, while satisfying specified financial, regulatory and project relationship requirements.

7.4.2.2 Network and Graph Problems

Many optimization problems can be represented by a network where a network (or graph) is defined by nodes and by arcs connecting those nodes. Many practical problems arise around physical networks such as city streets, highways, rail systems, communication networks, and integrated circuits. In addition, there are many problems which can be modeled as networks even when there is no underlying physical network. For example, one can think of the assignment problem where one wishes to assign a set of persons to some set of jobs in a way that minimizes the cost of the assignment. Here one set of nodes represents the people to be assigned, another set of nodes represents the possible jobs, and there is an arc connecting a person to a job if that person is capable of performing that job.

7.4.2.3 Space-Time Networks are Often Used in Scheduling Applications

Here one wishes to meet specific demands at different points in time. To model this problem, different nodes represent the same entity at different points in time. An example of the many scheduling problems that can be represented as a space-time network is the airline fleet assignment problem, which requires that one assign specific planes to pre-scheduled flights at minimum cost. Each flight must have one and only one plane assigned to it, and a plane can be assigned to a flight only if it is large enough to service that flight and only if it is on the ground at the appropriate airport, serviced and ready to depart when the flight is scheduled to take off. The nodes represent specific airports at various points in time and the arcs represent the flows of airplanes of a variety of types into and out of each airport. There are layover arcs that permit a plane to stay on the ground from one time period to the next, service arcs which force a plane to be out of duty for a specified amount of time, and connecting arcs which allow a plane to fly from one airport to another without passengers.

In addition, there are many graph-theoretic problems which examine the properties of the underlying graph or network. Such problems include the Chinese postman problem where one wishes to find a path (a connected sequence of edges) through the graph that starts and ends at the same node, that covers every edge of the graph at least once, and that has the shortest length possible. If one adds the restriction that each node must be visited exactly one time and drops the requirement that each edge be traversed, the problem becomes the notoriously difficult traveling salesman problem. Other graph problems include the vertex coloring problem, the object of which is to determine the minimum number of colors needed to color each vertex of the graph in order that no pair of adjacent nodes (nodes connected by an edge) share the same color; the edge coloring problem, whose object is to find a minimum total weight collection of edges such that each node is incident to at least one edge; the maximum clique problem, whose object is to find the largest subgraph of the original graph such that every node is connected to every other node in the subgraph; and the minimum cut problem, whose object is to find a minimum weight collection of edges which (if removed) would disconnect a set of nodes s from a set of nodes t.

Although these combinatorial optimization problems on graphs might appear, at first glance, to be interesting mathematically but have little application to the decision making in management or engineering, their domain of applicability is extraordinarily broad. The traveling salesman problem has applications in routing and

7.4 Combinatorial Optimization Problem   181

scheduling, in large-scale circuitry design and in strategic defense. The four-color problem (Can a map be colored in four colors or less?) is a special case of the vertex-coloring problem. Both the clique problem and the minimum cut problem have important implications for the reliability of large systems.

7.4.2.4 Scheduling Problems, Which are Rule-based

There are many problems where it is impossible to write down all of the restrictions in a mathematically "clean" way. Such problems often arise in scheduling where there are a myriad of labor restrictions, corporate scheduling preferences and other rules related to what constitutes a "feasible schedule." Such problems can be solved by generating all, or some reasonable large subset of the feasible schedules for each worker. One associates a matrix with such problems whose rows correspond to the tasks considered and whose columns correspond to individual workers, teams or crews. A column of the matrix has an entry of one in those rows that correspond to tasks that the worker will be assigned and a zero otherwise. Each "feasible" schedule defines one column of the constraint matrix and associated with each such schedule is a value. Thus the matrix of constraints consists of all zeroes and ones and the sense of the inequality indicates whether that job must be covered by exactly a specified number of people (called set partitioning), that it must be covered by at least a specific number (called set covering) or that it must be covered by not more that a specified number (called set packing). The optimization problem is then the problem of finding the best collection of columns, which satisfy these restrictions.

Apart from the above discussed, it can also be used to solve:

- Traveling salesman problem
- Minimum spanning tree problem
- Linear programming
- Eight queens puzzle

7.4.2.5 Solution Techniques for Integer Programming

Solving combinatorial optimization problems, i.e. finding an optimal solution to such problems, can be a difficult task. The difficulty arises from the fact that unlike linear programming, for example, whose feasible region is a convex set, in combinatorial problems, one must search a lattice of feasible points or, in the mixed-integer case, a set of disjoint half-lines or line segments to find an optimal solution. Thus, unlike linear programming where, due to the convexity of the problem, we can exploit the fact that any local solution is a global optimum, integer programming problems have many local optima and finding a global optimum to the problem requires one to prove that a particular solution dominates all feasible points by arguments other than the calculus-based derivative approaches of convex programming.

7.4.3 Methods

Heuristic search methods (metaheuristic algorithms) as those listed below have been used to solve combinatorial optimization problems.

- Local search
- Simulated annealing
- Quantum annealing
- GRASP
- Swarm intelligence
- Tabu search
- Genetic algorithms
- Quantum Based Genetic Algorithm
- Ant colony optimization , Reactive search

7.4.3.1 Quantum Based Genetic Algorithm to Solve N-Queens Problem

Genetic algorithms (GAs) are an approximated approach that has proved its efficiency for solving combinatorial optimization problems. Genetic algorithms use the biological Darwinian principal to optimize the solution obtained or measured. These algorithms work on a set of structures called chromosomes representing the solution to be optimized.

Another research field called Quantum Computing (QC) has appeared and induced intense researches in the last decade. This evolution that takes its origins from the quantum physics principles reduces remarkably the complexity. This is offered by the possibility of parallel computing. Such possibility of parallel computing can be exploited to solve combinatorial optimization problems, which use a great set of data. So quantum computing allows the possibility of designing very powerful algorithms. However, these algorithms may not be well exploited before developing powerful quantum machines. Awaiting the construction of such machines, the idea of simulating quantum algorithms on classical computers or to combine them to other conventional methods has appeared. At this prospect, lets study the genetic quantum hybridization and its contribution in solving combinatorial problems.

N-Queens Problem

The N-Queens problem is a classical artificial intelligence problem. It is a general case of the 8-Queens problem. This combinatorial optimization problem has been studied for more than a century. A chess player, Max Bezzel, first introduced the 8-Queens problem in 1848. The N-Queens problem, as generalization of the 8-Queens problem, was first proposed in 1867. Since 1850, the problem has attracted the attention of several famous mathematicians including Gauss, Polya, and Lucas.

The N-Queens problem can be defined as follows: place N queens on an NxN chessboard, one queen on each square, so that no queen captures any of the others, that is, a configuration in which there exists at most one queen on the

7.4 Combinatorial Optimization Problem

Fig. 7.1 Representation of the N-Queens problem where N = 8

same row, column or diagonal. A representation of N-Queens problem is given on a chessboard in Fig. 7.1.

During the last three decades, the problem has been discussed in the context of computer science and used as an example of backtrack algorithms, permutation generation, divide and conquer paradigm, program development methodology, constraint satisfaction problems, integer programming, specification and neural networks.

A common way to solve this problem consists in trying to put the queens on the board squares one after the other. If one queen threatens the newly introduced queen, we withdraw the queen and search for another position. If we cannot find a solution, we choose to remove a queen already positioned, assign it another position that has not yet been used, and start the search again. This last operation is called a backtrack, and the whole strategy is called a trial-and-error algorithm. It is known that for $n=8$, there are exactly 92 solutions, or less if we consider symmetric solutions as equal.

The number of solutions for n = 1, 2 ... 15, is 1, 0, 0, 2, 10, 4, 40, 92, 352, 724, 2680, 14200, 73712, 365596, 2279184.

Quantum Computing

In early 80s, Richard Feynman's observed that certain quantum mechanical effects cannot be simulated efficiently on a classical computer. His observation led to speculation that computation in general could be done more efficiently if it used this quantum effects. This speculation proved justified in 1994 when Peter Shor described a polynomial time quantum algorithm for factoring numbers. In quantum systems, the computational space increases exponentially with the size of the system, which enables exponential parallelism. This parallelism could lead to exponentially faster quantum algorithms than possible classically.

N-Queens Problem Solving

Conventional GAs operate on a set of individuals (chromosomes) forming a population. To be more representative this population must contains a fit number of chromosomes. This makes the solution space very large. So, the classical GAs are usually very costly. For reducing the number of chromosomes and consequently reducing the heaver computation time, we propose here an algorithm called Quantum Genetic Algorithm (QGA).

A QGA is a GA with quantum coding solutions. This representation will reduce the computation time by of increasing the number of chromosomes. Moreover we believe that it will give a better global solution. As in genetic algorithms, initial solutions are encoded in N chromosomes representing the initial population. The difference in a QGA is that each chromosome does not encode only one solution but all the possible solutions by putting them within a superposition

The Solution Modeling

Every queen on a checker square can reach the other squares that are located on the same horizontal, vertical, and diagonal line. So there can be at most one queen at each horizontal line, at most one queen at each vertical line, and at most one queen at each of the 4n-2 diagonal lines. Furthermore, since we want to place as many queens as possible, namely exactly n queens, there must be exactly one queen at each horizontal line and at each vertical line.

For a 1 × 1 board, there is one trivial solution:

For 2 × 2 and 3 × 3 boards, there are no solutions. For a 4 × 4 board, there are two:

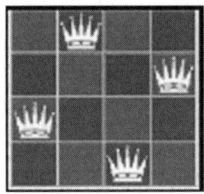 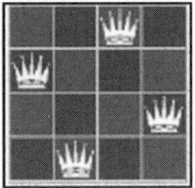

These are considered distinct solutions, even though the solutions are mirror images of each other. There is no quick and easy way to calculate the total number of NQP solutions for an NxN board. We can represent the NQP solution by an NxN matrix *A* containing only N ones and satisfying the constraint that only one 1 can be in a raw, in a column or in a diagonal.

For example the matrix below in in Fig. 7.2 represents a solution of the 4-Queens problem:

Fitness function

The penalty of one queen is equal to the number of queens she can check. The fitness of the configuration is equal the sum of all the queens penalties divided by two (deleting redundancy counting). For example the fitness of the solution presented in Fig. 7.2 is 0 and the fitness of the matrix solution in Fig. 7.3 is 8.

7.4 Combinatorial Optimization Problem

Fig. 7.2 A solution matrix of the 4-Queens problem

```
0 0 1 0
1 0 0 0
0 0 0 1
0 1 0 0
```

Fig. 7.3 A bad solution matrix of the 4-Queens problem

```
1 1 1 1
0 0 0 0
0 0 0 1
0 0 0 0
```

Quantum Representation

The solution representation given above can make the search space representation in a genetic algorithm very large. Because of this, we propose another representation of the solution (the chromosome). A quantum encoding offers a powerful mean to represent the solution space and reduces by the way the required number of chromosomes.

We have represented the solution by a quantum matrix which is equivalent to a chromosome in a conventional GA. For example the following matrix is a quantum matrix representing a 4*4 qubits, Such as every qubit represents a superposition of 1 and 0 states. So this matrix represents a superposition of all solutions including incorrect solutions.

The quantum solution matrix is a $(2*N)*N$ matrix which represents the superposition of all possible matrix (Fig. 7.4).

Quantum Genetic Algorithm for NQP solving

During the whole process we keep in memory the global best solution. The algorithm consists on applying cyclically the following quantum genetic operations:

The first operation is a quantum interference, which allows a shift of each qubit in all the direction of the corresponding bit value in the best solution. That is performed by applying a unitary quantum operator, which achieves a rotation whose angle is function of $ái$, $âi$ and the value of the corresponding bit in the best solution (Fig. 7.5). The second operation is a cross-over performed between each pair of chromosomes at a random position. Here is an example of a cross-over between two chromosomes (Fig. 7.6).

Fig. 7.4 A quantum solution matrix

```
0.4134  0.8435  0.8597  0.6633
0.9105  0.5371  0.5109  0.7484
0.9759  0.1819  0.6313  0.2524
0.2182  0.9833  0.7755  0.9676
0.5261  0.8530  0.7707  0.8326
0.8504  0.5219  0.6372  0.5539
0.5035  0.2984  0.2579  0.1440
0.8640  0.9544  0.9662  0.9896
```

Fig. 7.5 Quantum interference

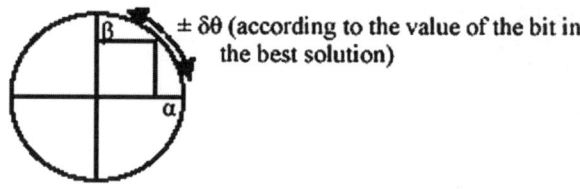

± δθ (according to the value of the bit in the best solution)

The third operation consists of a quantum mutation which will perform for some qubits, according to a probability, a permutation between their values *ái* and *âi*. That will invert the probabilities of having the values 0 and 1 by a measurement. An example is given in Fig. 7.7:

Finally, we perform a selection of *m* chromosomes among the *n* existing in the current generation. For this, we apply first a measurement on each chromosome to have from it one solution among all those present in superposition. But unlike pure quantum systems, the measurement here does not destroy the states' superposition. Since our algorithm operates on conventional computer and does not require the presence of a quantum machine, it is possible and in our interest to keep all the possible solutions in the superposition for the next iterations. For each measurement result, we extract a distribution of N-Queens. To evaluate the quality of a solution, we compute its fitness. The best solution is that having the minimal fitness (0 in the optimum).

Parents

0.0369	*0.4527*	*0.3390*	*0.0744*		*0.3778*	**0.1162**	**0.2699**	**0.7505**
0.9993	*0.8917*	*0.9408*	*0.9972*		*0.9259*	**0.9932**	**0.9629**	**0.6609**
0.0167	*0.5603*	*0.9884*	*0.2837*		*0.9956*	**0.1474**	**0.4823**	**0.4570**
0.9999	*0.8283*	*0.1518*	*0.9589*		*0.0939*	**0.9891**	**0.8760**	**0.8895**
0.9629	*0.3028*	*0.8884*	*0.1757*		*0.3332*	**0.1242**	**0.3013**	**0.3768**
0.2697	*0.9531*	*0.4591*	*0.9844*		*0.9428*	**0.9923**	**0.9535**	**0.9263**
0.4741	*0.7258*	*0.9857*	*0.3620*		*0.7492*	**0.9922**	**0.2232**	**0.7910**
0.8805	*0.6880*	*0.1686*	*0.9322*		*0.662*	**0.1250**	**0.9748**	**0.6119**
Parent 1.					Parent 2.			

Children

0.0369	0.1162	0.2699	0.7505
0.9993	0.9932	0.9629	0.6609
0.0167	0.1474	0.4823	0.4570
0.9999	0.9891	0.8760	0.8895
0.9629	0.1242	0.3013	0.3768
0.2697	0.9923	0.9535	0.9263
0.4741	0.9922	0.2232	0.7910
0.8805	0.1250	0.9748	0.6119

Fig. 7.6 Quantum cross-over

0.2383	0.1079	**0.5057**	0.8337
0.9712	0.9942	**0.8627**	0.5522

0.2383	0.1079	**0.8627**	0.8337
0.9712	0.9942	**0.5057**	0.5522

Fig. 7.7 Quantum mutation

7.5 Scheduling Problems

In today's complex manufacturing setting, with multiple lines of products, each requiring many different steps and machines for completion, the decision maker for the manufacturing plant must find a way to successfully manage resources in order to produce products in the most efficient way possible. The decision maker needs to design a production schedule that promotes on-time delivery, and minimizes objectives such as the flow time of a product. Out of these concerns grew an area of studies known as the scheduling problems.

Scheduling problems involve solving for the optimal schedule under various objectives, different machine environments and characteristics of the jobs. Some of the objectives of the scheduling problems include minimizing the makespan, or the last completion time of a job, minimizing the total completion time of all jobs, and minimizing the total "lateness" of jobs. The user can select any number of jobs and any number of parallel machines. Scheduling problems occurs almost everywhere in real-world scenario, especially in the industrial engineering world. Many scheduling problems from manufacturing process are quite complex in nature and very difficult to solve by conventional optimization techniques. They belong to NP-hard problems. This has paved way for the use of genetic algorithms to these types of problems. The various scheduling problems include:

- Job shop scheduling
- Multiprocessor scheduling
- Multitask scheduling
- Parallel Machine scheduling
- Group Job scheduling
- Resource constrained project scheduling
- Dynamic task scheduling and so on.

In the forthcoming section lets discuss the application of genetic algorithm to job shop scheduling problem.

7.5.1 Genetic Algorithm for Job Shop Scheduling Problems (JSSP)

Scheduling, especially job shop scheduling, has been studying for a long time. Because of its NP-Hard nature, there has not been found a global problem solver for this kind of problems. Recently, some meta-heuristics like Simulated Annealing (SA), Taboo Search (TS), and Genetic Algorithms (GA) have been implemented as pure methods and hybrid of different method, where the hybrid methods are superior over pure ones. The main problem is how to cope with local minima within a reasonable time. Among them, GA has been studied and implemented to like the other problems with success.

The JSSP consists of a number of machines, M, and a number of jobs, J. Each job consists of M tasks, each of fixed duration. Each task must be processed on a single specified machine, and each job visits each machine exactly once. There is a predefined ordering of the tasks within a job. A machine can process only one task at a time. There are no set-up times, no release dates and no due dates. The makespan is the time from the beginning of the first task to start to the end of the last task to finish. The aim is to find start times for each task such that the makespan is minimized. As a constraint problem, there are M*J variables, each taking positive integer values. The start time of *tth* task of the *jth* job will be denoted by *xjt*, and the duration of that task by *djt*. Each job introduces a set of *precedence* constraints on the tasks within that job: *xjt + djt _ xj(t+1)* for $t = 1$ to $M - 1$. Each machine imposes a set of *resource* constraints on the tasks processed by that machine: *xjt + djt _ xpq or xpq + dpq _ xjt*. The aim is to find values for the variables such that no constraint is violated. By defining an objective function on assignments (which simply takes the maximum of *xjt + djt*), and attempting to minimize the objective, we get a constraint optimization problem.

7.5.1.1 Types of Schedules

Schedules can be classified into one of following three types of schedules:

- *Semi-active schedule:* These feasible schedules are obtained by sequencing operations as early as possible. In a semi-active schedule, no operation can be started earlier without altering the processing sequences.
- *Active schedule:* These feasible schedules are schedules in which no operation could be started earlier without delaying some other operation or breaking a precedence constraint. Active schedules are also semi-active schedules. An optimal schedule is always active, so the search space can be safely limited to the set of all active schedules.
- *Non-delay schedule*: These feasible schedules are schedules in which no machine is kept idle when it could start processing some operation. Non-delay schedules are necessarily active and hence also necessarily semi-active.

7.5.1.2 The Genetic Algorithm Approach

The genetic algorithms (GA) mimic the evolution and improvement of life through reproduction, where each individual contributes with its own genetic information to build up new ones adapted to the environment with higher chances of survival. This is one of the main ideas behind genetic algorithms and genetic programming. Specialized Markov Chains underline the theoretical basis of GA in terms of change of states and search procedures. Each 'individual' of a generation represents a feasible solution as coded in a chromosome with distinct algorithms /parameters to be evaluated by a fitness function. GA operators are mutation (the change of a randomly chosen bit of the chromosome) and crossover (the exchange of randomly chosen slices of chromosome).

Figure 7.8 shows a generic cycle of GA where the best individuals are continuously being selected and operated by crossover and mutation. Following a number of generations, the population converges to the solution that performs better.

7.5 Scheduling Problems

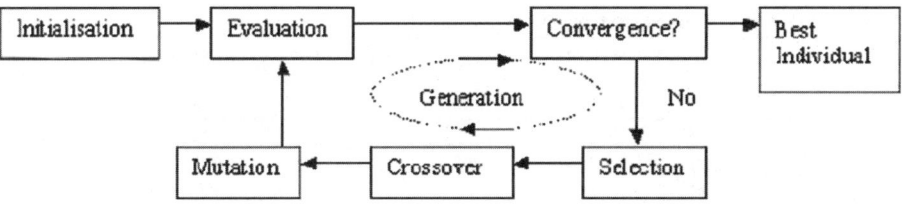

Fig. 7.8 Genetic algorithm: the sequence of operators and evaluation of each individual

GA applications for JSSP have special chromosome representation as well as genetic operators to be applied to feasible schedules. In our case, the chromosomes are coded as a list of sets of numerical values for each particular schedule.

7.5.1.3 Genetic Algorithms in JSSP

Schedules are generated in a particular way in which the chromosome will be feasible after performing genetic operators. The decision management in JSSP distributes the jobs for each machine, selecting sometimes one task among the other alternatives so as to have a better fitness. Chromosome is coded with M*J values between 0 and 1, one for each decision, which points for the job in the requesting jobs list that win the right to use the machine. Figure 7.9 shows an example of chromosome coding based in decision process. This approach allows using the same traditional GA operators to solve the problem because the chromosome contains a sequence of numbers, all representing feasible schedules. Two fitness functions can be applied to evaluate a solution: the makespan and the total idle time. The main problems with this approach are as follows:-

- the disruptive effect of crossover operator,
- the precocious convergence to some local minima that blinds the system to find the global one,
- eventually raising the inability of genetic operators to permute the solutions in a reasonable time,
- the lack of hill climbing in GA.

Job	Tasks			Job	Tasks			Job	Tasks		
0	1	0	2	0	1	0	2	0	1	0	2
1	0	1	2	1	1	2		1	1	2	
2	2	1	0	2	2	1	0	2	2	1	0
3	1	2	0	3	1	2	0	3	2	0	

Decision 1: machine 0 receives job 1, any value of the chromosome (only one candidate).

Decision 2: machine 1 could receive jobs 0, 1 or 3. If chromosome value is 0.7, job 3 begins

Decision 3: Machine 2 is required by jobs 2 and 3. If chromosome >0.5 then job 3 executes, otherwise job 2.

Fig. 7.9 Decisions sequence in JSSP

Table 7.1 Results obtained from genetic algorithms

	# Dec.	Opt.	GA makespan		GA Idle time	
			Time	%	Time	%
ABZ5	100	1234	1313	6.4	1314	6.4
ABZ6	100	943	994	5.4	982	4.1
LA18	100	848	940	10.8	897	5.7
LA17	100	784	872	11.2	854	8.9
LA20	100	902	979	8.5	1006	11.5
LA16	100	945	1031	9.1	1036	9.6
LA19	100	842	945	12.2	965	14.6
ORB01	100	1059	1230	16.1	1229	16.0
LA25	150	977	1129	15.5	1207	23.5
FT10	100	930	1032	10.9	1010	8.6
LA24	150	935	1091	16.6	1083	15.8
LA21	150	1046	1222	16.8	1260	20.4
LA27	200	1235	1507	22.0	1538	24.5
ABZ7	300	655	779	18.9	787	20.1
LA38	225	1196	1476	23.4	1485	24.1
LA40	225	1222	1569	28.3	1540	26.0
LA29	200	1130	1422	25.8	1501	32.8
ABZ9	300	656	795	21.1	843	28.5
ABZ8	300	638	823	28.9	827	29.6

The total size of the population is 100 individuals and the number of generations is number machines*number job *10, the crossover probability is 80% and mutation probability is 25%, and each decision is coded as 8 bits. The increase of population size does not change the results too much, but increase the processing time. For example, in LA20 problem using makespan fitness, the final result was 959 time units after 35 generations of 900 individuals (no better solution was founded until 149,500 generations) and 979 time units with 100 individuals. Table 7.1 shows the results for both fitness functions at columns "GA makespan" and "GA Idle time". Table 7.1 shows the results of several problems of Operations Research literature available in Internet. The columns "Time" contains the makespan and "%" contains the percentage of the optimal value (column "Opt.").

Thus, Genetic Algorithms can be applied to a wide range of scheduling problems of all kind.

7.6 Transportation Problems

The transportation problems include the determination of optimum transportation patterns, analysis of production scheduling problems including inventory control, transshipment problems and several other assignment problems. In this section lets discuss how genetic algorithm is applied to solve transportation problems.

7.6.1 Genetic Algorithm in Solving Transportation Location-Allocation Problems with Euclidean Distances

A transportation location-allocation problem is a problem in which both optimal source locations and the optimal amounts of shipments from sources to destinations are to be found.

7.6.1.1 Problem Description

Although the general transportation-location problem refers simply to "sources" and "destinations," for clarity, the algorithm will solve a particular example of a transportation-location problem, namely, identifying the optimal location of new powerplants to supply the new (or future) energy demands of a number of cities. The objective of this problem is to minimize the total power distribution cost. The power distribution cost is the sum of the products of the power distribution cost (per unit amount, per unit distance), the distance between the plant and the city, and the amount of power supplied from the plant to the city, for all plants and all cities. For each city, the total amount of energy supplied by all plants is made equal to the total demand of that city. And for each plant, the total amount of energy supplied by the plant is to be less than or equal to the total capacity of the plant.

The mathematical form of the problem can be written as,

$$Min, Cost(C) = \sum_{i=1}^{n} \sum_{j=1}^{m} \phi . \delta_{ij} . N_{ij}$$

$$subject\ to: \sum_{i=1}^{P} v_{ij} = d_j \quad for \quad j = 1, m$$

$$\sum_{j=1}^{C} v_{ij} \leq c_i \quad for \quad i = 1, n$$

Where

$\phi =$	transportation cost per unit amount per unit distance
$\delta_{ij} =$	distance from source i to destination j
$v_{ij} =$	amount supplied from source i to destination j
$n =$	number of plants
$m =$	number of cities
$x_i, y_i =$	X & Y coordinates of the source i
$x_j, y_j =$	X & Y coordinates of the destination j
$d_j \quad =$	demand of the destination
$c_i \quad =$	source capacity

(7.14)

It is noticed that the Euclidean distance term δ_{ij}, can be calculated using the equation given below.

$$\delta_{ij} = \sqrt{(x_i - x_j)^2 + (y_i - y_j)^2} \tag{7.15}$$

7.6.1.2 Genetic Algorithm Approach

A Two-Phase method is implemented to the solve location—allocation problem. Phase 1 involves the Genetic Algorithm technique, which is used to minimize the transportation cost by varying the source locations. Phase 2 includes a Linear Programming technique to allocate the power from the sources to the destinations in accordance with the constraints.

Phase 1

>   Step 1 : The locations and demands for each city; the lower and upper limits for the plant locations; the plant capacities; the population; and the number of generations are specified. The upper and lower limits are used to create the initial random population of the source locations.
>   Step 2: The objective function (7.15) is evaluated for the random population of plant locations by calling the phase 2 subroutine, which optimally allocates power from the plants to the demand points, and insures that the constraints are satisfied.
>   Step 3: The X and Y locations of all of the plants of the initial population are converted to base 10 integers and converted to their binary forms. From the objective function values, the probabilities and the cumulative probabilities for each individual in the population are calculated.
>   Step 4: Parent selection is made on the basis of fitness function. Individuals having higher fitness values are chosen more often. The greater the fitness value of an individual the more likely that the individual will be selected for recombination. The selection of mating parents is done by roulette wheel selection, in which a probability to each individual, i,
>
> $P_i = f_i/f_1 + f_2 + f_3 \ldots \ldots$ where P_i = Probability of individual,
> f_i = fitness values (7.16)
>
>   is computed. A parent is then randomly selected, based on this probability.
>   Step 5: The parents thus selected are made to mate using a single-point crossover method. The offspring thus obtained form a new population of plant locations. The binary version of the new population is converted to base-10 integers and then to real values.
>   Step 6: Steps 2–5 are repeated until the desired numbers of iterations have been performed.
>   Step 7: In order to maintain diversity in the population two operators, viz., mutation and elitism are included. Mutation is the random change of a gene from

7.6 Transportation Problems

0 to 1 (or 1 to 0). Elitism is the procedure by which the weakest individual of the current population is replaced by the fittest individual of the immediately preceding population. The mutation, and elitism operators offer the opportunity for new genetic material to be introduced into the population.

Step 8: The final cost and final (X and Y) location of the plants are reported.

Phase 2

In Phase 2 the random locations of the plants are received from Phase 1 and are solved as a linear transportation problem using the simplex algorithm. The Simplex Algorithm optimizes the cost for allocation of power from the plants to the cities, to a minimum. The optimal cost value, which is the objective function value in the Genetic Algorithm, is passed back to Phase 1.

A sample of 20 problems is solved using the above Genetic Algorithm and the results are displayed and analyzed. The method described above was applied to the sample problems given in Cooper (1972), and the efficiency of the Genetic Algorithm was observed.

In the case of quality of solution, it can be seen from Table 7.2, Genetic Algorithm converged to within 10% of the exact solutions. But for the large problems (19 & 20), the convergence is not more appropriate; as a result parallel genetic algorithms can be applied to improve the performance.

Table 7.2 Results obtained using GA approach

Problem No.	Source X destination	Exact solution	GA solution	Computation Cycles
1	2×7	50.450	50.465	15000
2	2×7	72.000	72.033	9000
3	2×7	38.323	38.334	12500
4	2×7	48.850	48.850	8000
5	2×7	38.033	38.398	8000
6	2×7	44.565	44.565	6500
7	2×7	59.716	59.921	15000
8	2×7	62.204	62.380	9000
9	2×4	54.14246	54.16013	12500
10	2×5	65.78167	66.83248	15000
11	2×6	68.28538	68.78933	12500
12	2×7	44.14334	44.17555	25000
13	2×8	93.65978	95.48586	20000
14	3×3	40.00267	40.28115	15000
15	3×4	40.00020	40.50941	10000
16	3×5	60.00000	60.74852	10000
17	3×6	54.14263	54.47150	15000
18	4×4	10.00000	11.06346	15000
19	8×16	216.549	502.9196	25000
20	12×16	160.000	444.1291	25000

7.6.2 Real-Coded Genetic Algorithm (RCGA) for Integer Linear Programming in Production-Transportation Problems with Flexible Transportation Cost

Among the various forms of linear programming problem, a popular and important type is the traditional transportation problem, in which the objective is to minimize the cost of transportation of various amounts of a single homogeneous commodity from different sources to different destinations. Generally, the traditional transportation problem (TP) is a minimization problem in which the total transportation cost for transporting the goods from source to the destination is minimized. However, due to the fierce competition resulting out of rapid changes of the Global economy and to maintain the quality of the item as well as the goodwill of the company, some manufacturing companies are forced to keep the following activities simultaneously under their own control:

1. Manufacturing and Marketing of the commodity.
2. Selling at different showrooms situated at different important markets / locations.
3. Transportation of commodities from different factories to different showrooms.

As a result, the overall objective of that manufacturing company is to maximize the profit of the system according to the prescribed demands of different markets as well as the capacity of different factories.

In the traditional transportation problem, it is assumed that the transportation cost (per unit) for transferring the commodities from a particular source to a particular destination is fixed. However, under real life situation when the number of transported units is above a certain limit, then one or more transport vehicle is generally hired to transport those units from a particular source to a particular destination, and this normally results in the lowering down of effective cost per unit. Also, the factors like road conditions, weather, etc can affect the unit transportation cost.

A genetic algorithm (GA) is a computerized stochastic search and optimization method which works by mimicking the evolutionary principles and chromosomal processing in natural genetics. It is based on Darwin's principles of "survival of the fittest". It is executed iteratively on the set of real / binary coded solution called population. Every solution is assigned a fitness, which is directly related to the objective function of the search and optimization problem. There after, applying three operators similar to natural genetic operators—selection / representation, crossover, and mutation, the population of solutions is modified to a new population. GA works iteratively by successively applying these three operators in each generation till a termination criterion is satisfied.

In this case, a realistic production-transportation model is developed under the assumption that a company is undergoing the following activities:

(i) Producing a single homogeneous product in different factories (situated in different places with different raw material cost, production cost and marketing cost per unit).

7.6 Transportation Problems

(ii) Transporting the product to different show-rooms (with different selling prices per unit). The unit transportation cost from a particular source to a particular destination is not fixed, but flexible. Generally, when the number of transported units is above a certain limit, then the transportation cost for full load of vehicle will be charged, otherwise transportation cost is charged per unit.

(iii) Selling the product in different markets provided that objective of the company is to maximize the total profit.

In order to solve the problem for discrete values of decision variables, a real coded genetic algorithm is developed for discrete values of decision variables with rank based selection, crossover and mutation.

7.6.2.1 Assumptions and Notations

The following assumptions and notations are used in developing the proposed model.

(i) A company has m factories F_i (producing a homogeneous product) with capacity a_i ($i=1, 2, \ldots, m$) and there are n showrooms in markets M_j with demand(requirement) bj ($j = 1, 2, \ldots, n$).

(ii) The transportation cost is constant for a transport vehicle of a given capacity (even if the quantity shipped is less than the full load capacity of that transport vehicle by some quantity).

(iii) The capacity of a transport vehicle is K units.

(iv) xij represents the unknown quantity to be transported from the factory Fi to the market Mj.

(v) Cij be the transportation cost for a full load of transport vehicle and $C'ij$ be the transportation cost per unit item from F_i to M_j.

(vi) Uij be the upper break point, some units less than K but more than Uij, the transportation cost for the whole quantity is Cij.
Hence $Uij = [Cij/C'ij] < K$ where $[Cij/C'ij]$ is the greatest integer value which is less than or equal to $Cij/C'ij$.

(vii) Cri and Cpi be the respective raw material and production costs per unit in the factory F_i ($i = 1, 2, \ldots, m$) of the company.

(viii) p_j be the selling price per unit in the market $(1, 2, \ldots,)$ M_j ($j = 1, 2, \ldots n$).

7.6.2.2 Model Formulation of the Problem

The Total Revenue TR of the company is given by,

$$TR = \sum_{j=1}^{n} \sum_{i=1}^{m} p_j x_{ij} \qquad (7.17)$$

and the total production cost including the raw material cost is,

$$\sum_{i=1}^{m}\sum_{j=1}^{n}(C_n + C_{pi})x_{ij} = \sum_{i=1}^{m}\sum_{j=1}^{n} c_u x_{ij} \quad \text{where } C_u = C_n + C_{pi} \quad (7.18)$$

Transportation Cost

When the transported quantity from the i th factory to the j th show-room is greater than one integral transport vehicle load, the transported quantity x_{ij} can be expressed as:

$$x_{ij} = n_{ij}K + k_i q_{ij} \quad \text{where } n_{ij} = 0 \text{ or any finite integer, } k_i = 0 \text{ or } 1 \text{ and } q_{ij} < K. \quad (7.19)$$

In this case, two situations may arise.

(i) $n_{ij}K < x_{ij} \leq n_{ij}K + U_{ij}$ \quad (ii) $n_{ij}K + U_{ij} < x_{ij} \leq m_{ij} + 1)K$ \quad (7.20)

Hence the transportation cost of xij units from the i th factory to the j th show-room is given by

$$TC_{ij} = n_{ij}C_{ij} + (x_{ij} - n_{ij}K).C_{ij}^l \quad \text{where } n_{ij}K < x_{ij} \leq n_{ij} + U_{ij}$$
$$= (n_{ij} + 1)C_{ij} \quad \text{where } n_{ij}K + U_{ij} < x_{ij} \leq (n_{ij} + 1)K \quad (7.21)$$

Now, the total profit of the company is given by

$$Z = <total\ revenue> - <production\ cost> - <row\ material\ cost>$$
$$- <transportation\ cost> \quad (7.22)$$

Again, the supply and demand constraints are as follows:

$$\sum_{j=1}^{n} x_{ij} = a_i, \quad i = 1, 2, \ldots, m \quad (7.23)$$

$$\sum_{i=1}^{m} x_{ij} = b_j, \quad j = 1, 2, \ldots, n \quad (7.24)$$

Hence the problem of the company is,

$$\text{Max } Z = \sum_{j=1}^{n}\sum_{i=1}^{m} x_{ij} p_j - \sum_{i=1}^{m}\sum_{j=1}^{n} C_u x_{ij} - \sum_{i=1}^{m}\sum_{j=1}^{n} TC_{ij} \quad (7.25)$$

7.6 Transportation Problems

subject to the constraints

$$\sum_{j=1}^{n} x_{ij} = a_i, \quad i = 1, 2, \ldots, m$$

$$\sum_{i=1}^{m} x_{ij} = b_j, \quad j = 1, 2, \ldots, n \tag{7.26}$$

$x_{ij} \geq 0$ and integers
In this situation, three cases may arise:

$$\text{Case-I}: \sum_{i=1}^{m} a_i = \sum_{j=1}^{n} b_j :$$

$$\text{Case-II}: \sum_{i=1}^{m} a_i > \sum_{j=1}^{n} b_j :$$

$$\text{Case-III}: \sum_{i=1}^{m} a_i < \sum_{j=1}^{n} b_j : \tag{7.27}$$

In Case-I, the above problem is a balanced problem, whereas in Case-II & Case-III, it is unbalanced. In Case-II, the total capacity of the source is greater than the total demand of the destination. Where as in Case-III, the total capacity is less than the total demand.

7.6.2.3 Implementation of GA

Now, we shall develop a GA with real value coding for solving the above constrained maximization problem involving m × n integer variables. The general working principle of GA is as follows:

Step-1: Initialize the parameters of Genetic Algorithm and different parameters of the transportation problem.
Step-2: $t = 0$ [t represents the number of current generation.]
Step-3: Initialize $P(t)$ [$P(t)$ represents the population at t-th generation]
Step-4: Evaluate $P(t)$.
Step-5: Find best result from $P(t)$.
Step-6: $t = t + 1$.
Step-7: If ($t >$ maximum generation number) go to step-14
Step-8: Select $P(t)$ from $P(t-1)$ by any selection process like roulette wheel selection, tournament selection, ranking selection etc.
Step-9: Alter $P(t)$ by crossover and mutation operation.

Step-10: Evaluate $P(t)$.
Step-11: Find best result from $P(t)$.
Step-12: Compare best results of $P(t)$ and $P(t-1)$ and store the better one.
Step-13: Go to step-6.
Step-14: Print final best result.
Step-15: Stop.

7.6.2.4 Representation of Chromosomes

For proper application of GA, the designing of an appropriate chromosome representation of solutions of the problem is an important task. In many situations including optimization problem with larger decision variables the classical binary coding is not well adopted. In this case, a chromosome is coded in the form of a matrix of real numbers, every component of chromosomes represents a variable of the function.

7.6.2.5 Evaluation Function

After getting a population of potential solutions, we need to see how good they are. Therefore, we have to calculate the fitness for each chromosome. In this problem, the value of the profit function for chromosome $V_j (j = 1, 2 \ldots POPSIZE)$ is taken as the fitness of V_j and it is denoted by $eval(V_j)$.

Consider an example, to solve the balanced production—transportation problem with the following values of different parameters:

$$m = 3, \ n = 4, \ [p_j] = [50.00, 40.0, 45.0, 35.00], \ [C_u] = [15.0, 22.00, 16.0],$$

$$[C_{ij}] = \begin{bmatrix} 60 & 90 & 105 & 75 \\ 120 & 48 & 130 & 150 \\ 110 & 65 & 80 & 100 \end{bmatrix} \ and \ [C'_{ij}] = \begin{bmatrix} 2.5 & 3.5 & 4.0 & 3.0 \\ 4.5 & 2.0 & 5.0 & 5.5 \\ 4.2 & 2.8 & 3.2 & 3.3 \end{bmatrix}$$

$$[a_i] = [60, 20, 25], \ [b_j] = [25, 30, 20, 30] \tag{7.28}$$

For different values of K, we have solved the balanced production—transportation problem by RCGA for discrete variables. The results are displayed in Table 7.3.

In this section, we have formulated and solved a production-transportation problem with the flexible transportation cost for transferring commodities from a

Table 7.3 Results of example

K	Values of decision variables	Profit of the company (Z)
20	$x_{11} = 25, x_{12} = 5, x_{14} = 30, x_{22} = 19, x_{23} = 1, x_{32} = 6,$ $x_{33} = 19$, all other decision variables are zero	$ 2344.40
25	$x_{11} = 25, x_{12} = 5, x_{14} = 30, x_{22} = 20, x_{32} = 5, x_{33} = 20,$ all other decision variables are zero	$2374.50
30	$x_{11} = 25, x_{12} = 5, x_{14} = 30, x_{22} = 20, x_{32} = 5, x_{33} = 20,$ all other decision variables are zero	$ 2389.50

particular factory to a particular show room. For this purpose, a real coded GA (RCGA) for discrete variables with rank based selection, whole crossover (applied for all genes of a chromosome) and a new type of mutation has been developed.

In real-life situation, transportation costs of goods are fixed for a finite capacity of a transport mode such as a truck, matador, etc. A fixed cost is charged against a certain amount of quantities (upper ceiling) or more when a truck is deployed whether it is utilized fully or partially. For the quantities less than that upper ceiling, a uniform rate per unit is charged. In this paper, we have considered the transportation cost explicitly considering the realistic situation.

The proposed production inventory problem is a constrained linear integer problem. To solve this problem, a real-coded GA for integer variables has been developed. However, in GA, there may arise a difficulty regarding the boundaries of the decision variables. In application problems, the selection of boundaries of the decision variables is a formidable task. We have overcome these difficulties by selecting the decision variables randomly and taking the minimum value of the corresponding capacity and requirement of the source and destination respectively. In this process, all constraints are satisfied automatically.

7.7 Network Design and Routing Problems

The great diversity of questions and problems that originate from today's network planning and design tasks requires a large number of algorithms, each of which specializes in a specific problem with specific constraints. Almost all of the optimization problems relevant in network design are NP-complete. For most problems, there is no known algorithm that could guarantee to find the global optimum in a polynomial amount of time. In many cases, sophisticated heuristics have to be developed to achieve satisfying results.

7.7.1 Planning of Passive Optical Networks

7.7.1.1 Problem Description

Passive Optical Networks (PON) provide a way to gradually introduce fiber optic technology into access networks while still deploying parts of the traditional copper line or coax-cable systems. PONs can be implemented in several topologies. One configuration of choice is a tree structure where the Optical Line Termination (OLT) in the central office can be seen as the root and the Optical Network Units (ONU) as the leaves of the tree. In the field between the OLT and the ONUs only passive elements—the fibers and optical splitters—are deployed. Customer access points are connected to the ONUs via traditional technologies like copper or coax lines. Figure 7.10 illustrates the PON tree structure.

When installing a new network in the access area, the majority of money has to be spent on digging the cable ducts. Thus, minimizing the total cost is mainly a matter

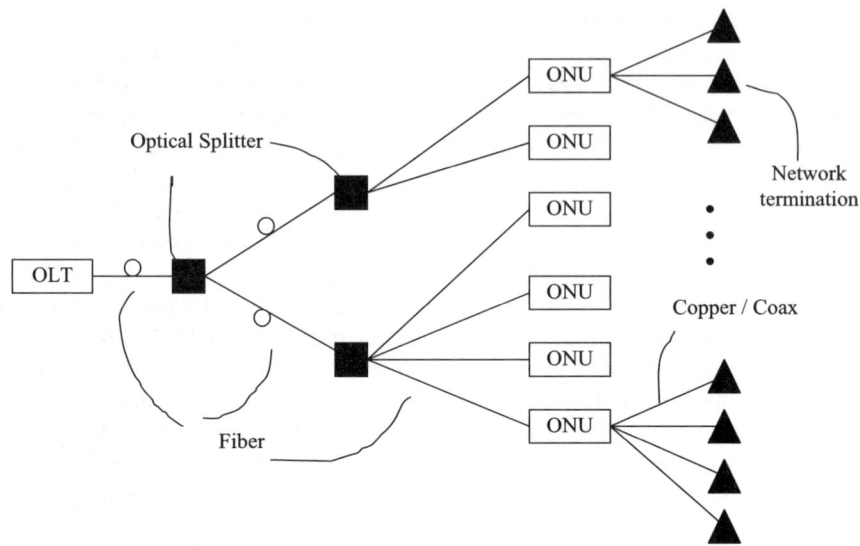

Fig. 7.10 Structure of passive optical networks

of finding the shortest street paths, which interconnect all ONUs with the OLT. A city map can be represented by a graph where the streets are the links, and the street junctions together with the ONUs and the OLT make up the nodes. The weights of the links are set to be proportional to the length of the respective streets. In some cases, for example, if some fiber lines exist or if some streets are preferred to be used as duct lines, special weight values can be assigned to theses edges. With this map representation, the optimization problem turns into the classical minimum Steiner tree problem. This means that we want to find a tree within a given graph, which spans a designated subset of nodes in such a way, that the sum of the costs of the selected edges becomes minimal. There already exist a number of algorithms that solve this problem exactly. Since the minimum Steiner tree problem is NP-complete, these algorithms have an exponential worst-case running time. Therefore, they are not applicable in the field of network planning where it is quite common to have a great number of nodes and edges.

7.7.1.2 Genetic Algorithm Approach

A simple genetic algorithm is applied to the design of passive optical networks. The genetic coding of a specific alternative of a Steiner tree consists of a string of integer values with one specific gene for every link in the graph. The integer value is assigned to the respective link as a pseudo link weight, which is not correlated to the real cost value of this edge. The pseudo link weights are only auxiliary parameters. Given a genetic string with a pseudo link weight for every edge in the graph, a minimum spanning tree is built over all nodes in this graph. From this minimum spanning tree all nodes and links are cut off which are not essential to connect

7.7 Network Design and Routing Problems

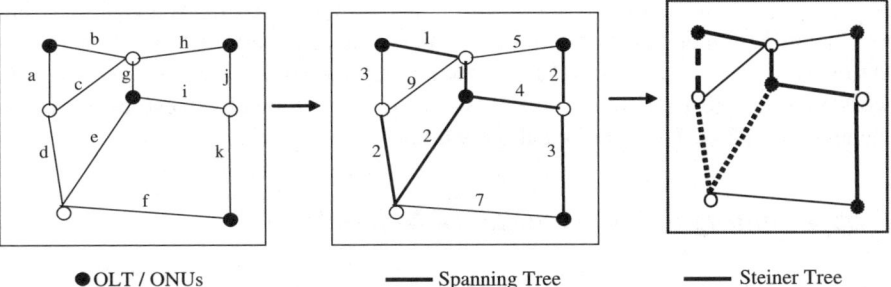

Fig. 7.11 Computation of a specific Steiner tree

the Steiner vertices with each other. The remaining tree is a specific Steiner tree solution that connects only the set of ONU and OLT nodes. Figure 7.11 illustrates the use of two simple heuristics to find a Steiner tree. In this example, the genetic string is 3-1-9-2-2-7-1-5-4-2-3 with the integers assigned to the edges a through k, respectively.

This type of genetic coding guarantees that a genetic string can represent all possible solution alternatives. To obtain one specific Steiner tree, one has to set the pseudo link weights of the Steiner edges smaller than the ones of the remaining edges. It is another advantage of this coding that the crossover and the mutation operators applied to genetic strings produce again valid strings.

The diagram in Fig. 7.12 shows the evolution of the costs for a typical run of the genetic algorithm for a network with 463 nodes, including 44 ONUs, and 709 edges. For every generation the maximum and the minimum cost values are shown. Starting from a random first generation with usually high cost values, it can be observed that the costs decrease gradually. This is true for the minimum as well as for the maximum values. It indicates that the principle of "survival of the fittest" also works in the area of network planning, and that a crossover of two good solutions has a high probability of creating another good or even better solution.

The running time of the genetic algorithm depends on the size of the network, the number of genetic strings per generation, and the total number of evolution cycles.

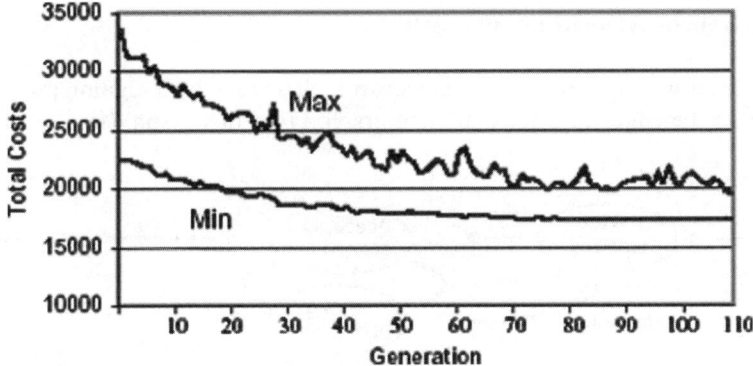

Fig. 7.12 Genetic algorithm cost evolution

A large portion of the time is spent within the method which maps a genetic string to a specific Steiner tree solution and calculates its costs. For our sample network, the evaluation of one string takes approximately 0.4 to 0.5 seconds on a SPARC workstation. With a generation size of 600 strings and a total of 110 evolution cycles, the test run of Fig. 7.12 required about 9 hours.

7.7.2 Planning of Packet Switched Networks

7.7.2.1 Problem Description

The design of packet switched networks requires the solution of different types of problems, e.g. node placement and link dimensioning, routing optimization, server specification, or address assignment. In our paper we consider only the link topology and routing optimization aspects. Our design problem can be formulated as follows: Given a set of node locations and a requirement matrix with traffic values for all node-to-node pairs, the link topology and the routing paths are optimized according to the costs, so the average end-to-end packet delay does not exceed a specified value. The packet arrival process is assumed to be a Poisson process, and the routers are modeled as independent M/M/1 waiting queues.

Since the three design aspects—link topology, capacity assignment, and routing optimization—are interdependent, they cannot be considered in isolation. Conventional planning algorithms usually handle this problem by repeatedly applying a sequence of methods to the different sub-problems as shown in Fig. 7.13. For a fixed topology and a certain routing scheme, the capacities of the links are optimized. Afterwards, the capacity values are kept constant while the routing strategy is improved. This is repeated until the results converge. If all constraints are met, the solution is returned and the algorithm exits. If one is not satisfied with this result, a different link topology is chosen, and the optimization procedure is started anew. Unlike this traditional approach, our genetic algorithm looks at the whole picture of a specific network solution and evaluates it.

7.7.2.2 Genetic Algorithm Approach

A genetic string represents a specific network alternative by assigning pseudo link weights to the edges of a fully meshed graph. From this graph, simple heuristics

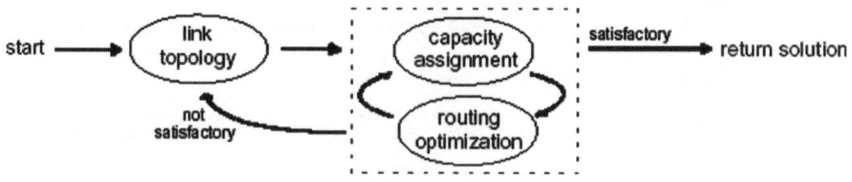

Fig. 7.13 Conventional optimization procedure

7.7 Network Design and Routing Problems

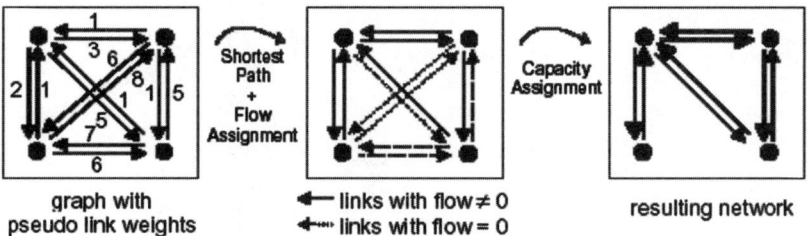

| graph with pseudo link weights | ← links with flow ≠ 0
←⋯ links with flow = 0 | resulting network |

Fig. 7.14 Computation of a specific network topology

create a valid network with a certain link topology, capacity assignment, and routing specification.

Figure 7.14 shows how a genetic string is converted to a specific network solution. Starting from the fully meshed graph with directed links, the shortest path for every node-to-node pair is determined according to the pseudo link weights, and the selected links are marked. Edges, which are not selected at all, are discarded. The shortest paths correspond to the packet routes between two end systems. Based on this routing pattern and on the given requirement matrix, it is now possible to compute the traffic load on every link and, furthermore, calculate the optimal capacities. In the case of discrete capacities, the continuous values are rounded up. Since current routing protocols like RIP and OSPF require bi-directional connectivity between neighbor gateways, an opposite edge is inserted for every link in the network if it does not already exist. The smallest possible capacity is assigned to these new edges, since only router messages will be transmitted over them. At the end, the costs for this specific network scenario are calculated and returned to the genetic framework.

The cost evolution resulting from the genetic algorithm for packet switched network planning looks similar to the one given in Fig. 7.15. Again, the minimum and maximum values decrease gradually until a convergence point is reached. This indicates that genetic algorithms can be used in the field of packet network planning. A sample network topology, which was computed by the genetic algorithm, is shown in Fig. 7.15. The structure strongly depends on the composition of the capacity costs. For high basic costs, which are only proportional to the length of a link, the network mesh degree will be low. For decreasing fixed link capacity costs, the number of links in the network increases.

Thus, the genetic algorithm is applied to the topology and routing optimization of packet switched networks.

7.7.3 Optimal Topological Design of All Terminal Networks

An important part of network design is to find the best way to layout the components (nodes and arcs) to minimize cost while meeting a performance criterion such as transmission delay, throughput or reliability. This design stage is called "Network

Fig. 7.15 Sample network topology

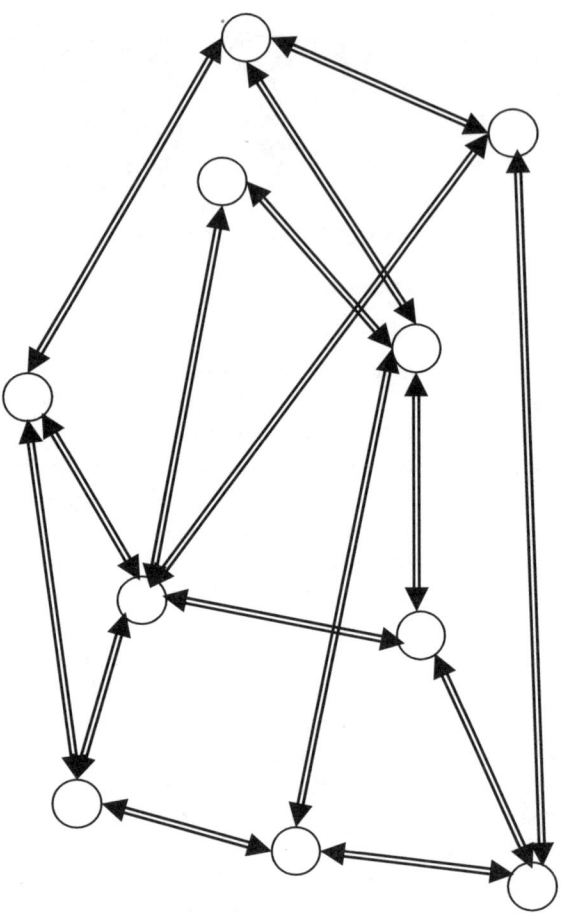

Topological Optimization". In a topological network design problem, a main concern is to design networks, which operate effectively and without interruption in the presence of component failures. Reliability is concerned with the ability of a network to carry out desired network operations.

Generally, a large-scale network has a multilevel, hierarchical structure consisting of a backbone network and several local access networks. Therefore, designing the topology of a large-scale network can be divided into two problems, the backbone network design and the local network design. This section is mainly interested in large-scale backbone network design. For backbone network design, an important connectivity measure is reliability. In a communication network, *all terminal* network reliability can be defined as the probability that every pair of nodes can communicate with each other.

Many studies have considered topological optimization with a network reliability criterion. For example, few researchers used a decomposition method based on branch and bound to minimize total network cost under a system reliability constraint. Their method can only solve small networks because as the number of arcs

7.7 Network Design and Routing Problems

increases, the number of possible layouts grows faster than exponentially. Because of this complexity, other existing methods are not computationally feasible for designing large-scale network topologies with very confining assumptions. Therefore, a heuristic search algorithm based on Genetic Algorithms (GAs) is developed to find a network topology, which has minimum cost, subject to a system reliability constraint.

A computer communication network can be modeled by a probabilistic graph $G = (N, L, p)$, in which N and L are the set of nodes and arcs that corresponds to the computer sites and communication links, respectively. The networks considered in this section are assumed to have bi-directional links and therefore are modeled by graphs with non-oriented edges. We further assume that the graph under discussion has no redundant arcs. Any graph $G = (N, L)$ is said to be connected if there is at least one path between every pair of nodes. A sub-graph G1 of G is a graph, of which all nodes and arcs are contained in G. i.e., $G1 = (N1, L1)$ where $N1 \subseteq N$ and $L1 \subseteq L$. If $N1 = N$, the sub-graph G1 is called a "spanning sub-graph". In a connected graph G of arcs and n nodes, a tree T is a spanning tree consisting of n-1 arcs. The deletion of any edge from a tree results in a disconnected graph. Therefore a connected graph should be at least a spanning tree with n-1 edges. A communication network topology should be at least a spanning tree and communication network reliability must be greater than the required system reliability value, p_0.

In addition to a simple network connectivity check (i.e., does a minimum spanning tree exist in the network), other researchers proposed a "2- connectivity" measure in the design of communication network topologies. "2- connectivity" means that there are at least two paths between each pair of nodes, rather than one. Originally, many studies considered this measure to be a reasonable constraint of reliability in the design of network topology. In this case, it is used to establish the initial population and to constrain subsequent populations. Therefore, the final network design will meet the system reliability constraint and contain at least two different paths between all pairs of nodes.

7.7.3.1 Problem Description

Under the following assumptions:

(1) the location of each network node is fixed and given,
(2) each c_{ij} and the p are fixed and known, where c_{ij} is the cost of link in the network between nodes i and j, and p, q are link reliability and unreliability (p + q = 1),
(3) each link is bi-directional,
(4) there are no redundant links in the network,
(5) the probability of failure of a link is independent of the states of the other links,

the main problem can be stated mathematically as follows:

$$\text{Minimize:} \quad z = \sum c_{ij} \, x_{ij}$$
$$\text{Subject to:} \quad f(x) \geq p_0 \quad\quad\quad (7.29)$$

$x_{ij} \in \{0, 1\}$ are the decision variables and f(x) is the network reliability. The all-terminal system reliability of a network is defined to be the probability that every pair of nodes can communicate with each other. At any instant of time, only some arcs of G may be operational. A state of G is a sub-graph (N, L') with $L' \in L$, where L' is the set of operational arcs. An operational state is generally called a pathset, and a minimal operational state is a min-path. A failed state L' is called $L \setminus L'$ (cutset) and when L' is a maximal failed state, $L \setminus L'$ is a min-cut. The reliability of G, $Rel_K(G)$, is the k-terminal reliability: If K = N, this is the all terminal reliability, Rel(G). It is easy to formulate a network in state $L' \subseteq L$, with reliability as follows:

$$\prod_{e \in L'} p_e \prod_{e \in (L \setminus L')} q_e \quad \text{where L' is the set of operational arcs.} \tag{7.30}$$

Summing this state occurrence probability over all operational states gives the network system reliability. There are basically two approaches to network system reliability calculation; simulation and analytic. All known analytic methods have worst-case computation times, which grow exponentially with the size of the network. Monte Carlo simulation methods, for which computation time grows only slightly faster than linear with network size, have been the method of choice for more than trivial sized networks. In this section, Monte Carlo simulation technique is used to predict the network reliability, which substantially reduces the variance of the estimator when compared to "crude" Monte Carlo. This reduced variance Monte Carlo is based on a two-tiered hierarchical approach to sampling, which makes use of how many arcs fail during a given simulation.

7.7.3.2 Genetic Algorithm Approach

A GA is developed as a solution methodology for network topological optimization with a reliability constraint. In GA, the search space is composed of possible solutions to the problem; each represented as some convenient data structure, referred to as the chromosome. Each chromosome has an associated objective function value, called the fitness value. A good chromosome is the one that has a high fitness value. A set of chromosomes together with their associated fitness values is called the population. This population, at a given stage of the GA, is referred to as a generation.

In a conventional GA, candidate solutions are represented by strings of numbers using a binary or non-binary alphabet. The present algorithm uses a binary coding structure for representing candidate solutions. A binary set is used to represent arcs, where the maximum number of non-redundant, undirected arcs for a network of n nodes is given by $(n-1)n/2$. For example, a simple network whose base graph consists of 5 nodes and 10 possible links can be represented by:

$$[\ 1 \ \ 1 \ \ 0 \ \ 1 \ \ 1 \ \ 0 \ \ 1 \ \ 1 \ \ 0 \ \ 1 \ ]$$
$$[\ x_{12}, x_{13}, x_{14}, x_{15}, x_{23}, x_{24}, x_{25}, x_{34}, x_{35}, x_{45} \ ]$$

7.7 Network Design and Routing Problems

where, x_{ij} represents a link connecting two nodes i and j. If x_{ij} is equal to 1, there is a connection between these two nodes. If x_{ij} is equal to 0, then there is no connection. The initial population, which consists of a set of feasible solutions (2- connected networks) is generated in a random fashion. For determining this initial population, a number of experiments were carried out. A candidate network consists of some randomly selected arcs between nodes. The selection of the probability values, which are used in deciding whether an arc exists or not was an important step to generate the initial population. In an experimental design with 10, 20 and 30 nodes, the following characteristics were systematically controlled.

- Arc probabilities between [0, 1], which determines the existence of an arc between nodes, are selected.
- The system reliability value of each connected network is estimated using Monte Carlo simulation.
- The probability values of the existence of arcs and the corresponding network reliability values are compiled.

The aim was to determine the intervals of the probability values, which result highly reliable networks. Any initial population can then be generated by using probabilities within these intervals. Table 7.4 shows the resulting probability intervals from the experiments described above which were used for the initial populations.

The choice of parameters for GAs can have a significant effect on performance of the algorithm. Parameter values were investigated by running the GA with different population sizes (10, 20, 30), crossover rates (0.55, 0.65, 0.75, 0.85, 0.95) and mutation rates (0.01, 0.05, 0.09, 0.10). It was found that the best results were: population size = 20, crossover rate = 0.95 and mutation rate = 0.05.

The objective function is the sum of the total cost for all arcs in the network plus a quadratic penalty function, which is applied when the network reliability prediction does not meet the network reliability requirement (i.e., infeasible). The objective of the penalty function is to lead the optimization algorithm to feasible solutions. It was important to allow infeasible solutions into the population because good solutions are often the result of breeding between feasible and infeasible solution. The objective function is,

$$Z = \sum_i \sum_j c_{ij} X_{ij} + \delta(\varepsilon(\text{Rel}(G)-p0))^2, \quad i=1,\ldots,n\text{-}1; \quad j=i+1,\ldots,n \tag{7.31}$$

where c_{ij}, x_{ij} and p0 were previously defined, Rel(G) equals f(x) (network reliability), ε is the maximum value of c_{ij} and $\delta = 0$ if Rel(G) is $\geq p_0$ and $\delta = 1$ if Rel(G) $< p_0$. The fitness is chosen to be (Zmax–Z) where Zmax is a constant, which is

Table 7.4 Probability values used to generate the initial population

Number of nodes (n)	Probability of an arc
10	(0.15–0.60)
20	(0.15–0.50)
30	(0.10–0.30)

the largest penalized cost of all networks in the current population. This subtraction translates the minimization problem to a maximization problem.

The reduced variance Monte Carlo estimation of system reliability is used to minimize computational effort. To further speed up the search, Jan's upper bound formulation (Jan, 1993) of network reliability is used. If this upper bound reliability value exceeds the required system reliability value, then the Monte Carlo simulation is used as a subroutine. Otherwise, the candidate network is considered to be infeasible. While it is possible that some networks, which are truly feasible, are discarded at this point, the probability of this occurring is very small. Use of the upper bound considerably reduces the number of network requiring simulation. Roulette wheel selection is used for each generation of our algorithm. In this mechanism, a candidate network is selected with probability equal to its relative fitness with respect to the whole population.

Classic crossover and mutation operators (Goldberg, 1989) are used to obtain the new candidate networks for the next population. After crossover and mutation, new candidate networks are checked for connectivity using the "Set Merging Algorithm". Then all new candidate networks replace their parents. Additionally, an elitist strategy appends the best performing candidate network of the previous generation to the current population. This strategy ensures that the candidate network with the best objective function value always survives to the next generation. A GA continues until a pre-determined stopping criterion has been met. The criterion is often based on the total number of generations. Our termination generation is determined according to the size of the network under study.

Thus, a heuristic search algorithm based on GAs was developed to solve network topology design with minimum cost subject to a reliability constraint. This can be applied to complex design problems.

7.8 Summary

Genetic algorithms (GAs) are a Meta heuristic searching techniques, which mimics the principles of evolution and natural genetics. These are a guided random search, which scans through the entire sample space, and therefore provide reasonable solutions in all situations like operations research, management science and engineering design. In recent years, genetic algorithms have received considerable attention regarding their potential as a novel optimization technique. Based on their simplicity, minimal requirements genetic algorithms have been widely applied in a variety of problems. This chapter provided a brief introduction to genetic optimization techniques and their applications.

Review Questions

1. Mention some of the areas where genetic algorithms can be applied.
2. How is fuzzy optimization performed?

3. In what way if-then rules are used for multiobjective optimization?
4. Write short note on Genetic Fuzzy Systems.
5. Explain in detail about Multiobjective reliability design problem.
6. Define Combinatorial Optimization
7. State how GA is applied to solve N-Queens Problem.
8. List some of scheduling problems where genetic algorithms can be applied.
9. Describe the application of Genetic Algorithm to Job Shop Scheduling Problem.
10. Give a short description on GA based transportation problems.
11. How is genetic algorithm concept applied to network planning and routing concept?
12. State the various advantages of Genetic Algorithm towards scheduling problems.

Exercise Problems

1. Write a MATLAB program to implement N-Queens problem.
2. Implement a parallel genetic algorithm to solve a network routing problem.
3. Develop a combinatorial optimization process for a bin-packing problem.
4. Make a study on constrained spanning tree problems with genetic algorithms
5. Develop a project for allocation and scheduling on multi-computers using genetic algorithms.
6. Apply Genetic Algorithm for VLSI Layout design.
7. Implement multiprocessor scheduling using Genetic Algorithm.
8. Implement vehicle routing problem using Genetic Algorithm.
9. Develop a C++ program for implementing multitask scheduling using GA approach
10. Implement multiobjective optimization for a reservoir management system.

Chapter 8
Genetic Algorithm Implementation Using Matlab

8.1 Introduction

MATLAB (**Mat**rix **Lab**oratory), a product of Mathworks, is a scientific software package designed to provide integrated numeric computation and graphics visualization in high-level programming language. Dr Cleve Moler, Chief scientist at MathWorks, Inc., originally wrote MATLAB, to provide easy access to matrix software developed in the LINPACK and EISPACK projects. The very first version was written in the late 1970s for use in courses in matrix theory, linear algebra, and numerical analysis. MATLAB is therefore built upon a foundation of sophisticated matrix software, in which the basic data element is a matrix that does not require predimensioning.

MATLAB has a wide variety of functions useful to the genetic algorithm practitioner and those wishing to experiment with the genetic algorithm for the first time. Given the versatility of MATLAB's high-level language, problems can be coded in m-files in a fraction of the time that it would take to create C or Fortran programs for the same purpose. Couple this with MATLAB's advanced data analysis, visualisation tools and special purpose application domain toolboxes and the user is presented with a uniform environment with which to explore the potential of genetic algorithms.

The Genetic Algorithm Toolbox uses MATLAB matrix functions to build a set of versatile tools for implementing a wide range of genetic algorithm methods. The Genetic Algorithm Toolbox is a collection of routines, written mostly in m-files, which implement the most important functions in genetic algorithms.

8.2 Data Structures

MATLAB essentially supports only one data type, a rectangular matrix of real or complex numeric elements. The main data structures in the Genetic Algorithm toolbox are:

- chromosomes
- objective function values
- fitness values

These data structures are discussed in the following subsections.

8.2.1 Chromosomes

The chromosome data structure stores an entire population in a single matrix of size Nind × Lind, where Nind is the number of individuals in the population and Lind is the length of the genotypic representation of those individuals. Each row corresponds to an individual's genotype, consisting of base-n, typically binary, values.

$$\text{Chrom} = \begin{bmatrix} g_{1,1} & g_{1,2} & g_{1,3} & \cdots & g_{1,\text{Lind}} \\ g_{2,1} & g_{2,2} & g_{2,3} & \cdots & g_{2,\text{Lind}} \\ g_{3,1} & g_{3,2} & g_{3,3} & \cdots & g_{3,\text{Lind}} \\ \cdot & \cdot & \cdot & \cdots & \cdot \\ g_{\text{Nind},1} & g_{\text{Nind},2} & g_{\text{Nind},3} & \cdots & g_{\text{Nind},\text{Lind}} \end{bmatrix} \begin{matrix} \text{individual 1} \\ \text{individual 2} \\ \text{individual 3} \\ \cdot \\ \text{individual Nind} \end{matrix}$$

This data representation does not force a structure on the chromosome structure, only requiring that all chromosomes are of equal length. Thus, structured populations or populations with varying genotypic bases may be used in the Genetic Algorithm Toolbox provided that a suitable decoding function, mapping chromosomes onto phenotypes, is employed.

8.2.2 Phenotypes

The decision variables, or phenotypes, in the genetic algorithm are obtained by applying some mapping from the chromosome representation into the decision variable space. Here, each string contained in the chromosome structure decodes to a row vector of order Nvar, according to the number of dimensions in the search space and corresponding to the decision variable vector value. The decision variables are stored in a numerical matrix of size $\text{Nind} \times \text{Nvar}$. Again, each row corresponds to a particular individual's phenotype. An example of the phenotype data structure is given below, where bin2real is used to represent an arbitrary function, possibly from the GA Toolbox, mapping the genotypes onto the phenotypes.

```
Phen = bin2real(Chrom)  % map genotype to phenotype
```

$$= \begin{bmatrix} x_{1,1} & x_{1,2} & x_{1,3} & \cdots & x_{1,\text{Nvar}} \\ x_{2,1} & x_{2,2} & x_{2,3} & \cdots & x_{2,\text{Nvar}} \\ x_{3,1} & x_{3,2} & x_{3,3} & \cdots & x_{3,\text{Nvar}} \\ \cdot & \cdot & \cdot & \cdots & \cdot \\ x_{\text{Nind},1} & x_{\text{Nind},2} & x_{\text{Nind},3} & \cdots & x_{\text{Nind},\text{Nvar}} \end{bmatrix} \begin{matrix} \text{individual 1} \\ \text{individual 2} \\ \text{individual 3} \\ \cdot \\ \text{individual Nind} \end{matrix}$$

The actual mapping between the chromosome representation and their phenotypic values depends upon the bin2real function used. It is perfectly feasible using this

8.2 Data Structures

representation to have vectors of decision variables of different types. For example, it is possible to mix integer, real-valued, and binary decision variables in the same Phen data structure.

8.2.3 Objective Function Values

An objective function is used to evaluate the performance of the phenotypes in the problem domain. Objective function values can be scalar or, in the case of multiobjective problems, vectorial. Note that objective function values are not necessarily the same as the fitness values. Objective function values are stored in a numerical matrix of size Nind × Nobj, where Nobj is the number of objectives. Each row corresponds to a particular individual's objective vector.

```
Objv = OBJFUN(Phen) % Objective Function
```

$$= \begin{bmatrix} y_{1,1} & y_{1,2} & y_{1,3} & \cdots & y_{1,\text{Nvar}} \\ y_{2,1} & y_{2,2} & y_{2,3} & \cdots & y_{2,\text{Nvar}} \\ y_{3,1} & y_{3,2} & y_{3,3} & \cdots & y_{3,\text{Nvar}} \\ \cdot & \cdot & \cdot & \cdots & \cdot \\ y_{\text{Nind},1} & y_{\text{Nind},2} & y_{\text{Nind},3} & \cdots & y_{\text{Nind},\text{Nvar}} \end{bmatrix} \begin{matrix} \text{individual 1} \\ \text{individual 2} \\ \text{individual 3} \\ \cdot \\ \text{individual Nind} \end{matrix}$$

8.2.4 Fitness Values

Fitness values are derived from objective function values through a scaling or ranking function. Fitnesses are non-negative scalars and are stored in column vectors of length Nind, an example of which is shown below. Again, Ranking is an arbitrary fitness function.

```
Fitn = ranking(ObjV) % fitness function
```

$$\text{Fitn} = \begin{matrix} f_1 \\ f_2 \\ f_3 \\ \cdots \\ f_{\text{Nind}} \end{matrix} \quad \begin{matrix} \text{individual 1} \\ \text{individual 2} \\ \text{individual 3} \\ \\ \text{individual Nind} \end{matrix}$$

8.2.5 Multiple Subpopulations

This toolbox supports the use of a single population divided into a number of subpopulations or demes by modifying the use of data structures so that subpopulations are stored in contiguous blocks within a single matrix. For example, the chromosome data structure, Chrom, composed of Subpop subpopulations, each of length N individuals Ind, is stored as:

$$\text{Chrom} = \begin{array}{c} \ldots \\ \text{Ind}_1 \; \text{Subpop}_1 \\ \text{Ind}_2 \; \text{Subpop}_1 \\ \ldots \\ \text{Ind}_N \; \text{Subpop}_1 \\ \text{Ind}_1 \; \text{Subpop}_2 \\ \text{Ind}_2 \; \text{Subpop}_2 \\ \ldots \\ \text{Ind}_N \; \text{Subpop}_2 \\ \ldots \\ \text{Ind}_1 \; \text{Subpop}_{subpop} \\ \text{Ind}_2 \; \text{Subpop}_{subpop} \\ \ldots \\ \text{Ind}_N \; \text{Subpop}_{subpop} \end{array}$$

This is known as the *regional model*, also called *migration* or *island model*.

8.3 Toolbox Functions

The Genetic Algorithm and Direct Search Toolbox is a collection of functions that extend the capabilities of the Optimization Toolbox and the MATLAB numeric computing environment. The Genetic Algorithm and Direct Search Toolbox includes routines for solving optimization problems using

- Genetic algorithm
- Direct search

These algorithms enable you to solve a variety of optimization problems that lie outside the scope of the Optimization Toolbox.

The genetic algorithm uses three main types of rules at each step to create the next generation from the current population:

- *Selection rules* select the individuals, called *parents*, that contribute to the population at the next generation.
- *Crossover rules* combine two parents to form children for the next generation.
- *Mutation rules* apply random changes to individual parents to form children.

8.3 Toolbox Functions

The genetic algorithm at the command line, call the genetic algorithm function ga with the syntax

    [x fval] = ga(@fitnessfun, nvars, options)

where

- @fitnessfun is a handle to the fitness function.
- nvars is the number of independent variables for the fitness function.
- options is a structure containing options for the genetic algorithm. If you do not pass in this argument, 'ga' uses its default options.

The results are given by

- x — Point at which the final value is attained
- fval — Final value of the fitness function

Toolboxes are set of standard library functions, which consists of predefined algorithms. The genetic algorithm and direct search toolbox of MATLAB consists of the following functions:

Solvers
- ga   - Genetic algorithm solver.
- gatool   - Genetic algorithm GUI.
- patternsearch   - Pattern search solver.
- psearchtool   - Pattern search GUI

Accessing options
- gaoptimset   - Create/modify a genetic algorithm options structure.
- gaoptimget   - Get options for genetic algorithm.
- psoptimset   - Create/modify a pattern search options structure.
- psoptimget   - Get options for pattern search.

Fitness scaling for genetic algorithm
- fitscalingshiftlinear   - Offset and scale fitness to desired range.
- fitscalingprop   - Proportional fitness scaling.
- fitscalingrank   - Rank based fitness scaling.
- fitscalingtop   - Top individuals reproduce equally.

Selection for genetic algorithm
- selectionremainder   - Remainder stochastic sampling without replacement.
- selectionroulette   - Choose parents using roulette wheel.
- selectionstochunif   - Choose parents using stochastic universal sampling (SUS).
- selectiontournament   - Each parent is the best of a random set.
- selectionuniform   - Choose parents at random.

Crossover (recombination) functions for genetic algorithm.
- crossoverheuristic   - Move from worst parent to slightly past best parent.
- crossoverintermediate   - Weighted average of the parents.

crossoverscattered	- Position independent crossover function.
crossoversinglepoint	- Single point crossover.
crossovertwopoint	- Two point crossover.

Mutation functions for genetic algorithm

mutationgaussian	- Gaussian mutation.
mutationuniform	- Uniform multi-point mutation.

Plot Functions for genetic algorithm

gaplotbestf	- Plots the best score and the mean score.
gaplotbestindiv	- Plots the best individual in every generation as a bar plot.
gaplotdistance	- Averages several samples of distances between individuals.
gaplotexpectation	- Plots raw scores vs the expected number of offspring.
gaplotgenealogy	- Plot the ancestors of every individual.
gaplotrange	- Plots the min, mean, and max of the scores.
gaplotscordiversity	- Plots a histogram of this generations scores.
gaplotscores	- Plots the scores of every member of the population.
gaplotselection	- A histogram of parents.
gaplotstopping	- Display stopping criteria levels.

Output Functions for genetic algorithm

gaoutputgen	- Displays generation number and best function value in a separate window.
gaoutputoptions	- Prints all of the non-default options settings.

Custom search functions for pattern search

searchlhs	- Implements latin hypercube sampling as a search method.
searchneldermead	- Implements nelder-mead simplex method (FMINSEARCH) to use as a search method.
searchga	- Implements genetic algorithm (GA) to use as a search method.
searchfcntemplate	- Template file for a custom search method.

Plot Functions for pattern search

psplotbestf	- Plots best function value.
psplotbestx	- Plots current point in every iteration as a bar plot.
psplotfuncount	- Plots the number of function evaluation in every iteration.
psplotmeshsize	- Plots mesh size used in every iteration.

Output functions for pattern search

psoutputhistory	- Displays iteration number, number of function evaluations, function value, mesh size and method used in every iteration in a separate window.
psoutputfcntemplate	- Template file for a custom output function.

Utility functions

allfeasible	- Filter infeasible points.
gacreationuniform	- Create the initial population for genetic algorithm.
gray2int	- Convert a gray code array to an integer.
lhspoint	- Generates latin hypercube design point.
nextpoint	- Return the best iterate assuming feasibility.

Help files for genetic algorithm

fitnessfunction	- Help on fitness functions.
fitnessscaling	- Help on fitness scaling

8.3 Toolbox Functions

These are the toolbox functions present in the MATLAB. There also exist Genetic and Evolutionary Algorithm Toolbox for use with MATLAB that contains a broad range of tools for solving real-world optimization problems. They not only cover pure optimization, but also the preparation of the problem to be solved, the visualization of the optimization process, the reporting and saving of results, and as well as some other special tools. The list of various functions using Genetic and Evolutionary Algorithm Toolbox for use with MATLAB is as follows:

Objective functions

initdopi	- INITialization function for DOuble Integrator objdopi
initfun1	- INITialization function for de jong's FUNction 1
mopfonseca1	- MultiObjective Problem: FONSECA's function 1
mopfonseca2	- MultiObjective Problem: FONSECA's function 1
moptest	- MultiObjective function TESTing
obj4wings	- OBJective function FOUR-WINGS.
objbran	- OBJective function for BRANin rcos function
objdopi	- OBJective function for DOuble Integrator
objeaso	- OBJective function for EASom function
objfletwell	- OBJective function after FLETcher and PoWELL
objfractal	- OBJective function Fractal Mandelbrot
objfun1	- OBJective function for de jong's FUNction 1
objfun10	- OBJective function for ackley's path FUNction 10
objfun11	- OBJective function for langermann's function 11
objfun12	- OBJective function for michalewicz's function 12
objfun1a	- OBJective function for axis parallel hyper-ellipsoid
objfun1b	- OBJective function for rotated hyper-ellipsoid
objfun1c	- OBJective function for moved axis parallel hyper ellipsoid 1c
objfun2	- OBJective function for rosenbrock's FUNction
objfun6	- OBJective function for rastrigins FUNction 6
objfun7	- OBJective function for schwefel's FUNction
objfun8	- OBJective function for griewangk's FUNction
objfun9	- OBJective function for sum of different power FUNction 9
objgold	- OBJective function for GOLDstein-price function
objharv	- OBJective function for HARVest problem
objint1	- OBJective function for INT function 1
objint2	- OBJective function for INT function 1
objint3	- OBJective function for INT function 3
objint4	- OBJective function for INT function 4
objlinq	- OBJective function for discrete LINear Quadratic problem
objlinq2	- OBJective function for LINear Quadratic problem 2
objone1	- OBJective function for ONEmax function 1
objpush	- OBJective function for PUSH-cart problem
objridge	- OBJective function RIDGE
objsixh	- OBJective function for SIX Hump camelback function
objsoland	- OBJective function for SOLAND function
objtsp1	- OBJective function for the traveling salesman example
objtsplib	- OBJective function for the traveling salesman library
plotdopi	- PLOTing of DO(Ppel)uble Integration results
plottsplib	- PLOTing of results of TSP optimization (TSPLIB examples)
simdopi1	- M-file description of the SIMULINK system named SIMDOPI1
simdopiv	- SIMulation Modell of DOPpelIntegrator, s-function, Vectorized

218   Genetic Algorithm Implementation Using Matlab

    simlinq1   - M-file description of the SIMULINK system named SIMLINQ1
    simlinq2   - Modell of Linear Quadratic Problem, s-function
    tsp_readlib   - TSP utility function, reads TSPLIB data files
    tsp_uscity   - TSP utility function, reads US City definitions

Conversion functions
    bin2int   - BINary string to INTeger string conversion
    bin2real   - BINary string to REAL vector conversion
    bindecod   - BINary DECODing to binary, integer or real numbers

Initialization functions
    initbp   - CReaTe an initial Binary Population
    initip   - CReaTe an initial (Integer value) Population
    initpop   - INITialization of POPulation (including innoculation)
    initpp   - Create an INITial Permutation Population
    initrp   - INITialize an Real value Population

Selection functions
    selection   - high level SELECTion function
    sellocal   - SELection in a LOCAL neighbourhood
    selrws   - SELection by Roulette Wheel Selection
    selsus   - SELection by Stochastic Universal Sampling
    seltour   - SELection by TOURnament
    seltrunc   - SELection by TRUNCation
    rankgoal   - perform goal preference calculation between multiple objective values
    ranking   - RANK-based fitness assignment, single and multi objective, linear
                               and nonlinear
    rankplt   - RANK two multi objective values Partially Less Than
    rankshare   - SHARing between individuals

Crossover functions
    recdis   - RECombination DIScrete
    recdp   - RECombination Double Point
    recdprs   - RECombination Double Point with Reduced Surrogate
    recgp   - RECombination Generalized Position
    recint   - RECombination extended INTermediate
    reclin   - RECombination extended LINe
    reclinex   - EXtended LINe RECombination
    recmp   - RECombination Multi-Point, low level function
    recombin   - high level RECOMBINation function
    recpm   - RECombination Partial Matching
    recsh   - RECombination SHuffle
    recshrs   - RECombination SHuffle with Reduced Surrogate
    recsp   - RECombination Single Point
    recsprs   - RECombination Single Point with Reduced Surrogate
    reins   - high-level RE-INSertion function
    reinsloc   - RE-INSertion of offspring in population replacing parents LOCal
    reinsreg   - REINSertion of offspring in REGional population model replac

Mutation functions
    mutate   - high level MUTATion function
    mutbin   - MUTation for BINary representation
    mutbmd   - real value Mutation like Discrete Breeder genetic algorithm
    mutcomb   - MUTation for combinatorial problems
    mutes1   - MUTation by Evolutionary Strategies 1, derandomized Self Adaption
    mutes2   - MUTation by Evolutionary Strategies 2, derandomized self adaption
    mutexch   - MUTation by eXCHange

mutint	- MUTation for INTeger representation
mutinvert	- MUTation by INVERTing variables
mutmove	- MUTation by MOVEing variables
mutrand	- MUTation RANDom
mutrandbin	- MUTation RANDom of binary variables
mutrandint	- MUTation RANDom of integer variables
mutrandperm	- MUTation RANDom of binary variables
mutrandreal	- MUTation RANDom of real variables
mutreal	- real value Mutation like Discrete Breeder genetic algorithm
mutswap	- MUTation by SWAPping variables
mutswaptyp	- MUTation by SWAPping variables of identical typ

Other functions

compdiv	- COMPute DIVerse things of GEA Toolbox
compdiv2	- COMPute DIVerse things of GEA Toolbox
compete	- COMPETition between subpopulations
comploc	- COMPute LOCal model things of toolbox
compplot	- COMPute PLOT things of GEA Toolbox
geamain2	- MAIN function for Genetic and Evolutionary Algorithm toolbox for matlab

Plot Functions

fitdistc	- FITness DISTance Correlation computation
meshvar	- create grafics of objective functions with plotmesh.
plotmesh	- PLOT of objective functions as MESH Plot
plotmop	- PLOT properties of MultiObjective functions
reslook	- LOOK at saved RESults
resplot	- RESult PLOTing of GEA Toolbox optimization
samdata	- sammon mapping: data examples
samgrad	- Sammon mapping gradient calculation
sammon	- Multidimensional scaling (SAMMON mapping)
samobj	- Sammon mapping objective function
samplot	- Plot function for Multidimensional scaling (SAMMON mapping)

8.4 Genetic Algorithm Graphical User Interface Toolbox

The Genetic Algorithm Tool is a graphical user interface that enables you to use the genetic algorithm without working at the command line. To open the Genetic Algorithm Tool, enter

    gatool

at the MATLAB command prompt.

This opens the tool as shown in the following Fig. 8.1

To use the Genetic Algorithm Tool, you must first enter the following information:

Fitness function – The objective function you want to minimize. Enter the fitness function in the form @fitnessfun, where fitnessfun.m is an M-file that computes the fitness function.

Number of Variables – The number of variables in the given fitness function should be given.

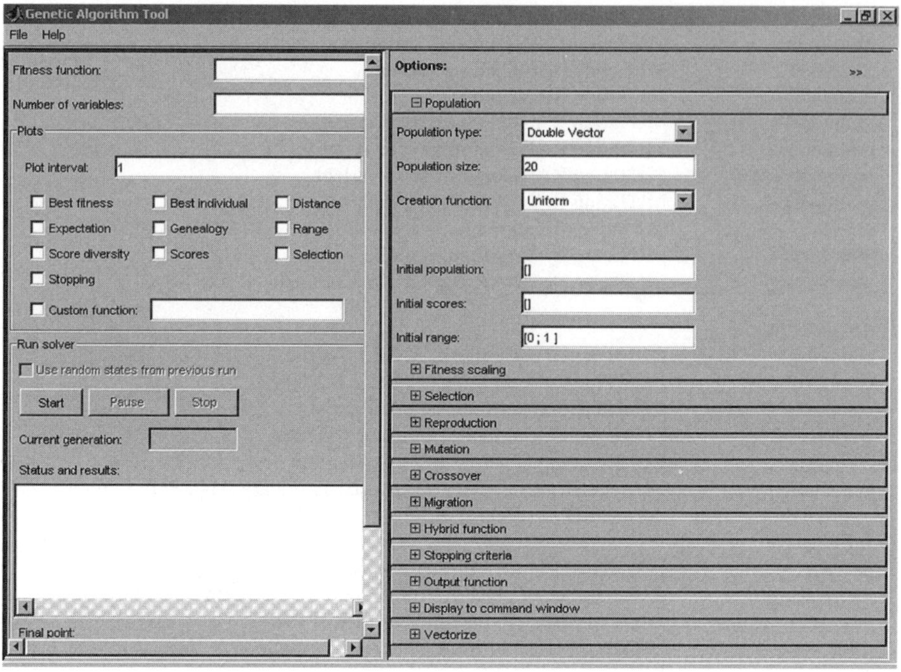

Fig. 8.1 Genetic Algorithm Tool

The **plot** options

1. Best fitness
2. Best individual
3. Distance
4. Expectation
5. Genealogy
6. Range
7. Score Diversity
8. Scores
9. Selection
10. Stopping

Based upon ones problem, custom function my also be built.

The various parameters essential for running Genetic algorithm tool should be specified appropriately. The parameters appear on the right hand side of the GA tool. The description is as follows:

1. **Population**

   In this case population type, population size and creation function may be selected. The initial population and initial score may be specified, if not, the "Ga tool" creates them. The initial range should be given.

2. **Fitness Scaling**

   The fitness scaling should be any of the following

8.4 Genetic Algorithm Graphical User Interface Toolbox

   a. Rank
   b. Proportional
   c. Top
   d. Shift Linear
   e. Custom

3. **Selection**

   The selection is made on any one of the following mentioned methods

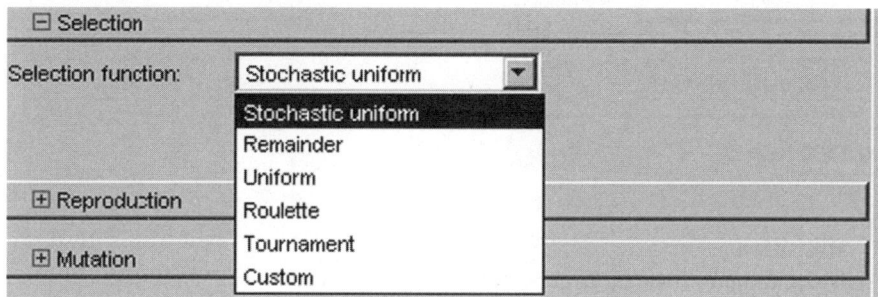

Fig. 8.2 Selection

4. **Reproduction**

   In reproduction the elite count and cross over fraction should be given. If elite count not specified, it is taken as 2.

Fig. 8.3 Reproduction

5. **Mutation**

   Generally Gaussian or Uniform Mutation is carried out. The user may define own customized mutation operation.

Fig. 8.4 Mutation

6. Crossover

The various crossover techniques are as follows:

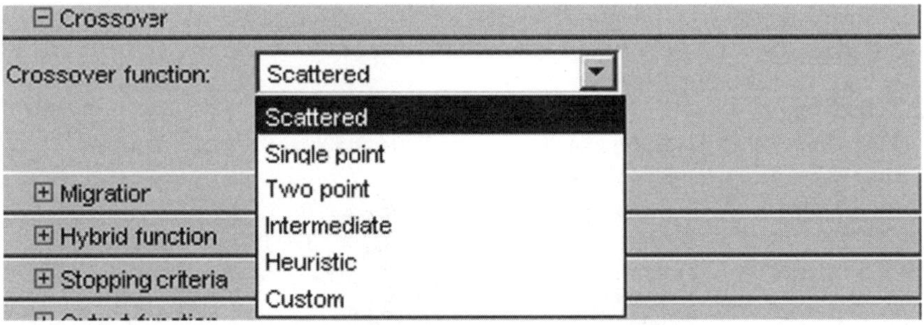

Fig. 8.5 Crossover

7. Migration

The parameter for migration should be defined as follows:

Fig. 8.6 Migration

8. Hybrid Function

Any one of the following hybrid functions may be selected,

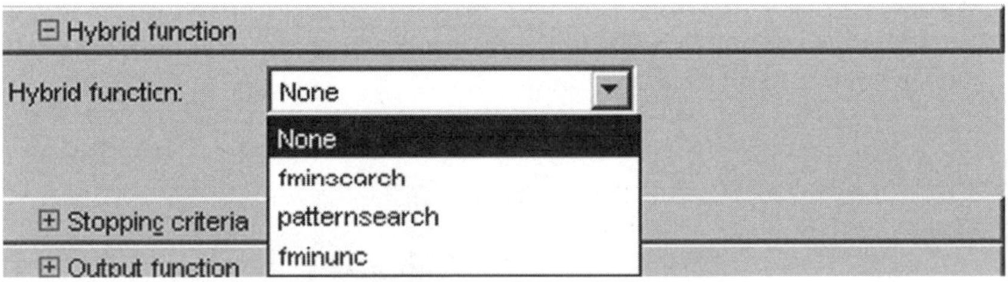

Fig. 8.7 Hybrid function

9. Stopping Criteria

The stopping criteria plays a major role in simulation. They are:

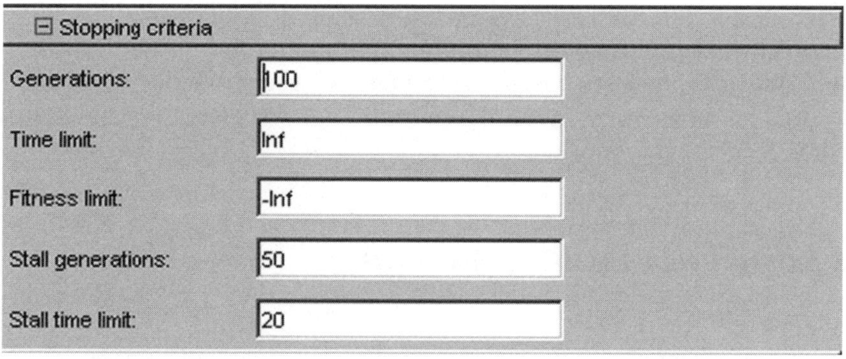

8.4 Genetic Algorithm Graphical User Interface Toolbox

Stopping criteria	
Generations:	100
Time limit:	Inf
Fitness limit:	-Inf
Stall generations:	50
Stall time limit:	20

Fig. 8.8 Stopping criteria

The other parameters **Output Function, Display to command window and Vectorize** may be suitably defined by the user.

10. **Running and Simulation**

The menu shown below helps the user for running the GA tool.

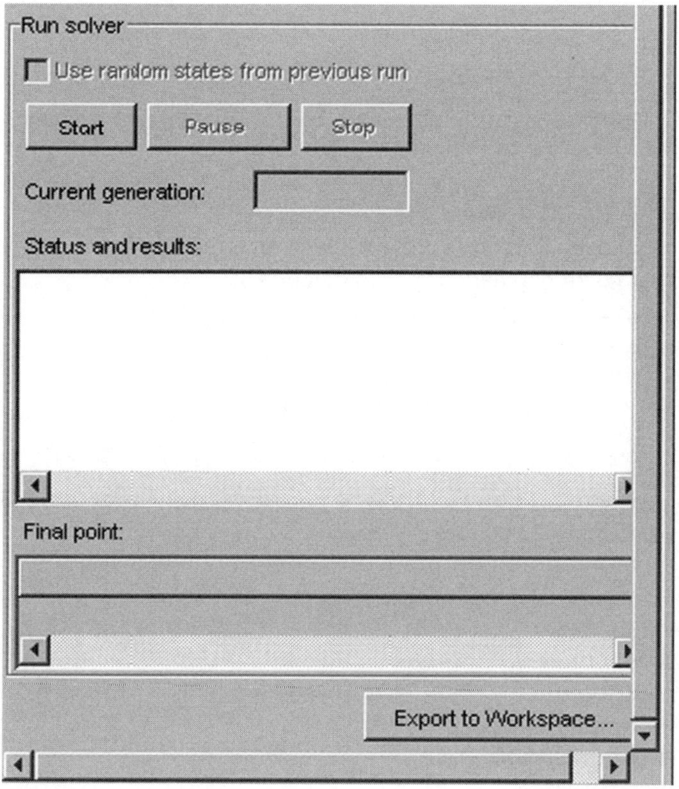

Fig. 8.9 Run solver

The running process may be temporarily stopped using "Pause" option and permanently stopped using "Stop" option. The "current generation" will be displayed during the iteration. Once the iterations are completed, the status and results will be displayed. Also the "final point" for the fitness function will be displayed.

8.5 Solved Problems using MATLAB

Problem 1

Write a MATLAB program for maximizing $f(x) = x^2$ using genetic algorithm, where x ranges from 0 to 31. Perform 4 iterations.

Note

In MATLAB % indicates comment statement.

Source Code

```
%program for Genetic algorithm to maximize the function f(x) =xsquare
clear all;
clc;
%x ranges from 0 to 31 2power5 = 32
%five bits are enough to represent x in binary representation
n=input('Enter no. of population in each iteration');
nit=input('Enter no. of iterations');
%Generate the initial population
[oldchrom]=initbp(n,5)
%The population in binary is converted to integer
FieldD=[5;0;31;0;0;1;1]
for i=1:nit
    phen=bindecod(oldchrom,FieldD,3); % phen gives the integer value of the
    binary population %obtain fitness value
    sqx=phen.^2;
    sumsqx=sum(sqx);
    avsqx=sumsqx/n;
    hsqx=max(sqx);
    pselect=sqx./sumsqx;
    sumpselect=sum(pselect);
    avpselect=sumpselect/n;
    hpselect=max(pselect);
    %apply roulette wheel selection
```

8.5 Solved Problems using MATLAB

```
FitnV=sqx;
Nsel=4;
newchrix=selrws(FitnV, Nsel);
newchrom=oldchrom(newchrix,:);
%Perform Crossover
crossoverrate=1;
newchromc=recsp(newchrom,crossoverrate); %new population after crossover
%Perform mutation
vlub=0:31;
mutationrate=0.001;
newchromm=mutrandbin(newchromc,vlub,mutationrate); %new population after
mutation disp('For iteration');
i
disp('Population');
oldchrom
disp('X');
phen
disp('f(X)');
sqx
oldchrom=newchromm;
end
```

Output

```
Enter no. of population in each iteration4
Enter no. of iterations4
oldchrom =
    1  0  0  1  0
    0  1  0  1  0
    0  0  1  1  0
    1  1  1  1  0
FieldD =
    5
    0
   31
    0
    0
    1
    1
For iteration
i =
    1
Population
oldchrom =
```

```
        1 0 0 1 0
        0 1 0 1 0
        0 0 1 1 0
        1 1 1 1 0
X
phen =
    18
    10
     6
    30
f(X)
sqx =
    324
    100
     36
    900
For iteration
i =
    2
Population
oldchrom =
        1 1 1 0 0
        0 1 1 0 1
        0 0 1 1 0
        1 0 1 0 1
X
phen =
    28
    13
     6
    21
f(X)
sqx =
    784
    169
     36
    441
For iteration
i =
    3
Population
oldchrom =
        0 0 0 0 1
        0 0 1 1 1
        0 0 0 0 1
        1 0 1 0 0
```

8.5 Solved Problems using MATLAB

X
phen =
   1
   7
   1
  20
f(X)
sqx =
    1
   49
    1
  400
For iteration
i =
   4
Population
oldchrom =
  1 0 0 0 0
  1 1 0 1 1
  1 0 0 1 1
  0 1 1 1 1
X
phen =
  16
  27
  19
  15
f(X)
sqx =
  256
  729
  361
  225

Problem 2

Find a minimum of a non-smooth objective function using the Genetic Algorithm (GA) function in the Genetic Algorithm and Direct Search Toolbox.

Description

Traditional derivative-based optimization methods, like those found in the Optimization Toolbox, are fast and accurate for many types of optimization problems.

These methods are designed to solve "smooth", i.e., continuous and differentiable, minimization problems, as they use derivatives to determine the direction of descent. While using derivatives makes these methods fast and accurate, they often are not effective when problems lack smoothness, e.g., problems with discontinuous, non-differentiable, or stochastic objective functions. When faced with solving such non-smooth problems, methods like the genetic algorithm or the more recently developed pattern search methods, both found in the Genetic Algorithm and Direct Search Toolbox, are effective alternatives.

Source Code

```
clear all; close all;format compact
Objfcn = @nonSmoothFcn;        %Handle to the objective function
X0 = [2 -2];                   % Starting point
range = [-6 6;-6 6];           %Range used to plot the objective function
rand('state',0);               %Reset the state of random number generators
randn('state',0);
type nonSmoothFcn.m            % Non-smooth Objective Function
showNonSmoothFcn(Objfcn,range);
set(gca,'CameraPosition',[-36.9991 62.6267 207.3622]);
set(gca,'CameraTarget',[0.1059 -1.8145 22.3668])
set(gca,'CameraViewAngle',6.0924)
%Plot of the starting point (used by the PATTERNSEARCH solver)
plot3(X0(1),X0(2),feval(Objfcn,X0),'or','MarkerSize',10,'MarkerFaceColor','r');
fig = gcf;
% Minimization Using The Genetic Algorithm
FitnessFcn = @nonSmoothFcn;
numberOfVariables = 2;
optionsGA = gaoptimset('PlotFcns',@gaplotbestfun,'PlotInterval',5, ...
'PopInitRange',[-5;5]);
% We run GA with the options 'optionsGA' as the third argument.
[Xga,Fga] = ga(FitnessFcn,numberOfVariables,optionsGA)
% Plot the final solution
figure(fig)
hold on;
plot3(Xga(1),Xga(2),Fga,'vm','MarkerSize',10,'MarkerFaceColor','m');
hold off;
fig = gcf;
```

% The optimum is at x* = (−4.7124, 0.0). GA found the point % (−4.7775,0.0481) near the optimum, but could not get closer with the default stopping criteria. By changing the stopping criteria, we might find a more accurate solution, but it may take many more function evaluations to reach x* = (−4.7124, 0.0). Instead, we can use a more efficient local search that starts where GA left off. The hybrid function field in GA provides this feature automatically.

8.5 Solved Problems using MATLAB

% Minimization Using A Hybrid Function

% Our choices are FMINSEARCH, PATTERNSEARCH, or FMINUNC. Since this optimization example is smooth near the optimizer, we can use the FMINUNC function from the Optimization toolbox as our hybrid function as this is likely to be the most efficient. Since FMINUNC has its own options structure, we provide it as an additional argument when specifying the hybrid function.

% Run GA-FMINUNC Hybrid
optHybrid = gaoptimset(optionsGA,'Generations',15, 'PlotInterval',1,...
            'HybridFcn',{ @fminunc,optimset('OutputFcn',@fminuncOut)});
[Xhybrid,Fhybrid] = ga(Objfcn,2,optHybrid);
% Plot the final solution
figure(fig);
hold on;
plot3(Xhybrid(1),Xhybrid(2),Fhybrid+1,' ^ c','MarkerSize',10,
'MarkerFaceColor','c');
hold off;
disp(['The norm of |Xga - Xhb| is ', num2str(norm(Xga-Xhybrid))]);
disp(['The difference in function values Fga and Fhb is ',
num2str(Fga - Fhybrid)]);

%% Minimization Using The Pattern Search Algorithm

% To minimize our objective function using the PATTERNSEARCH function, we need to pass in a function handle to the objective function as well as specifying a start point as the second argument.

ObjectiveFunction = @nonSmoothFcn;
X0 = [2 -2]; % Starting point
% Some plot functions are selected to monitor the performance of the solver.
optionsPS = psoptimset('PlotFcns',@psplotbestf);
% Run pattern search solver
[Xps,Fps] = patternsearch(Objfcn,X0,[],[],[],[],[],[],optionsPS)
% Plot the final solution
figure(fig)
hold on;
plot3(Xps(1),Xps(2),Fps+1,'*y','MarkerSize',14,'MarkerFaceColor','y');
hold off;

The various functions used in optimization of non-smooth function are a follows:

function [f, g] = **nonSmoothFcn**(x)
%NONSMOOTHFCN is a non-smooth objective function

```
for i = 1:size(x,1)
    if x(i,1) < -7
        f(i) = (x(i,1))^2 + (x(i,2))^2 ;
    elseif x(i,1) < -3
        f(i) = -2*sin(x(i,1)) - (x(i,1)*x(i,2)^2)/10 + 15 ;
    elseif x(i,1) < 0
        f(i) = 0.5*x(i,1)^2 + 20 + abs(x(i,2))+ patho(x(i,:));
    elseif x(i,1) >= 0
        f(i) = .3*sqrt(x(i,1)) + 25 +abs(x(i,2)) + patho(x(i,:));
    end
end
%Calculate gradient
g = NaN;
if x(i,1) < -7
    g = 2*[x(i,1); x(i,2)];
elseif x(i,1) < -3
    g = [-2*cos(x(i,1))-(x(i,2)^2)/10; -x(i,1)*x(i,2)/5];
elseif x(i,1) < 0
    [fp,gp] = patho(x(i,:));
    if x(i,2) > 0
        g = [x(i,1)+gp(1); 1+gp(2)];
    elseif x(i,2) < 0
        g = [x(i,1)+gp(1); -1+gp(2)];
    end
elseif x(i,1) >0
    [fp,gp] = patho(x(i,:));
    if x(i,2) > 0
        g = [.15/sqrt(x(i,1))+gp(1); 1+ gp(2)];
    elseif x(i,2) < 0
        g = [.15/sqrt(x(i,1))+gp(1); -1+ gp(2)];
    end
end
function [f,g] = patho(x)
Max = 500;
f = zeros(size(x,1),1);
g = zeros(size(x));
for k = 1:Max %k
    arg = sin(pi*k^2*x)/(pi*k^2);
    f = f + sum(arg,2);
    g = g + cos(pi*k^2*x);
end

function showNonSmoothFcn(fcn,range)
if(nargin == 0)
    fcn = @rastriginsfcn;
    range = [-5,5;-5,5];
```

8.5 Solved Problems using MATLAB 231

```
end
pts = 25;
span = diff(range')/(pts - 1);
x = range(1,1): span(1) : range(1,2);
y = range(2,1): span(2) : range(2,2);
pop = zeros(pts * pts,2);
k = 1;
for i = 1:pts
    for j = 1:pts
        pop(k,:) = [x(i),y(j)];
        k = k + 1;
    end
end
values = feval(fcn,pop);
values = reshape(values,pts,pts);
surf(x,y,values)
shading interp
light
lighting phong
hold on
contour(x,y,values)
rotate3d
view(37,60)
%Annotations
figure1 = gcf;
% Create arrow
annotation1 = annotation(figure1,'arrow',[0.5946 0.4196],[0.9024 0.6738]);
% Create textbox
annotation2 = annotation(...
  figure1,'textbox',...
  'Position',[0.575 0.9071 0.1571 0.07402],...
  'FitHeightToText','off',...
  'FontWeight','bold',...
  'String',{'Start point'});
% Create textarrow
annotation3 = annotation(...
  figure1,'textarrow',...
  [0.3679 0.4661],[0.1476 0.3214],...
  'String',{'Non-differentiable regions'},...
  'FontWeight','bold');
% Create arrow
annotation4 = annotation(figure1,'arrow',[0.1196 0.04107],[0.1381 0.5429]);
% Create textarrow
annotation5 = annotation(...
  figure1,'textarrow',...
  [0.7411 0.5321],[0.05476 0.1381],...
```

```
    'LineWidth',2,...
    'Color',[1 0 0],...
    'String',{'Smooth region'},...
    'FontWeight','bold',...
    'TextLineWidth',2,...
    'TextEdgeColor',[1 0 0]);
% Create arrow
annotation6 = annotation(...
    figure1,'arrow',...
    [0.8946 0.9179],[0.05714 0.531],...
    'Color',[1 0 0]);

function stop = fminuncOut(x,optimvalues, state)
persistent fig gaIter
stop = false;
switch state
    case 'init'
        fig = findobj(0,'type','figure','name','Genetic Algorithm');
        limits = get(gca,'XLim');
        gaIter = limits(2);
        hold on;
    case 'iter'
        set(gca,'Xlim', [1 optimvalues.iteration + gaIter]);
        fval = optimvalues.fval;
        iter = gaIter + optimvalues.iteration;
        plot(iter,fval,'dr')
        title(['Best function value: ',num2str(fval)],'interp','none')
    case 'done'
        fval = optimvalues.fval;
        iter = gaIter + optimvalues.iteration;
        title(['Best function value: ',num2str(fval)],'interp','none')
        % Create textarrow
        annotation1 = annotation(...
            gcf,'textarrow',...
            [0.6643 0.7286],[0.3833 0.119],...
            'String',{'Algorithm switch to FMINUNC'},...
            'FontWeight','bold');
        hold off
end
```

Output

Optimization terminated: maximum number of generations exceeded.

Xga =
    -4.7775 0.0481

8.5 Solved Problems using MATLAB

Fga =
 13.0053

Optimization terminated: maximum number of generations exceeded.
Switching to the hybrid optimization algorithm (FMINUNC).
In fminunc at 241
In ga at 268
In nonSmoothOpt at 101
Optimization terminated: relative infinity-norm of gradient less than options.TolFun.
The norm of |Xga - Xhb| is 0.08092
The difference in function values Fga and Fhb is 0.0053385
Optimization terminated: current mesh size 9.5367e-007 is less than 'TolMesh'.

Xps =
 -4.7124 0
Fps =
 13.0000

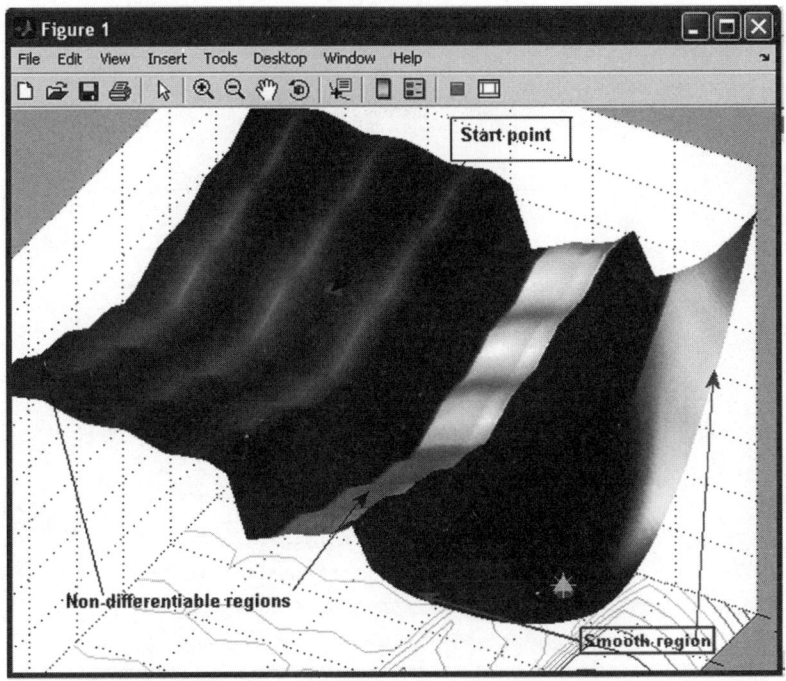

Fig. 8.10 Optimization of objective function using GA

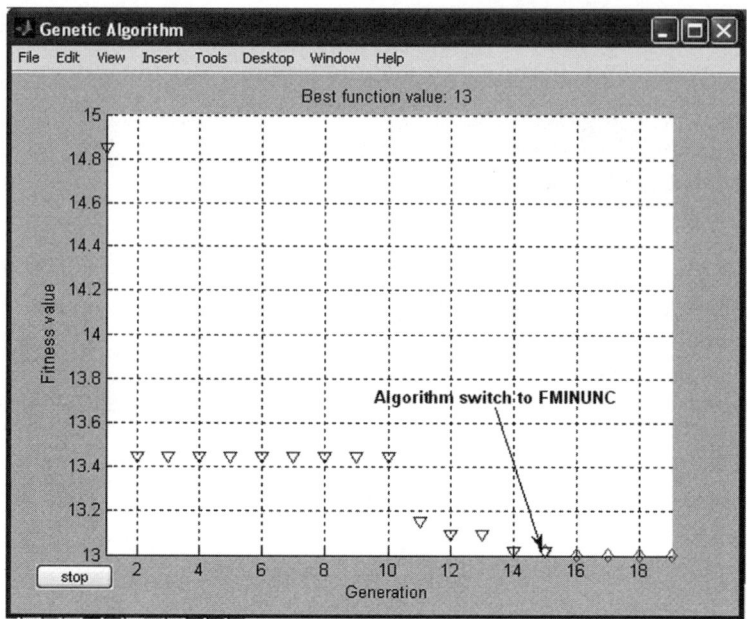

Fig. 8.11 Optimization using hybrid GA – FMINUNC

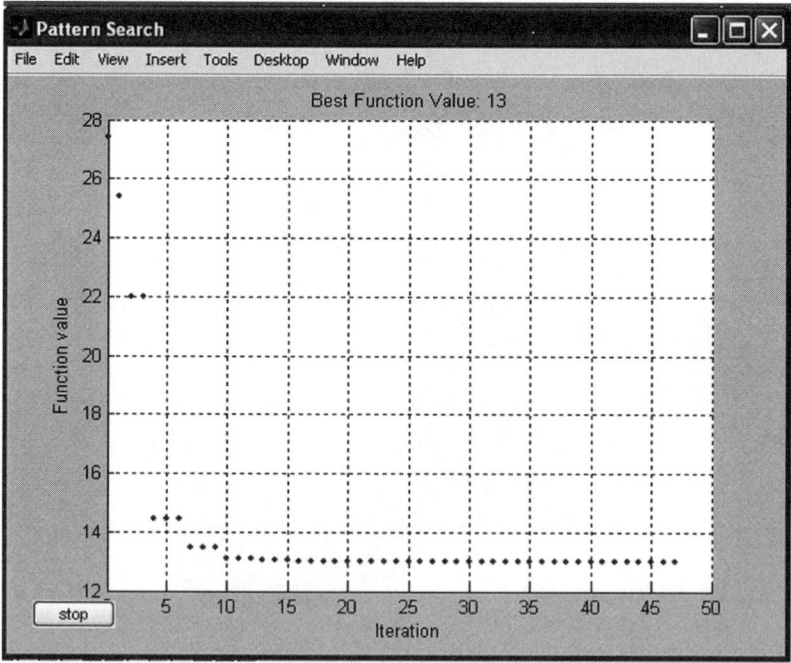

Fig. 8.12 Optimization using pattern search

8.5 Solved Problems using MATLAB

Problem 3

Find a minimum of a stochastic objective function using PATTERNSEARCH function in the Genetic Algorithm and Direct Search Toolbox.

Source Code

```
% Pattern search optimization solver
format compact
X0 = [2.5 -2.5];                          %Starting point.
LB = [-5 -5];                             %Lower bound
UB = [5 5];                               %Upper bound
range = [LB(1) UB(1); LB(2) UB(2)];
Objfcn = @smoothFcn;                      %Handle to the objective function.
% Plot the smooth objective function
clf;showSmoothFcn(Objfcn,range); hold on;
title('Smooth objective function')
plot3(X0(1),X0(2),feval(Objfcn,X0)+30,'om','MarkerSize',12, ...
  'MarkerFaceColor','r'); hold off;
set(gca,'CameraPosition',[-31.0391 -85.2792 -281.4265]);
set(gca,'CameraTarget',[0 0 -50])
set(gca,'CameraViewAngle',6.7937)
fig = gcf;


%% Run FMINCON on smooth objective function


% The objective function is smooth (twice continuously differentiable). Solving the
% optimization problem using FMINCON function from the Optimization Toolbox.
% FMINCON finds a constrained minimum of a function of several variables. This
% function has a unique minimum at the point x* = (-5.0, -5) where it has a function
% value f(x*) = -250.


% Set options to display iterative results.
options = optimset('Display','iter','OutputFcn',@fminuncOut1);
[Xop,Fop] = fmincon(Objfcn,X0,[],[],[],[],LB,UB,[],options)
figure(fig);
hold on;
%Plot the final point
plot3(Xop(1),Xop(2),Fop,'dm','MarkerSize',12,'MarkerFaceColor','m');
hold off;
% Stochastic objective function The objective function is same as the previous
function and some noise
```

```
% added to it.
%Reset the state of random number generators
randn('state',0);
noise = 8.5;
Objfcn = @(x) smoothFcn(x,noise); %Handle to the objective function.
%Plot the objective function (non-smooth)
figure;
for i = 1:6
    showSmoothFcn(Objfcn,range);
    title('Stochastic objective function')
    set(gca,'CameraPosition',[-31.0391 -85.2792 -281.4265]);
    set(gca,'CameraTarget',[0 0 -50])
    set(gca,'CameraViewAngle',6.7937)
    drawnow; pause(0.2)
end
fig = gcf;
%% Run FMINCON on stochastic objective function
options = optimset('Display','iter');
[Xop,Fop] = fmincon(Objfcn,X0,[],[],[],[],LB,UB,[],options)
figure(fig);
hold on;
plot3(X0(1),X0(2),feval(Objfcn,X0)+30,'om','MarkerSize',16,
'MarkerFaceColor','r');
plot3(Xop(1),Xop(2),Fop,'dm','MarkerSize',12,'MarkerFaceColor','m');
%% Run PATTERNSEARCH. A pattern search algorithm does not require any
derivative information of the objective function to find an optimal point.
PSoptions = psoptimset('Display','iter','OutputFcn',@psOut);
[Xps,Fps] = patternsearch(Objfcn,X0,[],[],[],[],LB,UB,PSoptions)
figure(fig);
hold on;
plot3(Xps(1),Xps(2),Fps,'pr','MarkerSize',18,'MarkerFaceColor','r');
hold off
```

The various functions used in the above program are as follows:

```
function y = smoothFcn(z,noise)
% Objective function
if nargin < 2
    noise = 0;
end
LB = [-5 -5];        %Lower bound
UB = [5 5];          %Upper bound
y = zeros(1,size(z,1));
for i = 1:size(z,1)
```

8.5 Solved Problems using MATLAB

```
      x = z(i,:);
      if any(x<LB) || any(x>UB)
         y(i) = Inf;
      else
      y(i) = x(1)^3 - x(2)^2 + ...
         100*x(2)/(10+x(1)) + noise*randn;
      end
   end
end


function showSmoothFcn(fcn,range)
pts = 100;
span = diff(range')/(pts - 1);
x = range(1,1): span(1) : range(1,2);
y = range(2,1): span(2) : range(2,2);
pop = zeros(pts * pts,2);
k = 1;
for i = 1:pts
   for j = 1:pts
      pop(k,:) = [x(i),y(j)];
      k = k + 1;
   end
end
values = feval(fcn,pop);
values = reshape(values,pts,pts);
clf;
surf(x,y,values)
shading interp
light
lighting phong
hold on
rotate3d
view(37,60)
set(gcf,'Renderer','opengl');
set(gca,'ZLimMode','manual');
function stop = fminuncOut1(X, optimvalues, state)
stop = false;
figure1 = gcf;
if strcmpi(state,'done')
annotation1 = annotation(figure1,'arrow',[0.7506 0.7014],[0.3797 0.625]);
   % Create textbox
   annotation2 = annotation(...
      figure1,'textbox',...
      'Position',[0.6857 0.2968 0.1482 0.0746],...
      'FontWeight','bold',...
```

```
        'String',{'start point'},...
        'FitHeightToText','on');
    % Create arrow
annotation3 = annotation(figure1,'arrow',[0.4738 0.3489],[0.1774 0.2358]);
    % Create textbox
    annotation4 = annotation(...
        figure1,'textbox',...
        'Position',[0.4732 0.1444 0.2411 0.06032],...
        'FitHeightToText','off',...
        'FontWeight','bold',...
        'String',{'FMINCON solution'});
end


function [stop options changed] = psOut(optimvalues,options,flag)
stop = false;
changed = false;
figure1 = gcf;
if strcmpi(flag,'done')
    % Create textbox
    annotation1 = annotation(...
        figure1,'textbox',...
        'Position',[0.4679 0.1357 0.3321 0.06667],...
        'FitHeightToText','off',...
        'FontWeight','bold',...
        'String',{'Pattern Search solution'});
    % Create textbox
    annotation2 = annotation(...
        figure1,'textbox',...
        'Position',[0.5625 0.2786 0.2321 0.06667],...
        'FitHeightToText','off',...
        'FontWeight','bold',...
        'String',{'FMINCON solution'});
    % Create textbox
    annotation3 = annotation(...
        figure1,'textbox',...
        'Position',[0.3714 0.6905 0.1571 0.06449],...
        'FitHeightToText','off',...
        'FontWeight','bold',...
        'String',{'Start point'});
    % Create arrow
    annotation4 = annotation(figure1,'arrow',[0.7161 0.6768],[0.3452 0.4732]);
    % Create arrow
    annotation5 = annotation(figure1,'arrow',[0.4697 0.35],[0.1673 0.2119]);
    % Create arrow
```

8.5 Solved Problems using MATLAB

```
    annotation6 = annotation(figure1,'arrow',[0.4523 0.6893],[0.6929 0.6]);
end
```

Output

```
    In fmincon at 260
    In PS at 33
Xop =
    -5 -5
Fop =
    -250
In fmincon at 260
    In PS at 70
Xop =
    1.2861 -4.8242
Fop =
    -86.0221
Xps =
    -5 -5
Fps =
    -247.3159
```

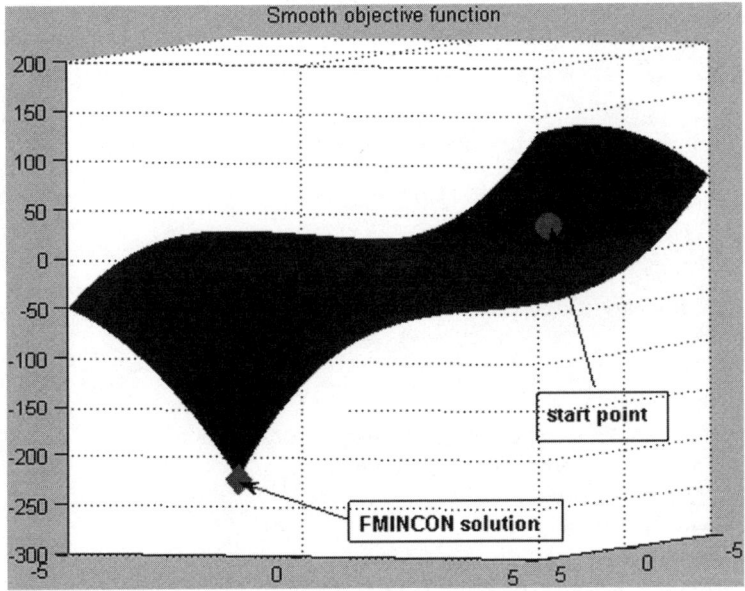

Fig. 8.13 Smooth objective function

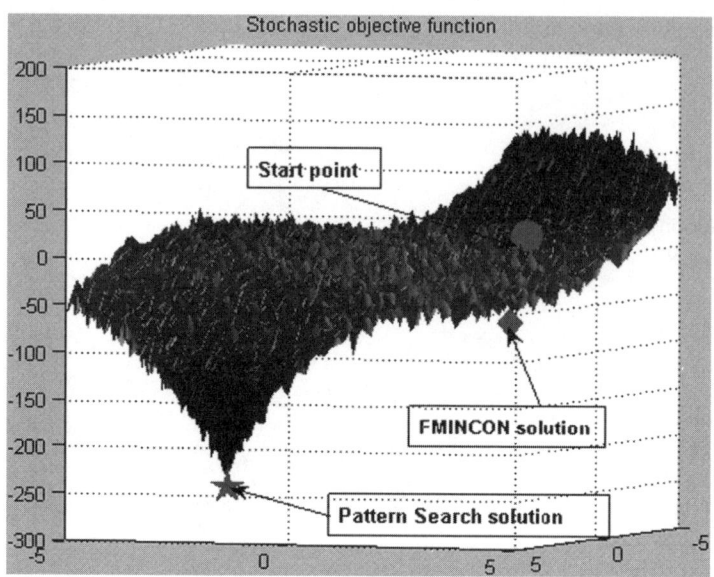

Fig. 8.14 Stochastic objective function

Pattern search algorithm is not affected by random noise in the objective functions. Pattern search requires only function value and not the derivatives, hence a noise (of some uniform kind) may not affect it.

Pattern search requires a lot more function evaluation to find the minima, a cost for not using the derivatives.

Problem 4

Write a Program to maximize sin(x) within the range 0<x<3.14

Source Code

```
%program for Genetic algorithm to maximize the function f(x) =sin(x)
clear all;
clc;
%x ranges from 0 to 3.14
%five bits are enough to represent x in binary representation
n=input('Enter no. of population in each iteration');
nit=input('Enter no. of iterations');
%Generate the initial population
[oldchrom]=initbp(n,5)
%The population in binary is converted to integer
FieldD=[5;0;3.14;0;0;1;1];
```

8.5 Solved Problems using MATLAB

```
for i=1:nit
    phen=bindecod(oldchrom,FieldD,3); % phen gives the integer value of the
    %binary population
    %obtain fitness value
    FitnV=sin(phen);
    %apply roulette wheel selection
    Nsel=4;
    newchrix=selrws(FitnV, Nsel);
    newchrom=oldchrom(newchrix,:);
    %Perform Crossover
    crossoverrate=1;
    newchromc=recsp(newchrom,crossoverrate); %new population after crossover
    %Perform mutation
    vlub=0:31;
    mutationrate=0.001;
    newchromm=mutrandbin(newchromc,vlub,mutationrate); %new population
    %after mutation
    disp('For iteration');
    i
    disp('Population');
    oldchrom
    disp('X');
    phen
    disp('f(X)');
    FitnV
    oldchrom=newchromm;
end
```

Output

Enter no. of population in each iteration5
Enter no. of iterations5
oldchrom =
    0 1 0 0 0
    0 1 0 0 1
    1 0 1 0 1
    1 0 0 1 1
    0 0 0 1 0
FieldD =
    5.0000
        0
    3.1400
        0
        0

    1.0000
    1.0000
For iteration
i =
    1
Population
oldchrom =
    0 1 0 0 0
    0 1 0 0 1
    1 0 1 0 1
    1 0 0 1 1
    0 0 0 1 0
X
phen =
    1
    1
    2
    2
    0
f(X)
FitnV =
    0.8415
    0.8415
    0.9093
    0.9093
         0
For iteration
i =
    2
Population
oldchrom =
    0 0 1 0 1
    1 0 1 1 0
    0 0 0 0 0
    1 0 0 0 1
X
phen =
    1
    2
    0
    2
f(X)
FitnV =
    0.8415
    0.9093
         0

8.5 Solved Problems using MATLAB

```
   0.9093
For iteration
i =
   3
Population
oldchrom =
   0 0 0 0 1
   0 1 1 0 1
   0 1 1 1 0
   1 0 0 0 0
X
phen =
   0
   1
   1
   2
f(X)
FitnV =
        0
   0.8415
   0.8415
   0.9093
For iteration
i =
   4
Population
oldchrm =
   1 1 0 1 1
   0 0 1 0 0
   0 0 1 1 1
   0 1 0 1 1
X
phen =
   3
   0
   1
   1
f(X)
FitnV =
   0.1411
        0
   0.8415
   0.8415
For iteration
i =
   5
```

Population
oldchrom =
   0 1 1 0 0
   1 0 0 0 1
   0 0 0 0 0
   1 0 0 1 1
X
phen =
   1
   2
   0
   2
f(X)
FitnV =
   0.8415
   0.9093
        0
   0.9093

Problem 5

Find the minimum of the quadratic equation $f(x)=x^2+5x+2$

Source Code

The minimizes of the given quadratic equation is done within a single command line function. The function used is "ga"

1. Define the given function $f(x) = x^2+5x+2$ in a separate m-file as shown below.

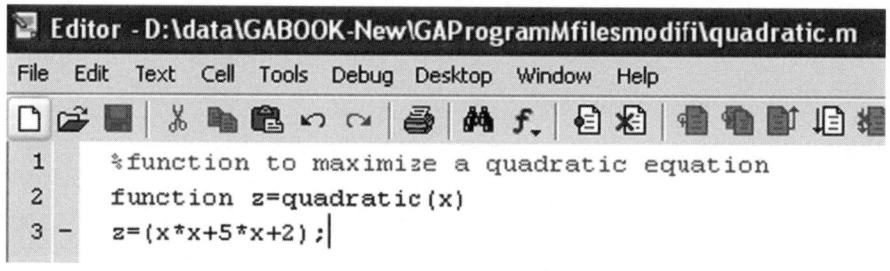

Fig. 8.15 Declaration of the quadratic function

2. Then use the command "ga" to obtained the minimized value of f(x). The format of command "ga" is,

8.5 Solved Problems using MATLAB

X = GA(FITNESSFCN,NVARS) finds the minimum of FITNESSFCN using GA. NVARS is the dimension (number of design variables) of the FITNESSFCN.

3. Thus the command for the given problem is given by,

    x=ga(@quadratic,1)

On running the above command, the output is obtained as given below.

Output

Optimization terminated: maximum number of generations exceeded.

x =
   -2.5002

In this case, all the operations are performed with the default setting of the command "ga".
In the above case, if the command is specified as,

[x, fval, reason, output, population, scores] = ga(@quadratic,1)

then the output obtained is,
Optimization terminated: maximum number of generations exceeded.

x =
   -2.5002
fval =
   -4.2500
reason =
Optimization terminated: maximum number of generations exceeded.
output =
    randstate: [35x1 double]
    randnstate: [2x1 double]
    generations: 100
     funccount: 2000
      message: 'Optimization terminated: maximum number of generations exceeded.'
population =
   -2.5002
   -2.5002
   -2.5002
   -2.4933
   -2.5002
   -2.5002
   -2.5002

-2.5002
-2.5002
-2.5002
-2.5002
-2.5097
-2.5002
-2.5002
-2.5173
-2.5002
-2.5002
-2.5002
-2.4855
-2.5002
scores =
-4.2500
-4.2500
-4.2500
-4.2500
-4.2500
-4.2500
-4.2500
-4.2500
-4.2500
-4.2500
-4.2500
-4.2499
-4.2500
-4.2500
-4.2497
-4.2500
-4.2500
-4.2500
-4.2498
-4.2500

Problem 6

Write a program to minimize Rastrigin's function. Also plot the best fitness value.

Description

For two independent variables, Rastrigin's function is defined as,

$$R(x) = 20 + x_1^2 + x_2^2 - 10(\cos 2\pi x_1 + \cos 2\pi x_2).$$

8.5 Solved Problems using MATLAB

The MATLAB toolbox contains an M-file, rastriginsfcn.m, that computes the values of Rastrigin's function.

Source Code

The function is defined as follows:

```
function Rasfun = rastriginsfcn(pop) %RASTRIGINSFCN Compute the
  %"Rastrigin" function.
  Rasfun = 10.0 * size(pop,2) + sum(pop .^2 - 10.0 * cos(2 * pi .* pop),2);
```

The program for minimizing this function is given as,

```
%Program to minimize Rastrigins Function
%Depending upon user's need Options can be specified using the command
   'gaoptimset'. %If
%Options not specified default options are chosen.
options=gaoptimset('CrossoverFcn',@crossoversinglepoint,...
      'MutationFcn',@mutationuniform,'Plotfcns',@gaplotbestf)
%Generating the genetic algorithm for 10 variables with the options specified
    %above
[x,fval,reason] = ga(@rastriginsFcn,10,options)
```

Output

```
options =
    PopulationType: 'doubleVector'
    PopInitRange: [2x1 double]
    PopulationSize: 20
    EliteCount: 2
  CrossoverFraction: 0.8000
  MigrationDirection: 'forward'
  MigrationInterval: 20
  MigrationFraction: 0.2000
     Generations: 100
       TimeLimit: Inf
    FitnessLimit: -Inf
    StallGenLimit: 50
    StallTimeLimit: 20
   InitialPopulation: [ ]
    InitialScores: [ ]
     PlotInterval: 1
       CreationFcn: @gacreationuniform
   FitnessScalingFcn: @fitscalingrank
```

       SelectionFcn: @selectionstochunif
       CrossoverFcn: @crossoversinglepoint
       MutationFcn: @mutationuniform
         HybridFcn: [ ]
         Display: 'final'
         PlotFcns: @gaplotbestf
         OutputFcns: [ ]
         Vectorized: 'off'
Optimization terminated: maximum number of generations exceeded.
x =
    Columns 1 through 9
       0.9840 0.8061 0.9698 0.0493 0.0070 0.0408 0.0671 0.0970 0.9412
    Column 10
       0.9848
fval =
    15.4207
reason =

    Optimization terminated: maximum number of generations exceeded.

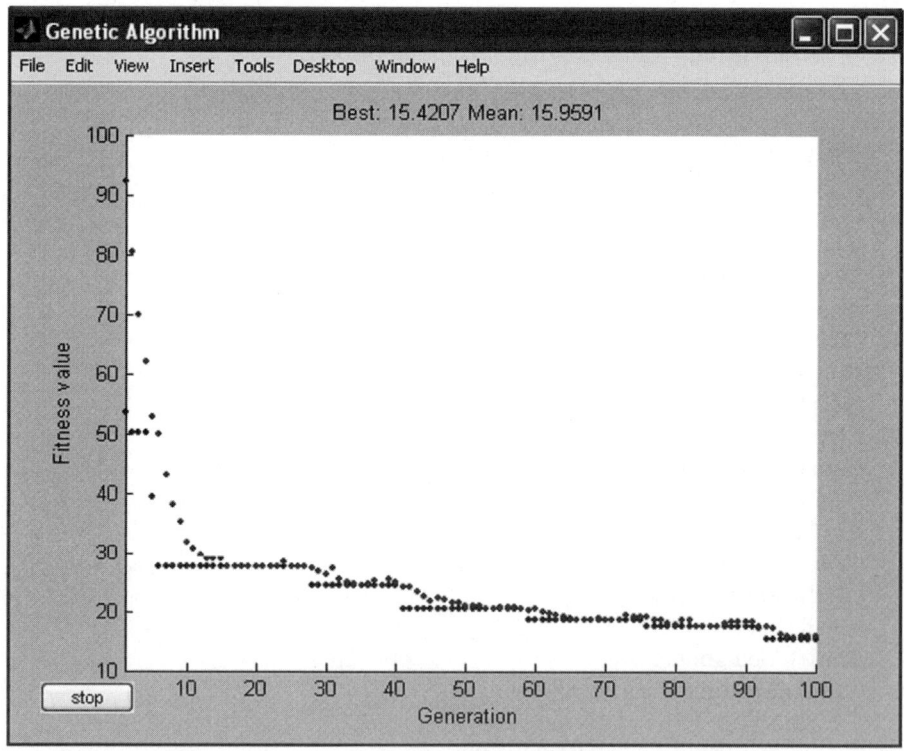

Fig. 8.16 Best fitness plot for minimization of Rastrigins function

8.5 Solved Problems using MATLAB

Problem 7

Write a program to optimize a function $f(x_1,x_2)=3x_1+9x_2$. Set suitable plot options.

Source Code

The function is defined as follows:

```
%function to be optimized
function z=twofunc(x)
z=(3*x(1)+9*x(2));
```

The program code for optimization process is as follows:

```
%Program to optimize a function with two variables
%Options are set using the command 'gaoptimset'.
clc;
clear all;
options = gaoptimset('PlotFcns',...
         {@gaplotbestf,@gaplotbestindiv,@gaplotexpectation,
         @gaplotstopping});
%Generating the genetic algorithm for 2 variables
[x,fval,reason] = ga(@twofunc,2,options)
```

Output

x =
  -14.2395 -24.4720
fval =
  -262.9662
reason =

Optimization terminated: maximum number of generations exceeded.

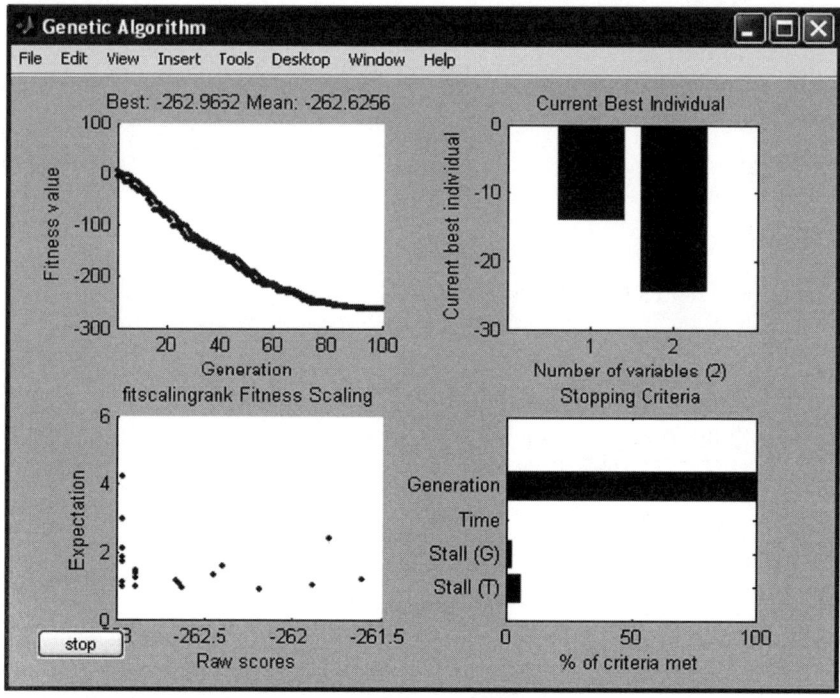

Fig. 8.17 Different Plots during optimization of the function f(x)=3x$_1$+9x$_2$

Problem 8

Obtain the best fitness value when the given linear function f(x$_1$,x$_2$,x$_3$)=−(3x$_1$+7x$_2$+6x$_3$) is minimized.

Source Code

The function is defined as follows:

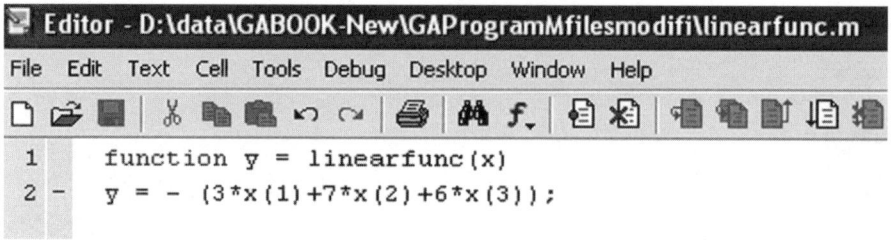

Fig. 8.18 Definition of the linear function

8.5 Solved Problems using MATLAB

The program code is given by,

```
%Program to optimize the linear function with 3 variables
clc;
clear all;
%Setting the required options
options = gaoptimset('PlotFcns',...
          {@gaplotbestf,@gaplotbestindiv});
%Generating the genetic algorithm
[x,fval]=ga(@linearfunc,3,options)
```

Output

Optimization terminated: maximum number of generations exceeded.

x =
  12.3074 23.0895 19.5800
fval =
  -316.0282

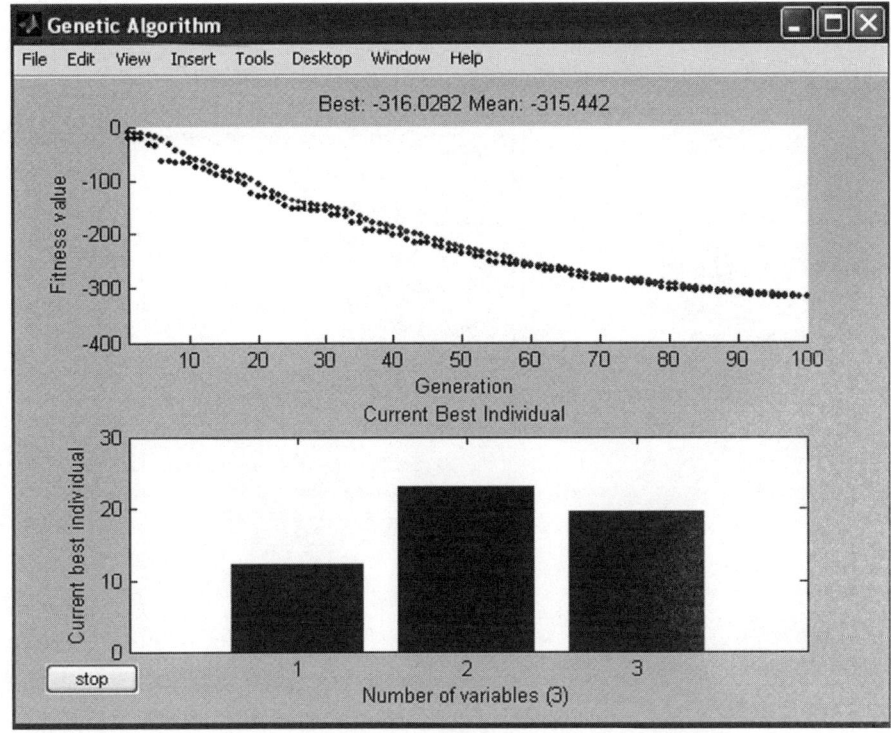

Fig. 8.19 Plot of the best fitness and best individual of the given linear function

Problem 9

Use **Gatool** and maximize the quadratic equation $f(x) = x^2+3x+2$ within the range $-6 \leq x \leq 0$.

Function Definition

Define the given function $f(x) = x^2+3x+2$ in a separate m-file as shown in Fig 8.20

```
%Function to maximize a quadratic function
function z=qudratic(x)
z=(x*x+3*x+2);
```

Fig. 8.20 M-file showing defined quadratic function

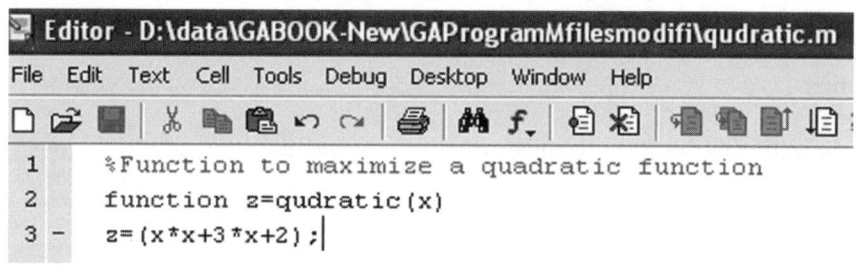

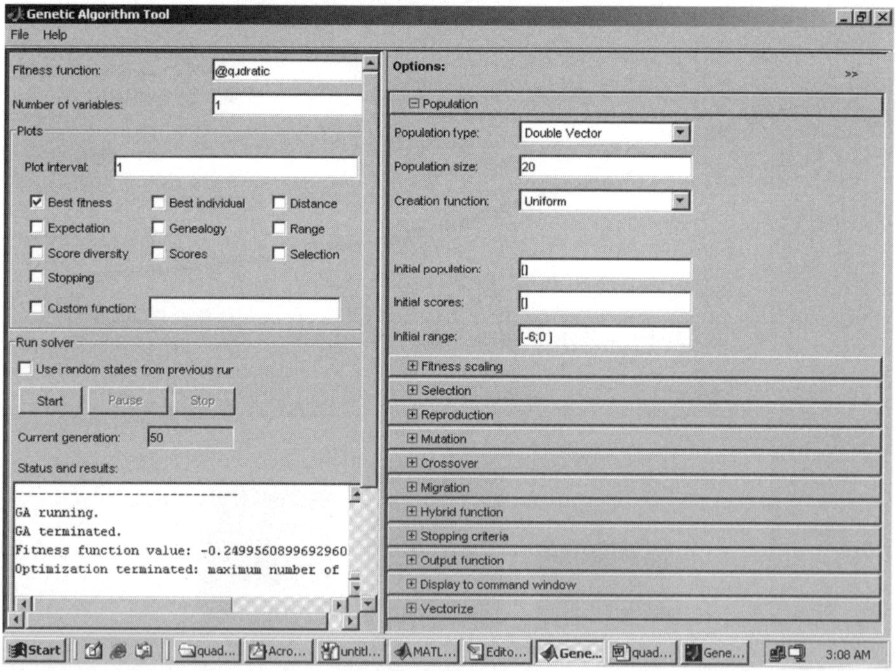

Fig. 8.21 Genetic algorithm tool for quadratic equation

8.5 Solved Problems using MATLAB

Creation of Gatool

On typing "gatool" in the command prompt, the GA tool box opens. In tool, for fitness value type **@qudratic** and mention the number of variables defined in the function. Select **best fitness** in plot and specify the other parameters as shown in Fig. 8.21

Output

The output showing the best fitness for 50 generations is shown in Fig, 8.22

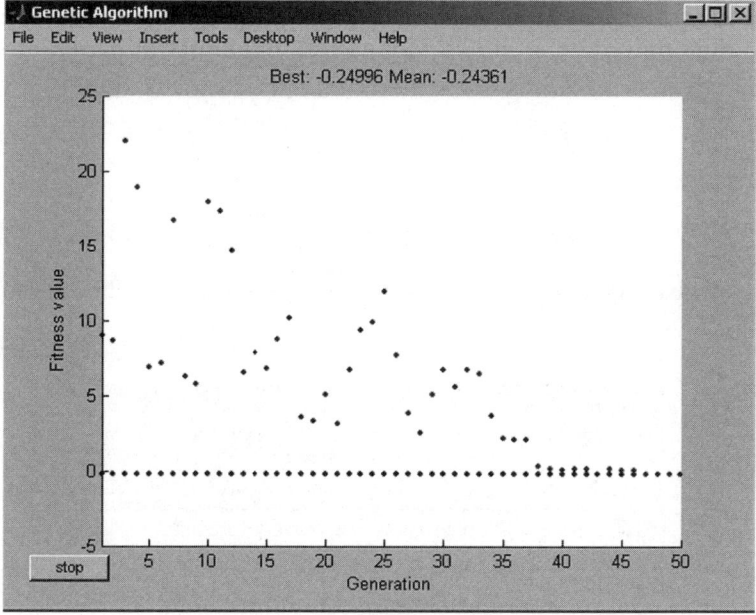

Fig. 8.22 Output response (Best fitness) for the function $f(x) = x^2+3x+2$

The status and results for this functions for 50 generations is as shown in Fig. 8.23

Problem 10

Create a **Gatool** to maximize the function $f(x1,x2)=4x_1+5x_2$ within the range 1 to 1.1

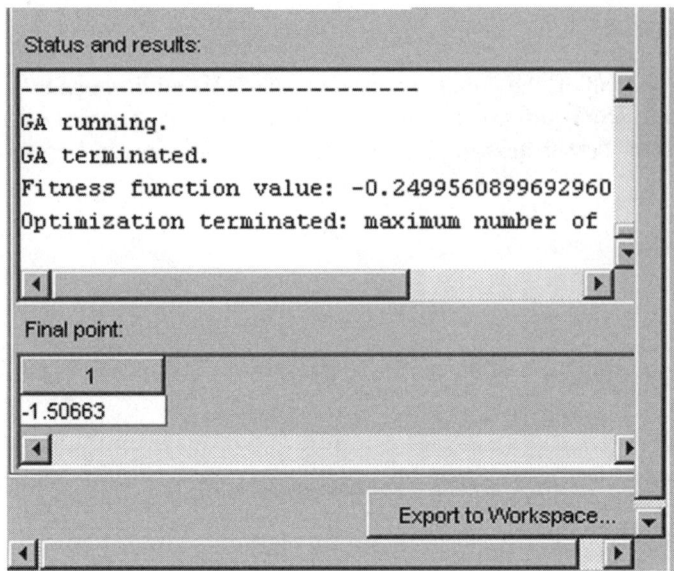

Fig. 8.23 Status and results for the function $f(x) = x^2+3x+2$

Function Definition

Define the given function $f(x1,x2)=4x_1+5x_2$ in a separate m-file as shown in Fig. 8.24

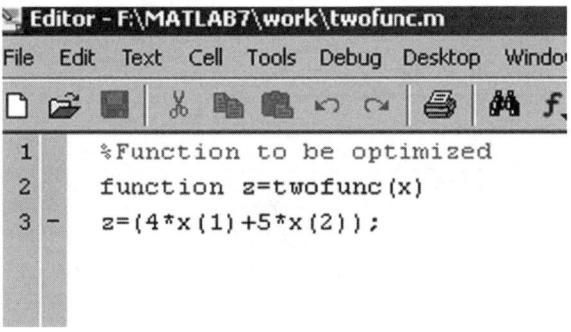

Fig. 8.24 M-file showing defined function

Creation of Gatool

On typing "gatool" in the command prompt, the GA toolbox opens. In tool, for fitness value type **@twofunc** and mention the number of variables defined in the

8.5 Solved Problems using MATLAB

function. Select **best fitness** and **best individual** in plot and specify the other parameters as shown in Fig. 8.25

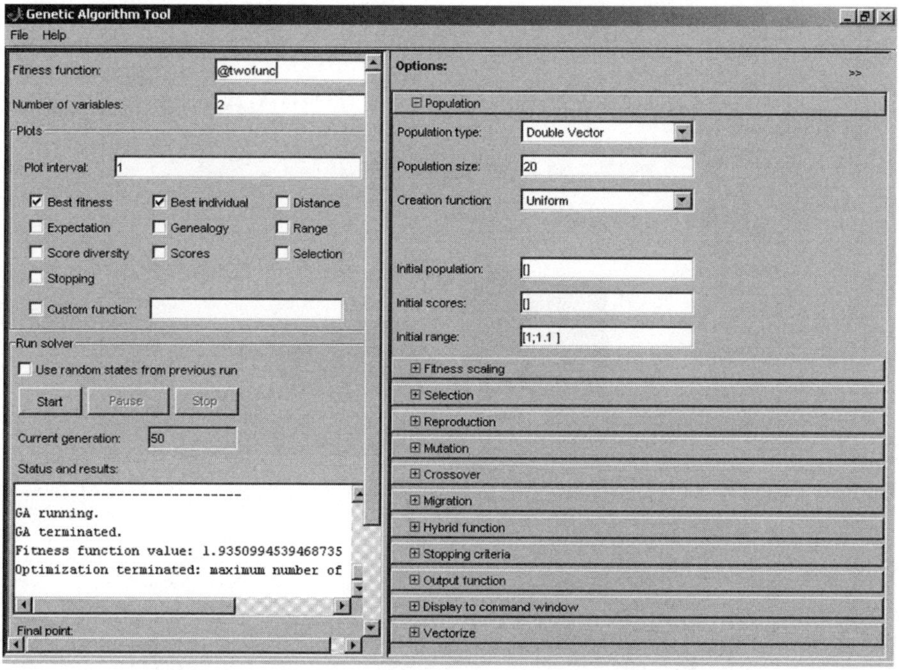

Fig. 8.25 Genetic algorithm tool for given function

Output

The output for 50 generations is as shown in Fig. 8.26 The output also shows the best inidvidual.
The status and results for this function is as shown in Fig. 8.27

Problem 11

Use **Gatool** and maximize the function

$$f(x1, x2, x3) = -5\sin(x1)\sin(x2)\sin(x3) + -\sin(5x1)\sin(5x2)\sin(x3)$$

where $0 <= xi <= pi$, for $1 <= i <= 3$.

256 Genetic Algorithm Implementation Using Matlab

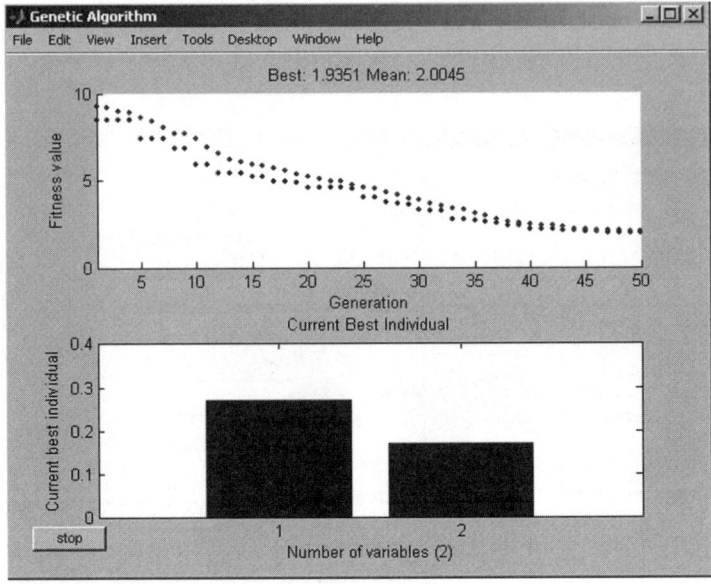

Fig. 8.26 Output response (Best fitness and best individual)

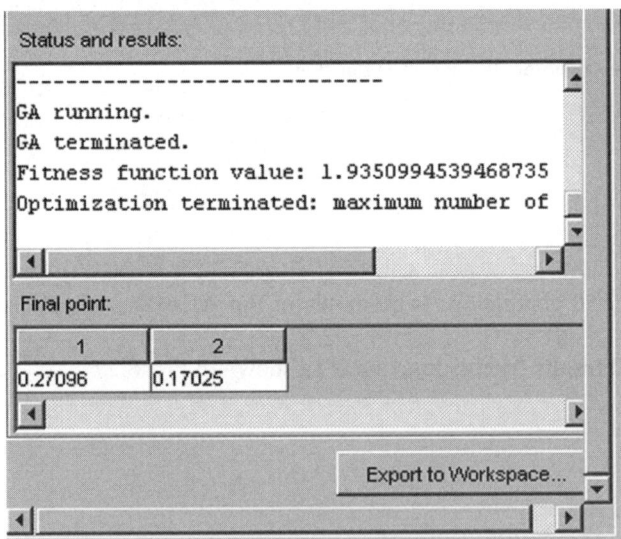

Fig. 8.27 Status and results for the function $f(x1,x2)=4x_1+5x_2$

8.5 Solved Problems using MATLAB

Function Definition

Define the given function

$$f(x1, x2, x3) = -5\sin(x1)\sin(x2)\sin(x3) + -\sin(5x1)\sin(5x2)\sin(x3)$$

in a separate m-file as shown in Fig. 8.28

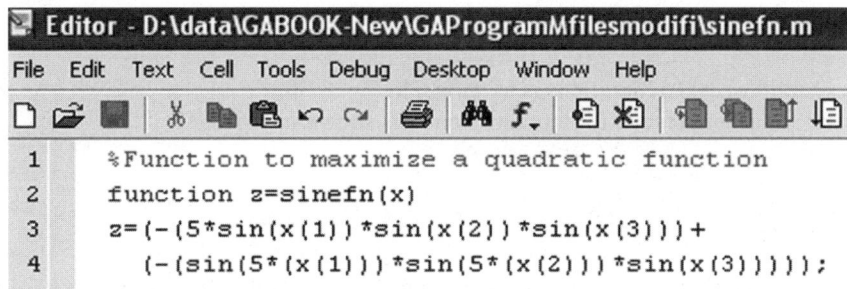

Fig. 8.28 M-file showing defined sine function

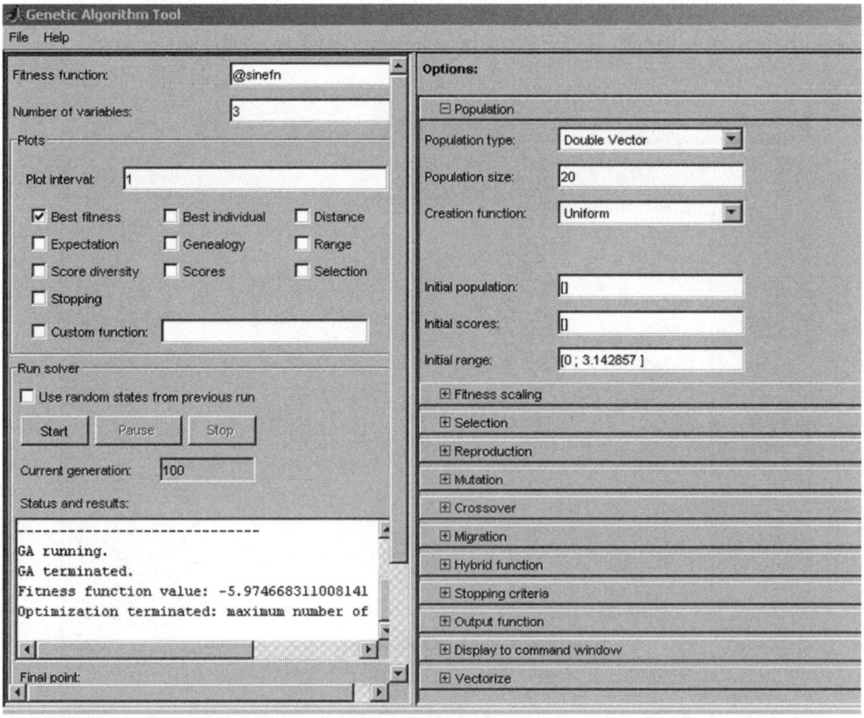

Fig. 8.29 Genetic algorithm tool for sine equation

Creation of Gatool

On typing "gatool" in the command prompt, the GA tool box opens. In tool, for fitness value type **@sinefn** and mention the number of variables defined in the function. Select **best fitness** in plot and specify the other parameters as shown in Fig. 8.29

Output

The output for 100 generations is as shown in Fig. 8.30.
  The status and results for this function is as shown in Fig. 8.31

Problem 12

Create a "gatool" to minimize the function $f(x) = \cos x$ within the range $0 \leq x \leq 3.14$

Function Definition

Define the given function $f(x) = \cos x$ in a separate m-file as shown below:

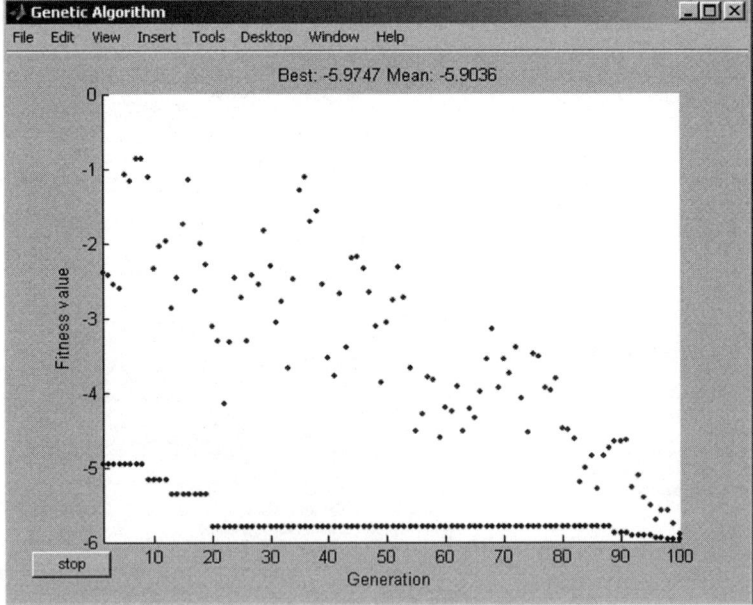

Fig. 8.30 Output response (Best fitness) for the function $f(x1, x2, x3) = -5\sin(x1)\sin(x2)\sin(x3)+\sin(5x1)\sin(5x2)\sin(x3)$

8.5 Solved Problems using MATLAB

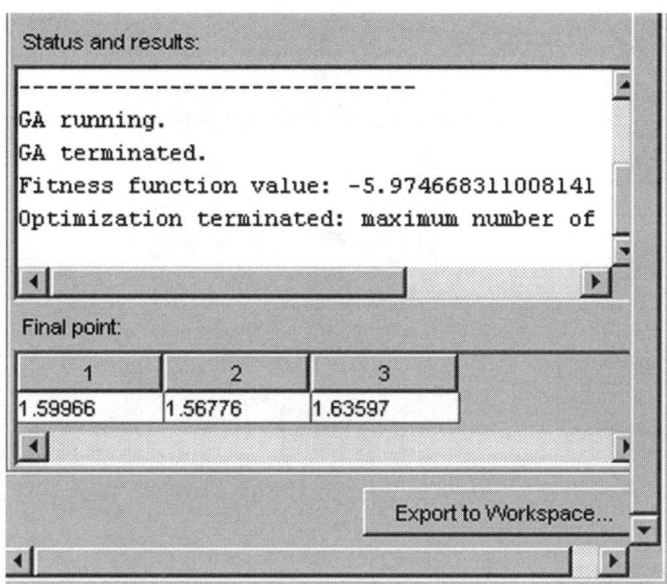

Fig. 8.31 Status and results for the function f(x1,x2,x3) = -5sin(x1)sin(x2)sin(x3)+-sin(5x1)sin(5x2)sin(x3)

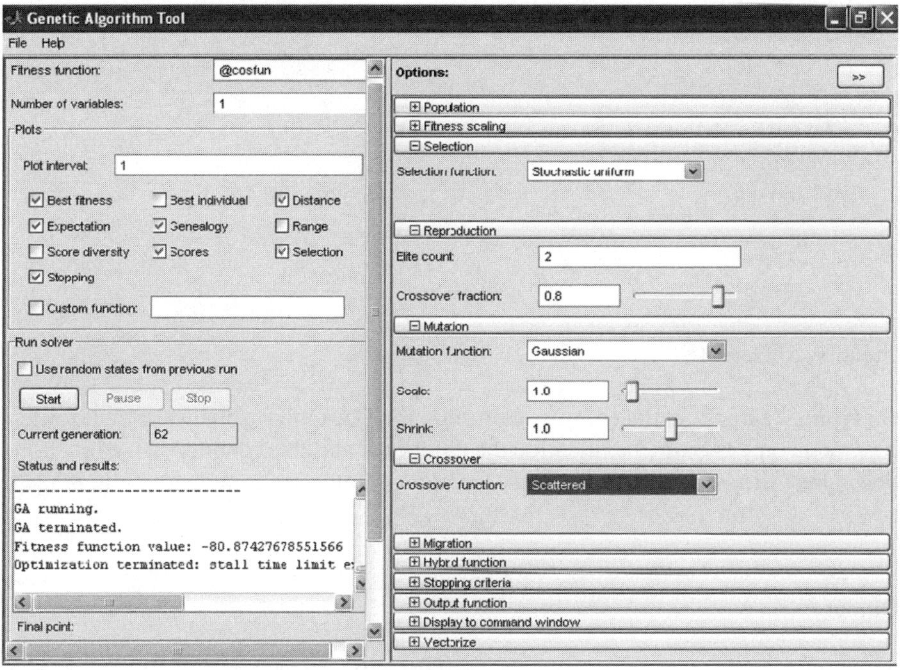

Fig. 8.32 Genetic algorithm tool for cosine function

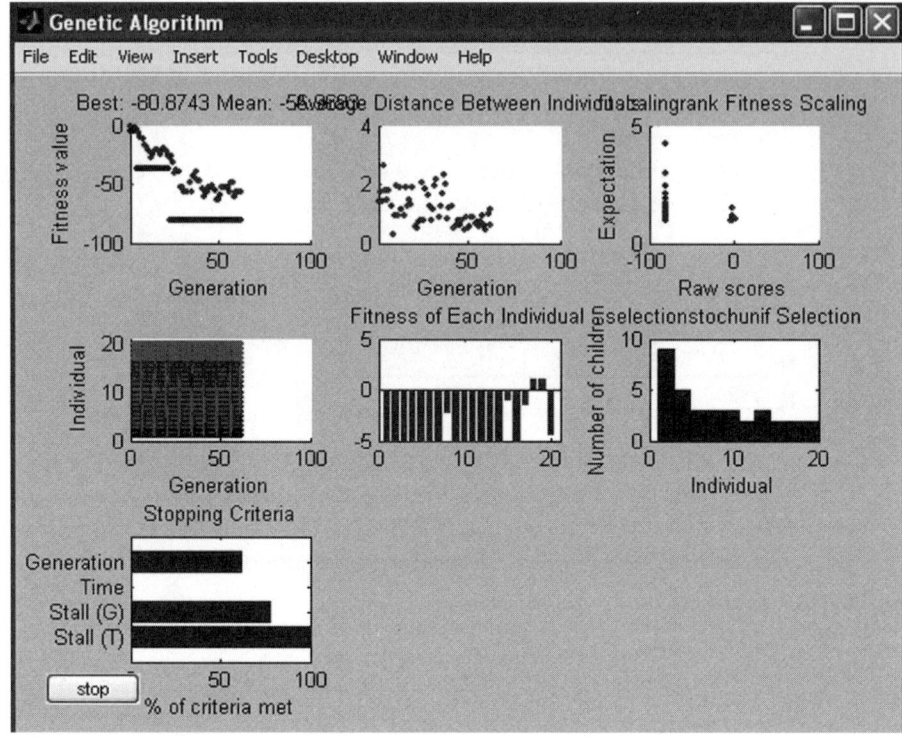

Fig. 8.33 Output response (Best fitness) for the function f(x) = cosx

```
%Function to minimize cosine function
function z=cosfun(x)
z=1/cos(x);
```

Creation of Gatool

On typing "gatool" in the command prompt, the GA toolbox opens. In tool, for fitness value type **@cosfn** and mention the number of variables defined in the function. Select **best fitness** in plot and specify the other parameters as shown in Fig. 8.32

Output

The output for 62 generations is as shown in Fig. 8.33
 The status and results for this function is shown in Fig. 8.34

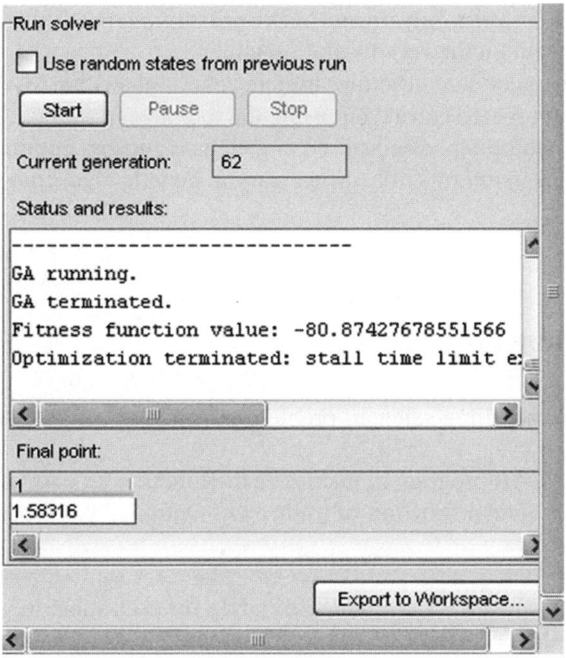

Fig. 8.34 Status and results for the function f(x) = cosx

8.6 Summary

In this chapter, the implementation of genetic algorithm using MATLAB software has been dealt. The various functions, which includes objective functions, crossover operations, mutation operations, plot functions, insertion operators, fitness scaling, utility functions and so on are listed for implementing the optimization process. One of the best tool of MATLAB is its Graphical user Interface (GUI) toolbox. The Genetic Algorithm GUI Toolbox plays a major role for obtaining an optimized solution and to find the best fitness value. This GUI tool gives us different plot related to best individual, best scores, distance, range, scorediversity, genealogy, stopping condition, best fitness value and generations. Few examples have been dealt using the MATLAB functions that are simulated and the outputs for easy reference.

Review Questions

1. Write note on the importance of MATLAB Software.
2. List the various functions used in MATLAB for performing crossover nd mutation operations.
3. Mention the different objective functions present in MATLAB Toolbox.
4. State the advantages of Graphicl User Interface toolbox.

5. Differentiate between forward and backward migration. What is migration rate?
6. Discuss in detail on the various plot functions.
7. What are the major data structures used in genetic algorithm MATLAB toolbox?
8. How is pattern search carried out using the toolboxes?
9. Mention few stopping criterias used in genetic algorithm optimization process.
10. What are conversion functions neccessary in genetic algorithm simulation process?

Exercise Problems

1. Implement a travelling saleman problem covering about 20 cities using MATLAB.
2. Write a MATLAB program to maximize the function $f(x)=4x^4+3x^3+2x^2+x+1$
3. Consider a binomial function of your own. Optimize the function to obtain a minimized solution.
4. Minimize the function $f(x,y,z)=x^2+y^2+z^2$, where x, y, and z are permitted to vary between -512 and $+512$. Use suitable coding for each substring.
5. Implement a MATLAB routine to perform mutation using an exponential distribution
6. Consider a Dejong's function, using MATLAB tool compare and contrast the alternative selection methods
7. Choose an application of your own, compare and contrast the output performance obtained using various crossover and mutation schemes
8. Create a "gatool" to minimize the function $f(x) = \sec x$ within the range $0 \leq x \leq 3.14$
9. Maximize the function $f(x1, x2, x3) = -10\sin(x1)\sin(x2)\sin(x3)+20\sin(5x1)\sin(5x2)\sin(x3)$ where $0 <= xi <= pi$, for $1 <= i <= 3$.
10. Obtain the best fitness value when the given linear function $f(x_1,x_2,x_3)=-(9x_1+7x_2+6x_3)$ is minimized

Chapter 9
Genetic Algorithm Optimization in C/C++

9.1 Introduction

C is a general-purpose structured programming language that is powerful, efficient, and compact. C combines the features of a high level language with the elements of the assembler and thus is, close to man and machine. Programs written in C are very efficient and fast. C++ is an object-oriented language that a C programmer can appreciate, especially who is an early age assembly language programmer. C++ orients towards execution performance and then towards flexibility. The name C++ signifies the evolutionary nature of the changes from C. Thus genetic algorithm being an approach based on natural evolution can be implemented using the structured programming and object programming languages. This chapter discusses few problems solved using genetic algorithm in C/C++.

9.2 Traveling Salesman Problem (TSP)

In traveling salesman problem, salesman travels n cities and returns to the starting city with the minimal cost, he is not allowed to cross the city more than once. In this problem we are taking the assumption that all the n cities are inter connected. The cost indicates the distance between two cities. To solve this problem we make use of genetic algorithm because the cities are randomly. Initial population for this problem is randomly selected cities. Fitness function is nothing but the minimum cost. Initially the fitness function is set to the maximum value and for each travel, the cost is calculated and compared with the fitness function. The new fitness value is assigned to the minimum cost. Initial population is randomly chosen and taken as the parent. For the next generation, the cyclic crossover is applied over the parent.
Cyclic Crossover

Let P1 and P2 are two parents

P1 : 2 8 0 1 3 4 5 7 9 6
P2 : 1 0 5 4 6 8 9 7 2 3

Select the first city of P1 make it as the first city of offspring1(O1)

`O1: 2 - - - - - - - - -`

To find the next city of offspring O1 search current city, which is selected from P1 in P2. Find the location of city in P2 and select the city which is in the same location in P1.

`O1: 2 - - - - - - 9 -`

Continue the same procedure, we will get O1 as

`O1: 2 8 0 1 - 4 5 - 9 -`

In the next step we will get the city 2 which is already present in O1 and then stop the procedure. Copy the cities from parent P2 in the corresponding locations

`O1: 2 8 0 1 6 4 5 7 9 3`

For the generation offspring O2 the initial selection is from the parent P2, and repeat the procedure with P1

`O2: 1 5 4 3 8 9 7 2 6`

If the initial population contain N parents it will generate N(N-1)/2 offsprings. The next generation the offsprings are considered as parent. The procedure is continued for N number of generation to find the minimum cost.

Source Code

```
#include<stdio.h>
#include<conio.h>
int tsp[10][10]={{999,10,3,2,5,6,7,2,5,4},
            {20,999,3,5,10,2,8,1,15,6},
            {10,5,999,7,8,3,11,12,3,2},
            {3,4,5,999,6,4,10,6,1,8},
            {1,2,3,4,999,5,10,20,11,2},
            {8,5,3,10,2,999,6,9,20,1},
            {3,8,5,2,20,21,999,3,5,6},
            {5,2,1,25,15,10,6,999,8,1},
            {10,11,6,8,3,4,2,15,999,1},
            {5,10,6,4,15,1,3,5,2,999}
            };
int pa[1000][10]= {{0,1,2,3,4,5,6,7,8,9},
        {9,8,6,3,2,1,0,4,5,7},
```

9.2 Traveling Salesman Problem (TSP)

```
                {2,3,5,0,1,4,9,8,6,7},
                {4,8,9,0,1,3,2,5,6,7}
                };


int i,j,k,l,m,y,loc,flag,row,col,it,x=3,y=3;
int count,row=0,res[1][10],row1,col1,z;
int numoff=4;
int offspring[1000][10];
int mincost=9999,mc;
main()
{
    int gen;
    clrscr();
    printf("Number of Generation : ");
    scanf("%d",&gen);
    offcal1(pa);
    offcal2(pa);
    printf(" \n\t\t First Generation\n");
    for(i=0;i<count;i++)
{
    for(j=0;j<10;j++)
       printf("%d ",offspring[i][j]);
       printf("\n");
}
for(y=1;y<=gen-1;y++)
{
        getch();
        clrscr();
        for(i=0;i<count;i++)
           for(j=0;j<10;j++)
              pa[i][j]=offspring[i][j];
        numoff=count;
        offcal1(pa);
        offcal2(pa);
        printf(" \n\t\t %d Generation\n",y+1);
        for(i=0;i<count;i++)
        {
           for(j=0;j<10;j++)
           printf("%d ",offspring[i][j]);
           printf("\n");
        }
        getch();
        clrscr();
    }
    printf("\n\nMinimum Cost Path\n");
```

```c
      for(z=0;z<10;z++)
         printf("%d ",res[0][z]);
      printf("\nMinimum Cost %d \n",mincost);
}
/* finding the offspring using cyclic crossover */
offcal1(pa)
int pa[1000][10];
{
      count=0;
      for(i=0;i<1000;i++)
      for(j=0;j<10;j++)
         offspring[i][j]=-1;

      for(k=0;k<numoff;k++)
      {
        for(l=k+1;l<numoff;l++)
        {
          offspring[row][0]=pa[k][0];
          loc=pa[l][0];
          flag=1;
          while(flag != 0)
          {
             for(j=0;j<10;j++)
             {
               if(pa[k][j] == loc )
               {
                 if (offspring[row][j]==-1)
                 {
                    offspring[row][j]=loc;
                    loc=pa[l][j];
                 }
                 else
                 flag=0;
               }
             }
          }/* end while*/
          for(m=0;m<10;m++)
          {
             if(offspring[row][m] == -1)
                offspring[row][m]=pa[l][m];
          }
          for(z=0;z<10;z++)
          {
             if(z<9)
             {
                row1=offspring[row][z];
```

9.2 Traveling Salesman Problem (TSP)

```
                    col1=offspring[row][z+1];
                    mc=mc+tsp[row1][col1];
                  }
                  else
                  {
                    row1=offspring[row][z];
                    col1=offspring[row][0];
                    mc=mc+tsp[row1][col1];
                  }
                }
                if(mc < mincost)
                {
                  for(z=0;z<10;z++)
                      res[0][z]=offspring[row][z];
                  mincost=mc;
                }
                count++;
                row++;
              }/* end l*/
          }
    }
    offcal2(pa)
    int pa[1000][10];
    {
        for(k=0;k<numoff;k++)
        {
            for(l=k+1;l<numoff;l++)
            {
              offspring[row][0]=pa[l][0];
              loc=pa[k][0];
              flag=1;
              while(flag != 0)
              {
                for(j=0;j<10;j++)
                {
                    if(pa[l][j] == loc )
                    {
                      if (offspring[row][j]==-1)
                      {
                        offspring[row][j]=loc;
                        loc=pa[k][j];
                      }
                      else
                          flag=0;
                }
              }
```

```
        }/* end while*/
                for(m=0;m<10;m++)
                {
                   if(offspring[row][m] == -1)
                      offspring[row][m]=pa[k][m];
                }
                for(z=0;z<10;z++)
                {
            if(z<9)
            {
                row1=offspring[row][z];
                col1=offspring[row][z+1];
                mc=mc+tsp[row1][col1];
            }
            else
            {
                row1=offspring[row][z];
                col1=offspring[row][0];
                mc=mc+tsp[row1][col1];
            }
        }
                row++;
                if(mc < mincost)
                {
                   for(z=0;z<10;z++)
                      res[0][z]=offspring[row][z];
                   mincost=mc;
                }
                count++;
                }/* end l*/
        }
}
```

Output

Number of Generation : 2
First Generation
```
0 8 2 3 4 1 6 7 5 9
0 1 2 3 4 5 9 8 6 7
0 1 2 3 4 5 6 7 8 9
9 8 5 3 2 1 0 4 6 7
9 8 6 0 1 3 2 4 5 7
2 3 5 0 1 4 9 8 6 7
9 1 6 3 2 5 0 4 8 7
```

9.2 Traveling Salesman Problem (TSP)

```
2 3 5 0 1 4 6 7 8 9
4 8 9 0 1 3 2 5 6 7
2 3 6 0 1 4 9 8 5 7
4 8 9 3 2 1 0 5 6 7
4 8 9 0 1 3 2 5 6 7
2 Generation
0 1 2 3 4 5 9 8 6 7
0 1 2 3 4 5 6 7 8 9
0 8 2 3 4 1 6 7 5 9
0 8 6 3 4 1 2 7 5 9
0 8 2 3 1 4 6 7 5 9
0 1 2 3 4 5 6 7 8 9
0 8 2 3 1 4 6 7 5 9
0 8 9 3 4 1 2 5 6 7
0 8 2 3 1 4 6 7 5 9
0 8 2 3 4 1 6 7 5 9
0 8 9 3 4 1 2 5 6 7
0 1 2 3 4 5 6 7 8 9
0 8 5 3 2 1 9 4 6 7
0 8 2 3 1 5 9 4 6 7
0 1 2 3 4 5 9 8 6 7
0 1 6 3 2 5 9 4 8 7
0 1 2 3 4 5 6 7 8 9
0 1 9 3 4 5 2 8 6 7
0 1 2 3 4 5 9 8 6 7
0 8 2 3 4 1 9 5 6 7
0 1 9 3 4 5 2 8 6 7
0 1 2 3 4 5 6 7 8 9
0 1 6 3 4 5 2 7 8 9
0 1 2 3 4 5 9 8 6 7
0 1 2 3 4 5 6 7 8 9
0 1 2 3 4 5 6 7 8 9
0 1 2 3 4 5 6 7 8 9
0 1 2 3 4 5 6 7 8 9
0 1 2 3 4 5 6 7 8 9
0 1 2 3 4 5 6 7 8 9
9 8 6 0 1 3 2 4 5 7
9 8 5 3 2 1 0 4 6 7
9 1 6 3 2 5 0 4 8 7
9 3 5 0 2 1 6 4 8 7
9 8 5 0 1 3 2 4 6 7
9 8 6 3 2 1 0 4 5 7
9 8 5 3 2 1 0 4 6 7
9 8 5 0 1 3 2 4 6 7
9 3 5 0 1 4 2 8 6 7
9 1 6 3 2 5 0 4 8 7
```

```
9 8 6 0 1 3 2 4 5 7
9 8 6 0 1 3 2 4 5 7
9 3 6 0 1 4 2 8 5 7
9 8 6 3 2 1 0 4 5 7
9 8 6 0 1 3 2 4 5 7
2 3 6 0 1 5 9 4 8 7
2 3 5 0 1 4 6 7 8 9
2 3 5 0 1 4 9 8 6 7
2 3 6 0 1 4 9 8 5 7
2 8 9 3 1 4 0 5 6 7
2 3 5 0 1 4 9 8 6 7
9 1 6 3 2 5 0 4 8 7
9 1 6 3 2 5 0 4 8 7
9 1 6 3 2 4 0 8 5 7
9 1 6 3 2 5 0 4 8 7
9 1 6 3 2 5 0 4 8 7
2 3 9 0 1 4 6 5 8 7
2 3 6 0 1 4 9 8 5 7
2 8 9 3 1 4 0 5 6 7
2 3 9 0 1 4 6 5 8 7
4 8 9 0 1 3 2 5 6 7
4 8 9 3 2 1 0 5 6 7
4 8 9 0 1 3 2 5 6 7
2 8 9 3 1 4 0 5 6 7
2 3 6 0 1 4 9 8 5 7
4 8 9 0 1 3 2 5 6 7
0 8 2 3 4 1 6 7 5 9
0 8 2 3 4 1 6 7 5 9
9 8 5 3 2 1 0 4 6 7
9 8 2 0 1 3 6 4 5 7
2 3 5 0 4 1 9 8 6 7
9 8 6 3 2 1 0 4 5 7
2 3 5 0 4 1 6 7 8 9
4 8 2 0 1 3 6 7 5 9
2 3 6 0 4 1 9 8 5 7
4 8 9 3 2 1 0 5 6 7
4 8 2 0 1 3 6 7 5 9
0 1 2 3 4 5 9 8 6 7
9 1 2 3 4 5 0 8 6 7
9 1 6 0 4 3 2 8 5 7
2 3 5 0 1 4 9 8 6 7
9 1 2 3 4 5 0 8 6 7
2 3 5 0 1 4 9 8 6 7
4 8 2 0 1 3 9 5 6 7
2 3 6 0 1 4 9 8 5 7
4 1 9 3 2 5 0 8 6 7
```

9.3 Word Matching Problem

```
4 8 2 0 1 3 9 5 6 7
9 8 5 3 2 1 0 4 6 7
9 8 2 0 1 3 6 4 5 7
2 3 5 0 1 4 6 7 8 9
9 1 6 3 2 5 0 4 8 7
2 3 5 0 1 4 6 7 8 9
4 8 9 0 1 3 2 5 6 7
2 3 6 0 1 4 9 8 5 7
4 8 9 3 2 1 0 5 6 7
4 8 9 0 1 3 2 5 6 7
9 8 5 3 2 1 0 4 6 7
2 3 5 0 1 4 9 8 6 7
9 8 5 3 2 1 0 4 6 7
2 8 5 3 1 4 0 7 6 9
4 8 9 3 2 1 0 5 6 7
2 3 5 0 1 4 9 8 6 7
4 8 9 3 2 1 0 5 6 7
4 8 9 3 2 1 0 5 6 7
2 8 6 0 1 3 9 4 5 7
9 8 6 0 1 3 2 4 5 7
2 3 5 0 1 4 6 7 8 9
4 8 9 0 1 3 2 5 6 7
2 8 6 0 1 3 9 4 5 7
4 8 9 0 1 3 2 5 6 7
4 8 9 0 1 3 2 5 6 7
9 1 5 3 2 4 0 8 6 7
2 3 5 0 1 4 9 8 6 7
4 8 9 0 1 3 2 5 6 7
2 3 5 0 1 4 9 8 6 7
4 3 5 0 2 1 9 8 6 7
4 8 9 0 1 3 2 5 6 7
2 3 5 0 1 4 6 7 8 9
4 8 9 0 1 3 2 5 6 7
2 3 6 0 1 5 9 4 8 7
Minimum Cost Path
0 8 2 3 4 1 6 7 5 9
           Minimum Cost 53
```

9.3 Word Matching Problem

Freeman stated this problem in his work *Simulating Neural Networks with Mathematics*. It is a nice example to show the power of genetic algorithms. The word-matching problem tries to evolve an expression of "to be or not to be" from the randomly generated lists of letters with genetic algorithm. Since there are 26 possible letters for each of 13 locations in the list, the probability that we get the correct phrase in a pure random way is $(1/26)^{13} = 4.03038*10^{-19}$, which is about two chances out of a billion.

We use a list of ASCII integers to encode the string of letters. The lower case letters in ASCII are represented by numbers in the range [97,122] in the decimal number system. For example, the string of letters tobeornottobe is converted into the following chromosome represented with ASCII integers:

[116,111,98,101,111,114,110,111,116,116,111,98,101]

Initial population is generated randomly. Fitness is calculated as the number of matched letters. Genetic operators are used to obtain the output. This problem is implemented in C and the output is obtained.

Source Code

```
#include<stdio.h>
#include<conio.h>
#include<stdlib.h>
#include<dos.h>
char input[15],parent[50][15],child[50][15],mating_pool[105]
[15], mutant[05][15];
int pfit[50],cfit[50],fit[105],mfit[05],gen=0;

void get_input()
{
        int i;
        clrscr();
        printf("\n\n\n\t\tWORD MATCHING PROBLEM");
        printf("\n\t *************************************");
        printf("\n\n\n\n\t\tENTER THE WORD TO BE MATCHED : ");
        scanf("%s",input);
        printf("\n\n\n\t THE ASCII EQUIVALENT OF THE LETTERS IN
                    THE ENTERED WORD");
        printf("\n\t- - - - - - - - - - - - - - - - - - - - - ");
        printf("\n\n LETTERS :");
        for(i=0;i<strlen(input);i++)
        {
                printf(" %c ",input[i]);
        }
        printf("\n ASCII :");
        for(i=0;i<strlen(input);i++)
        {
                printf(" %3d",input[i]);
        }
```

9.3 Word Matching Problem

```c
            getch();
}
void initial_pop()
{
         int i,j;
         randomize();
         for(i=0;i<50;i++)
         {
                  for(j=0;j<strlen(input);j++)
                  {
                           parent[i][j]=random(26)+97;
                           if(parent[i][j]==input[j])
                           {
                                    pfit[i]++;
                           }
                  }
         }
}
void display()
{
          int i,j,nexti;
         clrscr();
         printf("\n\n\t\t THE CHROMOSOMES OF PARENTS
                  AND CHILDREN");
         printf("\n\t - - - - - - - - - - - - - - - - - - -\n");
         printf("\n\t\t PREVIOUS GENERATION CHILDREN
                  CHROMOSOMES\n\n");
         for(i=0;i<50;i++)
         {
                  if(((i)%4)==0) printf("\n");
                  for(j=0;j<strlen(input);j++)
                  {
                           printf("%c",child[i][j]);
                  }
                  printf("% 2d ",cfit[i]);
         }
         printf("\n\t\t\tMUTANTS OF THIS GENERATION\n");
         for(i=0;i<05;i++)
         {
                  if (i==3) printf("\n");
                  for(j=0;j<strlen(input);j++)
                  {
                           printf("%c",mutant[i][j]);
                   }
                  printf("% 2d ",mfit[i]);
         }
```

```c
            getch();
            clrscr();
            printf("\n\n\t\t THE CHROMOSOMES OF PARENTS AND
                    CHILDREN");
            printf("\n\t - - - - - - - - - - - - - - - -\n");
            printf("\n\t\t NEXT GENERATION PARENTS
                    CHROMOSOMES\n\n");
            for(i=0;i<50;i++)
            {
                    if(((i)%4)==0) printf("\n");
                    for(j=0;j<strlen(input);j++)
                    {
                            printf("%c",parent[i][j]);
                    }
                    printf("% 2d ",pfit[i]);
            }
            getch();
}
void reproduction() //sorting_based_on_fitness()
{
            char tempc;
            int temp;
            int i,j,k;
            for(i=0;i<50;i++)
            {
                    for(j=0;j<strlen(input);j++)
                    {
                            mating_pool[i][j]=parent[i][j];
                            fit[i]=pfit[i];
                    }
            }
            for(i=50;i<100;i++)
            {
                    for(j=0;j<strlen(input);j++)
                    {
                            mating_pool[i][j]=child[i-50][j];
                            fit[i]=cfit[i-50];
                    }
            }
            for(i=100;i<105;i++)
            {
                    for(j=0;j<strlen(input);j++)
                    {
                            mating_pool[i][j]=mutant[i-100][j];
                            fit[i]=mfit[i-100];
                    }
```

9.3 Word Matching Problem

```
            }
             //sorting
            for(i=0;i<105;i++)
            {
                    for(j=i+1;j<105;j++)
                    {
                            if(fit[i]<fit[j])
                            {
                                    for(k=0;k<strlen(input);k++)
                                    {
                                       tempc=mating_pool[i][k];
                                       mating_pool[i][k]=
                                       mating_pool[j][k];
                                       mating_pool[j][k]=tempc;
                                       temp=fit[i];
                                       fit[i]=fit[j];
                                       fit[j]=temp;
                                    }
                            }
                    }
            }
            for(i=0;i<50;i++)
            {
                    for(j=0;j<strlen(input);j++)
                    {
                            parent[i][j]=mating_pool[i][j];
                            pfit[i]=fit[i];
                    }
            }
            for(i=50;i<100;i++)
            {
                    for(j=0;j<strlen(input);j++)
                    {
                            child[i-50][j]=mating_pool[i][j];
                            cfit[i-50]=fit[i];
                    }
            }
}
void crossover()
{
        int xover_pt;
        int i,j,k;
        for(i=0;i<50;i++)
        {
                xover_pt=random(strlen(input));
```

```c
                cfit[i]=0;
                cfit[i+1]=0;
                for(j=0;j<xover_pt;j++)
                {
                        child[i][j]=parent[i][j];
                        if (input[j]==child[i][j])
                                cfit[i]++;
                        child[i+1][j]=parent[i+1][j];
                        if(input[j]==child[i+1][j])
                                cfit[i+1]++;
                }
                for(j=xover_pt;j<strlen(input);j++)
                {
                        child[i][j]=parent[i+1][j];
                        if(input[j]==child[i][j])
                                cfit[i]++;
                        child[i+1][j]=parent[i][j];
                        if(input[j]==child[i+1][j])
                                cfit[i+1]++;
                }
                i++;
        }
}
void mutation()
{
        int i,mut_pt,j;
        char mut_val;
        randomize();
        for(i=0;i<05;i++)
        {
                mut_pt=random(strlen(input));
                mut_val=random(26)+97;
                mfit[i]=0;
                 for(j=0;j<mut_pt;j++)
                {
                        mutant[i][j]=parent[1][j];
                        if (mutant[i][j]==input[j])
                        {
                                mfit[i]++;
                        }
                }
                mutant[i][mut_pt]=mut_val;
                if (mutant[i][j]==input[j])
                {
                        mfit[i]++;
```

9.3 Word Matching Problem

```
                }
                for(j=mut_pt+1;j<strlen(input);j++)
                 {
                         mutant[i][j]=parent[1][j];
                         if (mutant[i][j]==input[j])
                         {
                                 mfit[i]++;
                         }
                 }
         }
}
void results()
{
        int i;
        clrscr();
        printf("\n\n\n\t\tWORD MATCHING PROBLEM ");
        printf("\n\t ****************************************");
        printf("\n\n\n\t\t THE MATCHING WORD FOR THE
           GIEN INPUT WORD");
        printf("\n\n\t\t OBTAINED USING GENETIC ALGORITHM");
        printf("\n\n\t\t\t ");
        for(i=0;i<strlen(input);i++)
          {
                  printf("%c",parent[0][i]);
          }
        printf("\n\t\t\t -");
        for(i=0;i<strlen(input);i++)
          {
                  printf("-");
          }
        printf("-\n\n\n\t\t USER INPUT : %s",input);
        printf("\n\n\n\t THE FITNESS OF THE GA GENERATED WORD
           AND THE USER'S INPUT");
        printf("\n\n\t\t\t %2d/%d",pfit[0],strlen(input));
        printf("\n\n\n\t\t GENERATIONS COUNT : %d",gen);
}
int input_choice()
{
        int choice,i;
        clrscr();
        printf("\n\n\n\n\t\t GENEREATION NUMBER : %d",gen);
        printf("\n\t\t ---------------");
        printf("\n\n\n\t\tTHE FITTEST INDIVIDUAL TILL THE PREVIOUS
           GENERATION\n\n\n\t\t\t\t");
        for(i=0;i<strlen(input);i++)
```

```c
        {
                printf("%c",parent[0][i]);
        }
        printf(" / ");
        for(i=0;i<strlen(input);i++)
        {
                printf("%c",input[i]);
        }
        printf("\n\n\n\t\t\t WITH A FITNESS OF %d/%d",pfit[0],
          strlen(input));
        printf("\n\n\n\n\t\tENTER YOUR CHOICE (TO CONTINUE 1 TO
        EXIT 0) : ");
        scanf("%d",&choice);
        return choice;
}
void main()
{
        int i,choice;
        clrscr();
        get_input();
        initial_pop();
        //display();
        reproduction(); //sorting_based_on_fitness();
        display();
        printf("\nENTER YOUR CHOICE (TO CONTINUE 1 TO
          EXIT 0) : ");
        scanf("%d",&choice);
        while((choice==1)&&(pfit[0]!=strlen(input)))
        {
                crossover();
                gen++;
                mutation();
                reproduction(); //sorting_based_on_fitness();
                display();
                choice=input_choice();
        }
        sound(1000);
        delay(200);
        nosound();
        delay(200);
        results();
        getch();
        sound(1000);
        delay(200);
        nosound();
}
```

9.3 Word Matching Problem

Sample input and output

```
                      WORD MATCHING PROBLEM
                      ***************************
              ENTER THE WORD TO BE MATCHED : tobeornottobe
          THE ASCII EQUIVALENT OF THE LETTERS IN THE ENTERED WORD
     ---------------------------------------------------------------
     LETTERS :  t    o    b    e    o    r    n    o    t    t    o    b    e
     ASCII   :  116  111  98   101  111  114  110  111  116  116  111  98   101
                    THE CHROMOSOMES OF PARENTS AND CHILDREN
                    ----------------------------------------
               PREVIOUS GENERATION CHILDREN CHROMOSOMES
     0                     0                   0                    0
     0                     0                   0                    0
     0                     0                   0                    0
     0                     0                   0                    0
     0                     0                   0                    0
     0                     0                   0                    0
     0                     0                   0                    0
     0                     0                   0                    0
     0                     0                   0                    0
     0                     0                   0                    0
     0                     0                   0                    0
     0                     0                   0                    0
     0                     0
                        MUTANTS OF THIS GENERATION
     0                     0                   0
     0                     0
                    THE CHROMOSOMES OF PARENTS AND CHILDREN
                    ----------------------------------------
                    NEXT GENERATION PARENTS CHROMOSOMES


     tsltzcmzthxsl    2     tcowkezlitact    2     xsbdunrdshtae   2     slfmkkbomaelg   1
     cdxtuhfmwyoyt    1     fngzovwqvolka    1     yqejrravxxqof   1     gdpdjfqoqzznk   1
     uopvjzpkbddjn    1     ptcvwpouxtuts    1     jkolwpdlokupe   1     gayzwkmolrsgl   1
     kluooeczryszy    1     kybhhqrczprmy    1     sgztvrynbuipg   1     knvgfcngzkvhv   1
     tpummmrcmyfzr    1     ifthifkdltxgi    1     kpgnnitdaxoxt   1     wfuxvgsomepei   1
     ujhibkmyceqvx    0     xmpjpmmblqpxj    0     vxcukefnzkhlw   0     ymcpxaompgfwg   0
     hbcljezgischs    0     pkjfmulmyruay    0     jjyypxbtembqn   0     aqmgdwujrffsy   0
     zmfutbqamdhft    0     uupbgudnszamz    0     sbojtvrvbzkca   0     vsodsjigqmdaa   0
     wliomvmcmrhom    0     adwmduybimmhq    0     qevisjzvslsio   0     lywbccjywshtl   0
     mjsxyzajcdnof    0     hdldclhcuxcmu    0     sztgsxivreiso   0     dmnzunxrcngpa   0
     giuaadyylobdf    0     wkhsygjervjkj    0     djipdxywsykcn   0     acpsqjinomujh   0
     ienwboqpvmdmr    0     unnsbufppoqfq    0     amswufzjmkspz   0     dblbjpflwyepr   0
     jfazusrxmxlnj    0     nlthcxxremeri    0
     ENTER YOUR CHOICE (TO CONTINUE 1 TO EXIT 0) : 1
```

THE CHROMOSOMES OF PARENTS AND CHILDREN

PREVIOUS GENERATION CHILDREN CHROMOSOMES

ujhibkmyceqvx	0	xmpjpmmblqpxj	0	vxcukefnzkhlw	0	ymcpxaompgfwg	0
cdxtuhwqvolka	0	hbcljezgischs	0	pkjfmulmyruay	0	jjyypxbtembqn	0
aqmgdwujrffsy	0	ptcvjzpkbddjn	0	zmfutbqamdhft	0	uupbgudnszamz	0
sbojtvrvbzkca	0	vsodsjigqmdaa	0	wliomvmcmrhom	0	adwmduybimmhq	0
qevisjzvslsio	0	ifthifkdmyfzr	0	lywbccjywshtl	0	mjsxyzajcdnof	0
ujhipmmblqpxj	0	xmpjbkmyceqvx	0	vxcpxaompgfwg	0	ymcukefnzkhlw	0
hbjfmulmyruay	0	pkcljezgischs	0	jjyypxujrffsy	0	aqmgdwbtembqn	0
zmfbgudnszamz	0	uuputbqamdhft	0	sbojsjigqmdaa	0	vsodtvrvbzkca	0
wliomvmcmrhoq	0	adwmduybimmhm	0	qywbccjywshtl	0	levisjzvslsio	0
mjsxyzajcdnou	0	hdldclhcuxcmf	0	sztgsxxrcngpa	0	dmnzunivreiso	0
giuaadyylobdj	0	wkhsygjervjkf	0	djipdxinomujh	0	acpsqjywsykcn	0
iensbufppoqfq	0	unnwboqpvmdmr	0	dblbjpflwyepr	0	amswufzjmkspz	0
jfazusxremeri	0	nlthcxrxmxlnj	0				

MUTANTS OF THIS GENERATION

tcowkezlitace	3	tcowkenlitact	3	tcbwkezlitact	3
tcowkezoitact	3	tcoekezlitact	3		

THE CHROMOSOMES OF PARENTS AND CHILDREN

NEXT GENERATION PARENTS CHROMOSOMES

tcowkezlitace	3	tcowkenlitact	3	tcbwkezlitact	3	tcowkezoitact	3
tcoekezlitact	3	slfmkkboshtae	2	fngzovfmwyoyt	2	uopvwpouxtuts	2
tpummmrcltxgi	2	tsltzcmzthxsl	2	tcowkezlitact	2	xsbdunrdshtae	2
tsltzcmzthxst	2	tcowkezlitacl	2	sgztvrynbuipg	1	knvgfcngzkvhv	1
tpummmrcmyfzr	1	ifthifkdltxgi	1	kpgnnitdaxoxt	1	wfuxvgsomepei	1
slfmkkbomaelg	1	cdxtuhfmwyoyt	1	xsbdunrdmaelg	1	fngzovwqvolka	1
yqejrravxxqof	1	yqejjfqoqzznk	1	gdpdrravxxqof	1	gdpdjfqoqzznk	1
jkolwpdolrsgl	1	gayzwkmlokupe	1	kybhhqrczprmy	1	kluooeczryszy	1
sgztvrynzkvhv	1	knvgfcngbuipg	1	uopvjzpkbddjn	1	kpgnvgsomepei	1
wfuxnitdaxoxt	1	ptcvwpouxtuts	1	jkolwpdlokupe	1	gayzwkmolrsgl	1
kluooeczryszy	1	kybhhqrczprmy	1	djipdxywsykcn	0	acpsqjinomujh	0
ienwboqpvmdmr	0	unnsbufppoqfq	0	amswufzjmkspz	0	dblbjpflwyepr	0
jfazusrxmxlnj	0	nlthcxxremeri	0				

GENEREATION NUMBER : 2

THE FITTEST INDIVIDUAL TILL THE PREVIOUS GENERATION
tcowkenlitace / tobeornottobe
WITH A FITNESS OF 4/13
ENTER YOUR CHOICE (TO CONTINUE 1 TO EXIT 0): 0

9.4 Prisoner's Dilemma

Cooperation is usually analyzed in game theory by means of a non-zero-sum game called the "Prisoner's Dilemma". The two players in the game can choose between two moves, either "cooperate" or "defect". The idea is that each player gains when both cooperate, but if only one of them cooperates, the other one, who defects, will gain more. If both defect, both lose (or gain very little) but not as much as the

9.4 Prisoner's Dilemma

"cheated" cooperator whose cooperation is not returned. The whole game situation and its different outcomes can be summarized by table below, where hypothetical "points" are given as an example of how the differences in result might be quantified.

Action of A / Action B	Cooperate	Defect
Cooperate	Fairly good [+5]	Bad [-10]
Defect	Good [+10]	Mediocre [0]

The type of crossover that is performed is a "single point crossover" where the point of crossover is randomly selected. The mutation is expected to happen every two thousand generation. It is easy to change the mutation as it is implemented as a separate function.

Source Code

```
#include<stdlib.h>
#include<stdio.h>
#include<conio.h>

int calculate(int*);
int* select(int *);
void crossover(int*,int*);
void sort_select(void);
//THESE ARE SOME GLOBAL VARIABLE USED
int best_score[20];
int score[9];
int index[6];

void main()
{
        int a[10][70],select_string[5][70];
        int best_string[20][70],max,ind=0;
        int p,counter=1;
        int i,n,j,temp[10];
        randomize();
        clrscr();
        for(j=0;j<10;j++)
        for(i=0;i<70;i++)
                a[j][i]=random(2);
        //THE NUMBER OF GENERATION TO BE SCANED IN
        printf(" Enter the no of generation ");
        scanf("%d",&n);
        for(i=0;i<10;i++)
```

```
                score[i]=calculate(&a[i][0]);
//function for sorting the score array and finding the index
of best score
        sort_select();
        for(i=0;i<7;i++)
        {
                p=index[i]; //THE ORDER OF BEST SCORE STORED IN
                    INDEX.
                for(j=0;j<70;j++)
                        select_string[i][j]=a[p][j];
        }
        best_score[0]=score[0];
        for(i=0;i<70;i++)
                best_string[0][i]=select_string[0][i];

        while(counter < n)
        {
                for(i=0;i<7;i=i+2)
                        crossover(&a[i][0],&a[i+1][0]);
                        for(i=0;i<9;i++)
                                score[i]=0;
                        for(i=0;i<7;i++)
                                score[i]=calculate(&a[i][0]);
//CALCULATE FUNCTION RETURNS SCORE OF EACH STRING
                        sort_select();
                        best_score[counter]=score[0];
                        p=index[0];
                    for(j=0;j<70;j++)
                            best_string[counter][j]=a[p][j];
                    counter++;
        }

        //OUTPUT THE BEST SCORES.
        for(p=0;p<n;p++)
        {
                printf("The best score in the generation
                    %d :",p+1);
                printf(" %d \n", best_score[p]);
        }
        //OUTPUT THE BEST STRINGS.
        for(i=0;i<n;i++)
        {
                printf("\n\nTHE BEST STRNG IN GENERATION %d
                    :\n\n", i+1);
                for(j=0;j<70;j++)
                {
```

9.4 Prisoner's Dilemma

```
                    if(j%2==0&&j!=0)
                    printf(" ");
                    if(best_string[i][j] ==1)
                            printf("d");
                    //COVERTING 1'S AND 0'S TO d AND c
                    else
                            printf("c");

            }
       }
//CALCULATING THE BEST OF THE BEST
       for(i=0;i<n;i++)
            temp[i]=best_score[i];
       max=temp[0];
       for(i=1;i<n;i++)
       {
            if(max<temp[i])
            {
                    max=temp[i];
                    ind=i;
            }
       }
//CALCULATING THE BEST FROM THE SELECTED.
       printf("\n\n");
       printf("\nTHE BEST STRING IN ALL GENERATION IS \n\n");
       for(i=0;i<70;i++)
       {
            if(i%2==0&&i!=0)
            printf(" ");
            if(best_string[ind][i]==1)
                    printf("d");
            else
                    printf("c");
       }
printf("\n\nTHE CORRESPONDING BEST SCORE IS: %d ",best
_score[ind]);
getch();
}

int calculate(int* ptr)

{
       int *a;
       int p1,p2,i;
       a=ptr;
       p1=0; p2=0;
```

```
        for(i=0;i<70;i=i+2) //calculating the values according
        to truth table.
        {
                if(a[i]==1 && a[i+1]==1)
                {
                        p1=p1+3; p2=p2+3;
                }
                if(a[i]==1 && a[i+1]==0)
                {
                        p1=p1+5; p2=p2+0;
                }
                if(a[i]==0 && a[i+1]==1)
                {
                        p1=p1+0; p2=p2+5;
                }
                if(a[i]==0 && a[i+1]==0)
                {
                        p1=p1+1; p2=p2+1;
                }
        }
        return(p1+p2); //RETRUN THE TOTAL SCORE OF THE STRING.
}
void sort_select() //ORDINARY SORTING PROCEDURE
{
        int temp[9],i,j,t;
        for(i=0;i<10;i++)
        temp[i]=score[i];

        for(i=0;i<10;i++)
                for(j=9;j>=i;j-)
                {
                        if(temp[i]<temp[j]) //USUSAL SWAPPING
                           PROCEDURE.
                        {
                                t=temp[j];
                                temp[j]=temp[i];
                                temp[i]=t;
                        }
                }
        for(i=0;i<7;i++)
        for(j=0;j<10;j++)
                if(temp[i]==score[j])
                        index[i]=j;
                score[0]=temp[0];
```

9.4 Prisoner's Dilemma

```
}

void crossover(int *ptr1,int *ptr2)
{
        int temp,i,j;
        int ind=random(60); //RANDOM POINT OF CROSSOVER

        for(i=ind;i<70;i++)
        {
                temp=ptr1[i];
                ptr1[i]=ptr2[i];
                ptr2[i]=temp;
        }
}
```

Output

Enter the no of generation 5

```
The best score in the generation 1: 171
The best score in the generation 2: 160
The best score in the generation 3: 170
The best score in the generation 4: 166
The best score in the generation 5: 169

The best string in generation 1:
dd dc cd dc dc cd cd dd cc dc dc dc dd dc dd cd dc cd
   cd cc dc cd cc cd dd cd cd dd cd dc cc cd dc dd dd

The best string in generation 2:
cd cc cd cc cd cd dd dc cd cc dc cc dd cd dd dd cc cc dc dd
   dc cd cd dd dc dd dd cc cd dd dc dc cd dc cc

The best string in generation 3:
cd cc cd cc cd cd dc dd cd dc dd cc cd cd cc dd cd dd dc cd
   dc dc dd cd dc dc dc cd cd cd dc dc dd dc dd

The best string in generation 4:
cd cc cd cc cd cd dc dd cd dc dd cc cd cd cc dd cd dd dc cd
   dc dc dd dd dc dd dd cc cd dd dc dc cd dc cc

The best string in generation 5:
Cd dd cc cd dd dc cd cc dd cd dd dd dc cd cd cc dc cd cd dc
   cc dd dd dc dc dc dd dc dc cd dc cc dc cd dd
```

The best string in all generation is
dd dc cd dc dc cd cd dd cc dc dc dc dd dc dd cd dc
   cd cd cc dc cd cc cd dd cd cd dd cd dc cc cd dc dd dd

The corresponding best score is : 171

9.5 Maximize $f(x) = x^2$

A C++ Program for maximizing $f(x) = x^2$ using genetic algorithm, where x is ranges from 0 to 31.

1. Generate Initial four populations of binary string with 5 bits length.
2. Calculate corresponding x and fitness value $f(x) = x^2$.
3. Use the tournament selection method to generate new four populations.
4. Apply cross-over operator to the new four populations and generate new populations.
5. Apply mutation operator for each population.
6. Repeat the step 2 to 5 for some 20 iterations.
7. Finally print the result.

Source Code

```
#include<stdio.h>
#include<iostream.h>
#include<conio.h>
#include<stdlib.h>
#include<math.h>
#include<time.h>
int pop[10][10],npop[10][10],tpop[10][10],x[10],fx[10],m_max=1,
   ico=0,ico1,it=0;
void iter(int [10][10],int,int);
int u_rand(int);
void tour_sel(int,int);
void cross_ov(int,int);
void mutat(int,int);
void main()
{
int k,m,j,i,p[10],n=0,a[10],nit;

//time_t t;

clrscr();
//srand((unsigned) time(&t));
```

9.5 Maximize f(x) = x^2

```
randomize();
/*while (n<4)
{
k=0;
p[n]=u_rand(32);
a[n]=p[n];
for(i=0;i<=n;i++)
{
if (p[n]==p[i] && n!=i)
        k++;
}
if (k==0)
        {
        n++;
        }
} */
cout<<"Enter the number of Population in each iteration is : ";
cin>>n;
cout<<"Enter the number of iteration is : ";
cin>>nit;

m=5;
for(i=0;i<n;i++)
{
for(j=m-1;j>=0;j-)
{
/*if (a[i]==0)
        pop[i][j]=0;
else
        {*/
        pop[i][j]=u_rand(2);
//      a[i]=a[i]/2;
//      }
}
}
cout<<"\nIteration "<<it<<" is :\n";
iter(pop,n,m);
it++;
getch();
do
{
it++;
cout<<"\nIteration "<<it<<" is :\n";
tour_sel(n,m);
iter(pop,n,m);
getch();
```

```
}while(it<nit);
cout<<"\n\nAfter the "<<ico1<<" Iteration, the Maximum Value
   is : "<<(int)
sqrt(m_max);
getch();
}

void iter(int pp[10][10],int o, int p)
{
int i,j,sum,avg,max=1;
for(i=0;i<o;i++)
{
x[i]=0;
for(j=0;j<p;j++)
{
x[i]=x[i]+(pp[i][j]*pow(2,p-1-j));
}
fx[i]=x[i]*x[i];
sum=sum+fx[i];
if (max<=fx[i])
        max=fx[i];
}
avg=sum/o;
cout<<"\n\nS.No.\tPopulation\tX\tf(X)\n\n";
for(i=0;i<o;i++)
{
cout<<ico<<"\t";
ico++;
for(j=0;j<p;j++)
        cout<<pp[i][j];
cout<<"\t\t"<<x[i]<<"\t"<<fx[i]<<"\n";
}
cout<<"\n\t Sum : "<<sum<<"\tAverage : "<<avg<<"\tMaximum :
   "<<max<<"\n";
if (m_max<max)
        {
        m_max=max;
        ico1=it;
        }
}
int u_rand(int x)
{
int y;
y=rand()%x;
return(y);
}
```

9.5 Maximize f(x) = x^2

```
void tour_sel(int np,int mb)
{
int i,j,k,l,co=0,cc;
//time_t t;
//srand((unsigned) time(&t));
do
{
k=u_rand(np);
do
{
cc=0;
l=u_rand(np);
if (k==l)
        cc++;
}while(cc!=0);

if (fx[k]>fx[l])
{
        for(j=0;j<mb;j++)
                npop[co][j]=pop[k][j];
}
else if (fx[k]<fx[l])
        {
        for(j=0;j<mb;j++)
                npop[co][j]=pop[l][j];

        }
co++;
}while(co<np);

getch();
cross_ov(np,mb);
getch();
}
void cross_ov(int np1,int mb1)
{
int i,j,k,l,co,temp;
//time_t t;
//srand((unsigned) time(&t));

i=0;
do
{
k=rand()%2;
do
{
```

```
co=0;
l=u_rand(mb1);
if (((k==0) && (l==0)) || ((k==1) && (l==mb1)))
        co++;
}while(co!=0);

if ((k==0) && (l!=0))
{
        for(j=0;j<l;j++)
        {
                temp=npop[i][j];
                npop[i][j]=npop[i+1][j];
                npop[i+1][j]=temp;
        }
}
else if ((k==1) && (l!=mb1))
        {
                for(j=l;j<mb1;j++)
                {
                        temp=npop[i][j];
                        npop[i][j]=npop[i+1][j];
                        npop[i+1][j]=temp;
                }
        }
i=i+2;
}while(i<np1);

for(i=0;i<np1;i++)
{
for(j=0;j<mb1;j++)
{
tpop[i][j]=npop[i][j];
//pop[i][j]=tpop[i][j];
}
}
mutat(np1,mb1);

}

void mutat(int np2,int mb2)
{
int i,j,r,temp,k,z;
i=0;
do
{
        for(k=0;k<np2;k++)
```

9.5 Maximize f(x) = x^2

```
            {
            r=0;
            if (i!=k)
            {
            for(j=0;j<mb2;j++)
            {
            if (tpop[i][j]==tpop[k][j])
                    r++;
            }
            if (r!=mb2-1)
            {
                    z=u_rand(mb2);
                    if (tpop[i][z]==0)
                            tpop[i][z]=1;
                    else
                            tpop[i][z]=0;
                    if (npop[k][u_rand(mb2)]==0)
                            npop[k][u_rand(mb2)]=1;
                    else
                            npop[k][u_rand(mb2)]=0;
                    mutat(k,mb2);
            }
            }
            }
i++;
}while(i<np2);

for(i=0;i<np2;i++)
{
for(j=0;j<mb2;j++)
{
pop[i][j]=tpop[i][j];
}
}
}
```

Output

Enter the number of Population in each iteration is : 4
Enter the number of iteration is : 5
Iteration 0 is :

S.No.	Population	X	f(X)
0	01001	9	81
1	01110	14	196
2	11101	29	841
3	01111	15	225

Sum : 1620   Average : 405   Maximum : 841

Iteration 1 is :

S.No.	Population	X	f(X)
4	00000	0	0
5	11001	25	625
6	00001	1	1
7	11011	27	729

Sum : 1358   Average : 339   Maximum : 729

Iteration 2 is :

S.No.	Population	X	f(X)
8	10010	18	324
9	00010	2	4
10	11000	24	576
11	11001	25	625

Sum : 1531   Average : 382   Maximum : 625

Iteration 3 is :

S.No.	Population	X	f(X)
12	11110	30	900
13	11010	26	676
14	11110	30	900
15	11101	29	841

Sum : 3320   Average : 830   Maximum : 900

Iteration 4 is :

S.No.	Population	X	f(X)
16	11111	31	961
17	11101	29	841
18	01011	11	121
19	11000	24	576

Sum : 2501   Average : 625   Maximum : 961

Iteration 5 is :

S.No.	Population	X	f(X)
20	11110	30	900
21	01010	10	100
22	11010	26	676
23	11011	27	729

Sum : 2408   Average : 602   Maximum : 900

After the 5 Iteration, the Maximum Value is : 30

9.6 Minimization a Sine Function with Constraints

Many practical problems contain one or more constraints that must be satisfied. Here we consider incorporation of constraints into genetic algorithm search.

Constraints are usually classified as equality or inequality relations. Since equality constraints may be subsumed into a system model, we are only concerned with inequality constraints. A genetic algorithm generates a sequence of parameters to be tested using the system model, objective function and the constraints. We run the model, evaluate the objective function, and check to see if any constraints are violated. If not, the parameters set is assigned the fitness value corresponding to the objective function evaluation. If the constraints are violated, the solution is infeasible and thus has no fitness. We usually want to get some information out of infeasible solutions, by degrading their fitness ranking in relation to the degree of constraint violation. This method is called as the penalty method.

In this method, a constrained problem in optimization is transformed to an unconstrained problem by associating the cost or penalty with all constraint violations. The cost is included in the objective function evaluation.

9.6.1 Problem Description

Minimize $f(x1, x2, x3, x4, x5) = -5\sin(x1)\sin(x2)\sin(x3)\sin(x4)\sin(x5) +$
$- \sin(5x1)\sin(5x2)\sin(x3)\sin(5x4)\sin(5x5)$,

where $0 <= xi <= pi$, for $1 <= i <= 5$.

For the above mentioned problem, the known global solution is
$(x1, x2, x3, x4, x5) = (\Pi/2, \Pi/2, \Pi/2, \Pi/2, \Pi/2)$ and $f(\Pi/2, \Pi/2, \Pi/2, \Pi/2, \Pi/2) = -6$

To solve this problem we have chosen the heuristic crossover technique. This operator is a unique cross over for the following reasons :

1. It uses values of the objective function in determining the direction of the search.
2. It produces only one offspring and
3. It may produce no offspring at all.

The operator generates a single offspring x3 from two parents x1 and x2 according to the following rule:

$$X3 = r.(x2 - x1) + x2,$$

where r is a random number between 0 and 1, and the parent x2 is not worst than x2. i.e. $f(x2) <= f(x1)2$ for minimization problems.

It is possible for this operator to generate an offspring vector which is not feasible. In such a case another random value r is generated and another offspring created. If after w attempts no new solution meeting the constraints is found, the operator gives up and produces no offsprings.

The heuristic crossover contributes to the precision of the solution found; its major responsibilities are: fine tuning and search in the most promising direction.

After the experiment, it is found that the average value of the best point run was, **-5.986343** after 50 generations.

Source Code

```c
#include<stdio.h>
#include<conio.h>
#include<math.h>
#include<stdlib.h>

#define POPPA 20
#define POPCH 10
#define VAR 5
void main()
 {
  int par[POPPA][VAR];
  float chrom[POPPA][VAR];
  float newchrom[POPCH][VAR];
  float temp[1][5];
  float fit[POPPA],pi,fitch[POPCH],fitemp[1],fitchtemp[1];
  int gen=0;
  int i,j,k;
  int ri1,ri2,rj1,rj2;
  FILE *fp1;
  clrscr();
  fp1=fopen("inp.dat","rt");
  // Reading the initial population
  for(i=0;i<POPPA;i++)
   {
    for(j=0;j<VAR;j++)
     {
      fscanf(fp1,"%d",&par[i][j]);
      if(par[i][j]<=180 && par[i][j]>=0)
      {
      printf("%d ",par[i][j]);
      }
      else
      {
      printf("\nCONSTRAINT VIOLATED!\n 0<=xi<=180");
      exit(1);
      }
     }
     printf("\n");
   }
    fclose(fp1);
  // Parent Displaying
  for(i=0;i<POPPA;i++)
   {
    for(j=0;j<VAR;j++)
```

9.6 Minimization a Sine Function with Constraints

```c
      {
        chrom[i][j]=(float)par[i][j];
        printf("%f ",chrom[i][j]);
      }
      printf("\n");
    }
   pi = 4.0*atan(1.0);
   clrscr();
   while(gen!=50)
   {
   printf("Next Generation: %d\n",gen);
   for(i=0;i<POPPA;i++)
     {
   fit[i]=-
   5*(sin(pi*chrom[i][0]/180))*sin(pi*chrom[i][1]/180)
       *sin(pi*chrom[i][2]/180)
   *sin(pi*chrom[i][3]/180)*sin(pi*chrom[i][4]/180)-
   sin(5*pi*chrom[i][0]/180)*sin(5*pi*chrom[i][1]/180)
       *sin(5*pi*chrom[i][2]/180)
   *sin(5*pi*chrom[i][3]/180)*sin(5*pi*
   chrom[i][4]/180);
     printf("%f %f %f %f %f   %f\n",chrom[i][0],chrom[i][1],
        chrom[i][2],chrom[i][3],chrom[i][4],fit[i]);
    }
   printf("\n");
   //Crossover
   for(i=0,k=10;i<POPCH;i++,k++)
      {
      for(j=0;j<VAR;j++)
      {
        if(chrom[i][j]<=chrom[k][j])
        newchrom[i][j]=((random(10)/10)*(chrom[k][j]-chrom[i][j]))
           +chrom[k][j];
        else
        newchrom[i][j]=((random(10)/10)*(chrom[i][j]-chrom[k][j]))
           +chrom[i][j];
      }
      }

   //calculate fitness for children
   printf("child fitness:\n");
   for(i=0;i<POPCH;i++)
    {
     fitch[i]=-
   5*(sin(pi*newchrom[i][0]/180))*sin(pi*newchrom[i][1]/180)
   *sin(pi*newchrom[i][2]/180)*sin(pi*newchrom[i][3]/180)
```

```
         *sin(pi*newchrom[i][4]/180)-
sin(5*pi*newchrom[i][0]/180)*sin(5*pi*newchrom[i][1]/180)
*sin(5*pi*newchrom[i][2]/180)*sin(5*pi*newchrom[i][3]/180)
   *sin(5*pi*newchrom[i][4]/180);
    printf("%f\n",fitch[i]);
  }
  printf("\n\n\n");
  //sort based on fittness
  //parent
  for(i=0;i<POPPA;i++)
   {
     for(j=i+1;j<POPPA;j++)
       {
         if(fit[i]>fit[j])
         {
           for(k=0;k<VAR;k++)
             {
               temp[0][k]=chrom[j][k];
               fitemp[0]=fit[j];
             }
           for(k=0;k<VAR;k++)
             {
               chrom[j][k]=chrom[i][k];
               fit[j]=fit[i];
             }
           for(k=0;k<VAR;k++)
             {
               chrom[i][k]=temp[0][k];
               fit[i]=fitemp[0];
             }
         }
       }
    }
  //child
  for(i=0;i<POPCH;i++)
   {
     for(j=i+1;j<POPCH;j++)
       {
         if(fitch[i]>fitch[j])
          {
           for(k=0;k<VAR;k++)
             {
               temp[0][k]=newchrom[j][k];
               fitchtemp[0]=fitch[j];
             }
         for(k=0;k<VAR;k++)
```

9.6 Minimization a Sine Function with Constraints

```
              {
              newchrom[j][k]=newchrom[i][k];
              fitch[j]=fitch[i];
              }
            for(k=0;k<VAR;k++)
              {
              newchrom[i][k]=temp[0][k];
              fitch[i]=fitchtemp[0];
              }
            }
          }
        }
        //testing
        printf("\n");
        for(i=0;i<POPCH;i++)
      {
        printf("%f\n",fitch[i]);
       }
        printf("\n\n");
//Selecting fittest parent
  for(i=10,k=0;i<POPPA;i++,k++)
    {
     for(j=0;j<VAR;j++)
       {
        chrom[i][j]=newchrom[k][j];
       }
    }
//Mutation
if(gen%10==0)
  {
randomize();
    i=random(4);
    j=random(4);
randomize();
    ri1=random(4);
    rj1=random(4);
randomize();
    ri2=random(4);
    rj2=random(4);
    chrom[i][j]=(chrom[ri1][rj1]+chrom[ri2][rj2])/2;
  }
 gen++;
}//end of while
clrscr();
printf("***************************************************\n");
printf("\n\t\t\t OPTIMIZATION\n\n");
```

```
    printf("************************************************\n");
    printf("\n\nMinimize f(x1,x2,x3,x4,x5)=-5*sinx1*sinx2*sinx3
      *sinx4*sinx5
    +(-Sin(5x1)*sin(5x2)*sin(5x3)*sin(5x4)*sin(5x5))\n\n\n\n");
    // Displaying the last generation
    //Fitness
    for(i=0;i<POPCH;i++)
      {
    fitch[i]=-
    5*(sin(pi*chrom[i][0]/180))*sin(pi*chrom[i][1]/180)
       *sin(pi*chrom[i][2]/180)*sin(pi*chrom[i][3]/180)*sin(pi*
    chrom[i][4]/180)-
    sin(5*pi*chrom[i][0]/180)*sin(5*pi*chrom[i][1]/180)
       *sin(5*pi*chrom[i][2]/180)*sin(5*pi*chrom[i][3]/180)*si
    n(5*pi*chrom[i][4]/180);
       printf("%f\n",fitch[i]);
      }
    printf("The Last Generation:\n");
    for(i=0;i<POPCH;i++)
      {
      printf("%f %f %f%f%f%f\n",chrom[i][0],chrom[i][1],
         chrom[i][2],chrom[i][3],chrom[i][4],fitch[i]);
      }
    printf("\n");
    printf("The Solution is : %f",fitch[0]);
    getch();
    }
    // end of main
```

Output

```
Next Generation: 0
56.000000    89.000000    65.000000    90.000000    45.000000    -2.258168
23.000000    55.000000    120.000000   56.000000    89.000000    -0.381728
32.000000    56.000000    78.000000    51.000000    62.000000    -1.349719
98.000000    5.000000     63.000000    60.000000    90.000000    -3.193594
90.000000    80.000000    70.000000    40.000000    30.000000    -1.506204
32.000000    65.000000    98.000000    45.000000    12.000000    -0.441627
56.000000    90.000000    98.000000    23.000000    150.000000   -0.460083
100.000000   110.000000   90.000000    60.000000    51.000000    -3.020786
23.000000    45.000000    90.000000    67.000000    12.000000    -0.498938
89.000000    85.000000    90.000000    45.000000    62.000000    -3.598395
45.000000    45.000000    62.000000    21.000000    89.000000    -0.422367
12.000000    20.000000    60.000000    50.000000    40.000000    0.085765
78.000000    56.000000    89.000000    23.000000    12.000000    0.055674
```

9.6 Minimization a Sine Function with Constraints

```
45.000000    78.000000    65.000000    30.000000    20.000000    -0.635845
10.000000    20.000000    12.000000    32.000000    52.000000     0.194277
10.000000    52.000000    80.000000    89.000000    74.000000    -0.563703
45.000000    78.000000    60.000000    32.000000    21.000000    -0.669913
54.000000    98.000000    65.000000    32.000000    65.000000    -1.657383
78.000000    54.000000    65.000000    20.000000    32.000000    -0.746531
90.000000    25.000000    32.000000    54.000000    65.000000    -0.981731
```

child fitness:
-4.316244
-0.381728
-2.418798
-3.991571
-2.306131
-2.259517
-0.958608
-3.576286
-2.002319
-4.171978

-4.316244
-4.171978
-3.991571
-3.576286
-2.418798
-2.306131
-2.259517
-2.002319
-0.958608
-0.381728

Next Generation: 1
```
 89.000000    85.000000    90.000000    45.000000    62.000000    -3.598395
 98.000000    65.000000    63.000000    60.000000    90.000000    -3.193594
100.000000   110.000000    90.000000    60.000000    51.000000    -3.020786
 56.000000    89.000000    65.000000    90.000000    45.000000    -2.258168
 54.000000    98.000000    65.000000    32.000000    65.000000    -1.657383
 90.000000    80.000000    70.000000    40.000000    30.000000    -1.506204
 32.000000    56.000000    78.000000    51.000000    62.000000    -1.349719
 90.000000    25.000000    32.000000    54.000000    65.000000    -0.981731
 78.000000    54.000000    65.000000    20.000000    32.000000    -0.746531
 45.000000    78.000000    60.000000    32.000000    21.000000    -0.669913
 56.000000    89.000000    65.000000    90.000000    89.000000    -4.316244
 90.000000    85.000000    90.000000    54.000000    65.000000    -4.171978
 98.000000    78.000000    65.000000    60.000000    90.000000    -3.991571
```

100.000000	110.000000	90.000000	60.000000	65.000000	-3.576286
78.000000	56.000000	89.000000	51.000000	62.000000	-2.418798
90.000000	80.000000	70.000000	40.000000	52.000000	-2.306131
32.000000	65.000000	98.000000	89.000000	74.000000	-2.259517
78.000000	54.000000	90.000000	67.000000	32.000000	-2.002319
56.000000	90.000000	98.000000	32.000000	150.000000	-0.958608
23.000000	55.000000	120.000000	56.000000	89.000000	-0.381728

```
child fitness:
-5.986343
-3.670412
-4.103836
-4.257583
-3.622066
-2.306131
-2.259517
-3.132252
-1.348738
-2.182163
-5.986343
-4.257583
-4.103836
-3.670412
-3.622066
-3.132252
-2.306131
-2.259517
-2.182163
-1.348738
.....................{ ........convergence..........}
Next Generation: 49
```

89.000000	89.000000	90.000000	90.000000	89.000000	-5.986343
89.000000	89.000000	90.000000	90.000000	89.000000	-5.986343
89.000000	89.000000	90.000000	90.000000	89.000000	-5.986343
89.000000	89.000000	90.000000	90.000000	89.000000	-5.986343
89.000000	89.000000	90.000000	90.000000	89.000000	-5.986343
89.000000	89.000000	90.000000	90.000000	89.000000	-5.986343
89.000000	89.000000	90.000000	90.000000	89.000000	-5.986343
89.000000	89.000000	90.000000	90.000000	89.000000	-5.986343
89.000000	89.000000	90.000000	90.000000	89.000000	-5.986343
89.000000	89.000000	90.000000	90.000000	89.000000	-5.986343
89.000000	89.000000	90.000000	90.000000	89.000000	-5.986343
98.000000	89.000000	90.000000	90.000000	90.000000	-5.713716
98.000000	89.000000	98.000000	90.000000	90.000000	-5.486998
98.000000	89.000000	98.000000	90.000000	90.000000	-5.486998
100.000000	110.000000	90.000000	90.000000	90.000000	-4.515464

9.6 Minimization a Sine Function with Constraints

```
100.000000   110.000000    90.000000   90.000000    90.000000   -4.515464
100.000000   110.000000    98.000000   90.000000    90.000000   -4.496547
100.000000   110.000000    98.000000   90.000000    90.000000   -4.496547
100.000000   110.000000   120.000000   90.000000    90.000000   -4.103836
100.000000   110.000000    98.000000   90.000000   150.000000   -2.248273
```

child fitness:
-5.986343
-5.713716
-5.486998
-5.486998
-4.515464
-4.515464
-4.496547
-4.496547

-4.103836
-2.248273

-5.986343
-5.713716
-5.486998
-5.486998
-4.515464
-4.515464
-4.496547
-4.496547
-4.103836
-2.248273


```
*****************************************************************
                           OPTIMIZATION
*****************************************************************
```
Minimize f(x1,x2,x3,x4,x5)=-5*sinx1*sinx2*sinx3*sinx4*sinx5
  +(-Sin(5x1)*sin(5x2)
*sin(5x3)*sin(5x4)*sin(5x5))

-5.986343
-5.986343
-5.986343
-5.986343
-5.986343
-5.986343
-5.986343
-5.986343

-5.986343
-5.986343

The Last Generation:

89.000000	89.000000	90.000000	90.000000	89.000000	-5.986343
89.000000	89.000000	90.000000	90.000000	89.000000	-5.986343
89.000000	89.000000	90.000000	90.000000	89.000000	-5.986343
89.000000	89.000000	90.000000	90.000000	89.000000	-5.986343
89.000000	89.000000	90.000000	90.000000	89.000000	-5.986343
89.000000	89.000000	90.000000	90.000000	89.000000	-5.986343
89.000000	89.000000	90.000000	90.000000	89.000000	-5.986343
89.000000	89.000000	90.000000	90.000000	89.000000	-5.986343
89.000000	89.000000	90.000000	90.000000	89.000000	-5.986343
89.000000	89.000000	90.000000	90.000000	89.000000	-5.986343

The Solution is: -5.986343

9.7 Maximizing the Function $f(x) = x^*\sin(10^*\Pi^*x) + 10$

To find the solution of the function Max $F(x) = x^* \sin(10^*\Pi^*x) + 10$ with the constraint $-1 < x < 2$ by using genetic algorithm.

Steps involved

Step1 : Generate the random number as n
Step2 : Initialize i, j to n and m respectively
Step3 : Max ← 1, x[i] ← 0, sum ← 0, m_max ← 1
Step4 : Compute x[i] ← x[i]+(pp[i][j]/pow(10,j-1)) and
  fx[i]=x[i]*sin(10*Π*x[i])+10
  sum=sum+fx[i];
Step5 : if (max>=fx[i])
Step6 : max ← fx[i];
Step7 : until m_max>max
Step8 : Compute Maximum value

Source Code

```
#include<stdio.h>
#include<iostream.h>
#include<conio.h>
```

9.7 Maximizing the Function f(x) = x*sin(10*Π*x) + 10   303

```
#include<stdlib.h>
#include<math.h>
#include<time.h>

int ico=0,ico1,it=0;
long int pop[10][10],npop[10][10],tpop[10][10];
float x[10],fx[10],m_max=1.0;
void iter(long int [10][10],int,int);
int u_rand(int);
void tour_sel(int,int);
void cross_ov(int,int);
void mutat(int,int);
void main()
{
int k,m,j,i,p[10],n=0,a[10],nit;

//time_t t;

clrscr();
//srand((unsigned) time(&t));
randomize();
/*while (n<4)
{
k=0;
p[n]=u_rand(32);
a[n]=p[n];
for(i=0;i<=n;i++)
{
if (p[n]==p[i] && n!=i)
        k++;
}
if (k==0)
        {
        n++;
        }
} */
cout<<"Enter the number of Population in each iteration is : ";
cin>>n;
cout<<"Enter the number of iteration is : ";
cin>>nit;

m=7;
for(i=0;i<n;i++)
{
for(j=0;j<m;j++)
{
```

```
/*if (a[i]==0)
        pop[i][j]=0;
else
        {*/
if ((j==0) || (j==1))
        pop[i][j]=u_rand(2);
if (j>1)
        pop[i][j]=u_rand(10);
//      a[i]=a[i]/2;
//      }
}
}
cout<<"\nIteration "<<it<<" is :\n";
iter(pop,n,m);
it++;
getch();
do
{
it++;
cout<<"\nIteration "<<it<<" is :\n";
tour_sel(n,m);
iter(pop,n,m);
getch();
}while(it<nit);
cout<<"\n\nAfter the "<<ico1<<" Iteration, the Maximum Value
    is :
"<<m_max;
getch();
}

void iter(long int pp[10][10],int o, int p)
{
int i,j;
float sum,avg,max=1.0;
for(i=0;i<o;i++)
{
x[i]=0;
for(j=1;j<p;j++)
{
if (j==1)
        x[i]=x[i]+pp[i][j];
if (j>1)
        x[i]=x[i]+(pp[i][j]/pow(10,j-1));
}
j=0;
if (pp[i][j]==0)
```

9.7 Maximizing the Function f(x) = x*sin(10*Π*x) + 10

```
            x[i]=-x[i];

fx[i]=x[i]*sin(10*3.14*x[i])+10;
sum=sum+fx[i];
if (max<=fx[i])
         max=fx[i];

}
avg=sum/o;
cout<<"\n\nS.No.\tPopulation\tX\t\tf(X)\n\n";
for(i=0;i<o;i++)
{
cout<<ico<<"\t";
ico++;
for(j=0;j<p;j++)
         cout<<pp[i][j];
cout<<"\t\t"<<x[i]<<"\t\t"<<fx[i]<<"\n";
}
cout<<"\n\t Sum : "<<sum<<"\tAverage : "<<avg<<"\tMaximum : 
   "<<max<<"\n";
if (m_max<max)
         {
         m_max=max;
         ico1=it;
         }
}

int u_rand(int x)
{
int y;
y=rand()%x;
return(y);
}

void tour_sel(int np,int mb)
{
int i,j,k,l,co=0,cc;
//time_t t;
//srand((unsigned) time(&t));

do
{
k=u_rand(np);
do
{
cc=0;
```

```
l=u_rand(np);
if (k==l)
        cc++;
}while(cc!=0);

if (fx[k]>fx[l])
{
        for(j=0;j<mb;j++)
                npop[co][j]=pop[k][j];
}
else if (fx[k]<fx[l])
        {
        for(j=0;j<mb;j++)
                npop[co][j]=pop[l][j];
        }

co++;
}while(co<np);

getch();
cross_ov(np,mb);
getch();
}

void cross_ov(int np1,int mb1)
{
int i,j,k,l,co,temp;
//time_t t;
//srand((unsigned) time(&t));

i=0;
do
{
k=rand()%2;
do
{
co=0;
l=u_rand(mb1);
if (((k==0) && (l==0)) || ((k==1) && (l==mb1)))
        co++;
}while(co!=0);

if ((k==0) && (l!=0))
{
        for(j=0;j<l;j++)
        {
```

9.7 Maximizing the Function f(x) = x*sin(10*Π*x) + 10

```
                    temp=npop[i][j];
                    npop[i][j]=npop[i+1][j];
                    npop[i+1][j]=temp;
           }
}
else if ((k==1) && (l!=mb1))
           {
                    for(j=1;j<mb1;j++)
                    {
                             temp=npop[i][j];
                             npop[i][j]=npop[i+1][j];
                             npop[i+1][j]=temp;
                    }
           }
i=i+2;
}while(i<np1);

for(i=0;i<np1;i++)
{
for(j=0;j<mb1;j++)
{
tpop[i][j]=npop[i][j];
//pop[i][j]=tpop[i][j];
}
}
mutat(np1,mb1);

}

void mutat(int np2,int mb2)
{
int i,j,r,temp,k,z;

i=0;
do
{
          for(k=0;k<np2;k++)
          {
          r=0;
          if (i!=k)
          {
          for(j=0;j<mb2;j++)
          {
          if (tpop[i][j]==tpop[k][j])
                   r++;
          }
```

```
                if (r!=mb2-1)
                {
                        z=u_rand(mb2);
                        if ((tpop[i][z]==0) && ((z==0) || (z==1)))
                                tpop[i][z]=u_rand(2);
                        else if ((tpop[i][z]!=0) && ((z==0) || (z==1)))
                                tpop[i][z]=u_rand(2);
                        else
                                tpop[i][z]=u_rand(10);

                        if ((npop[k][u_rand(mb2)]==0) && ((z==0) ||
                          (z==1)))
                                npop[k][u_rand(mb2)]=u_rand(2);
                        else if ((npop[k][u_rand(mb2)]!=0) && ((z==0) ||
                          (z==1)))
                                npop[k][u_rand(mb2)]=u_rand(2);
                        else
                                npop[k][u_rand(mb2)]=u_rand(10);

                        mutat(k,mb2);
                }
            }
        }
i++;
}while(i<np2);

for(i=0;i<np2;i++)
{
for(j=0;j<mb2;j++)
{
pop[i][j]=tpop[i][j];
}
}
}
```

Output

Enter the number of Population in each iteration is : 5
   Enter the number of iteration is : 5
      Iteration 1 is :

9.7 Maximizing the Function f(x) = x*sin(10*Π*x) + 10

S.No.	Population	X	f(X)
0	0040269	-0.40269	10.031417
1	1182511	1.82511	11.257072
2	1103802	1.03802	10.958928
3	1025375	0.25375	10.25211
4	1038920	0.3892	9.868195

Sum : 52.367722   Average : 10.473544   Maximum : 11.257072

Iteration 2 is :

S.No.	Population	X	f(X)
5	1153524	1.53524	8.644072
6	0031433	-0.31433	9.864642
7	0137630	-1.3763	9.045399
8	0008313	-0.08313	10.042119
9	1074001	0.74001	9.298878

Sum : 46.895107   Average : 9.379022   Maximum : 10.042119

Iteration 3 is :

S.No.	Population	X	f(X)
10	1186753	1.86753	11.619774
11	0080292	-0.80292	10.063322
12	0158525	-1.58525	9.255855
13	0091516	-0.91516	9.592331
14	0103803	-1.03803	10.959062

Sum : 51.490341   Average : 10.298068   Maximum : 11.619774

Iteration 4 is :

S.No.	Population	X	f(X)
15	1173828	1.73828	8.396161
16	0064429	-0.64429	10.632739
17	1099734	0.99734	9.900976
18	0130327	-1.30327	9.893287
19	1088392	0.88392	10.438623

Sum : 49.261787   Average : 9.852358   Maximum : 10.632739

Iteration 5 is :

S.No.	Population	X	f(X)
20	1149948	1.49948	10.060295
21	0010010	-0.1001	9.999846
22	0130283	-1.30283	9.911272
23	0088675	-0.88675	10.36997
24	0017727	-0.17727	9.883524

Sum : 50.224907   Average : 10.044981   Maximum : 10.36997
After the 5 Iterations, the Maximum Value is: 11.619774

9.8 Quadratic Equation Solving

To find the roots of the quadratic equation using genetic algorithm. To solve the above problem for the quadratic equation- $x*x + 5*x + 6$ using following procedure. It could be used for solving any quadratic equation by changing fitness function- f(x) and changing length of chromosome.

Steps involved

> Step1: Initial population size is 10 and chromosome length is set to 5. Selecting initial population. i.e. random approximate solution to the problem, which are ten different 5-bit binary strings. Here initial population consists of ten chromosomes. Chromosomes are generated by using random number generator.
> Step 2: Converting the chromosome's genotypes to its phenotype (i.e.) Binary string into decimal value. In the binary string the most significant bit is sign bit. It's weight is $-2*(n - 1)$ and other bits are magnitude bits their weights are $2*(n - 1)$.
> Step 3: Evaluate the objective function $f(x) = x*x + 5*x + 6$. For each chromosome
>
>> 1) Convert the value of the objective function into fitness. Here for this problem fitness is simply equal to the value of the objective function.
>> 2) If f(x) = = 0 for a particular chromosome, that chromosome is required accurate solution , now display the value of chromosome and stop. Otherwise perform next generation by continuing following steps.
>
> Step 4: Implementation of selection operation. For this problem the tournament selection is adopted.
> The tournament selection is implemented as follows: Take any two chromosomes randomly and select one with min. Fitness for next generation. This process has to be repeated till we get ten chromosomes.
> Step 5:Implementation of crossover operation on new population.
> Take chromosome - 1 &2 randomly fix the cut-point position and randomly decide left or right crossover and interchange the bits and the resulting chromosomes are used in the next generation.
> Repeat the above process for chromosome pair (3,4), pair (5,6), Pair (7,8) and pair (9,10).
> This crossover operation generates ten new chromosomes for the next generation.
> Step 6: Jump to step-2. (i.e.) Perform next generation.

Source Code

```
#include <stdio.h>
#include <conio.h>
```

9.8 Quadratic Equation Solving

```c
#include <dos.h>
#include <math.h>
#include <stdlib.h>
#include <time.h>
int f(int);

void main()
{
 struct c{
   int chromosome[5];
   int decimal_val;
   int fittness;
 };
struct c ipop[10], newpop[10];
int i,j,cut,gen,t,flag,num,s1,s2;
clrscr();

/* generating Initial population */
randomize();
for(i=0;i<10; ++i)
 for(j=0; j<5; ++j)
   ipop[i].chromosome[j] = rand()%2;

/* start of the next generation */
gen=1;
while(1)
{
/* Converting a binary string into decimal value */
for(i=0;i<10; ++i)
{
   num=0;
   for(j=0;j<4;++j)
      num = num+ (ipop[i].chromosome[j] * pow(2,j));
   num = num-(ipop[i].chromosome[4]*pow(2,4));
   ipop[i].decimal_val = num;
}

/* Calculating fittness value */
for(i=0;i<10;++i)
 ipop[i].fittness = f(ipop[i].decimal_val);
printf("Generation- %1d\n", gen);
printf("Initial population- output\n");
for(i=0;i<10;++i)
{
 for(j=4; j>=0; -j)
   printf("%1d", ipop[i].chromosome[j]);
```

```c
      printf(" %d", ipop[i].decimal_val);
      printf(" %d", ipop[i].fittness);
      printf("\n");
}
for(i=0;i<10; ++i)
{
   if(ipop[i].fittness ==0)
   {
          printf("stop generations\n");
          printf("result = %d\n", ipop[i].decimal_val);
          goto l1;
       }
 }

/* tournament selection */
printf("tournament selection\n ");
i=0;
while(i<=9)
{
 s1 = rand()%10;
 s2 = rand()%10;
 printf("%d %d %d %d\n", s1,s2,ipop[s1].fittness, ipop[s2].
 fittness);
 getche();
   if( ipop[s1].fittness < ipop[s2].fittness)
   {
      for(j=0;j<5;++j)
      newpop[i].chromosome[j] = ipop[s1].chromosome[j];
   }
   else
   {
      for(j=0;j<5;++j)
      newpop[i].chromosome[j] = ipop[s2].chromosome[j];
   }
   i++;

}
getche();
printf("new population -output\n");
for(i=0;i<10;++i)
{
 for(j=4; j>=0; -j)
   printf("%1d", newpop[i].chromosome[j]);
 printf("\n");
}
getche();
```

9.8 Quadratic Equation Solving

```c
/*crossover operation */
printf("crossover operation\n");
printf("left/right cut-point position\n");
for(i=0;i<=4;++i)
{
 flag= rand()%2;
 cut= rand()%5;
 printf("%1d    %1d\n", flag, cut);
 if(flag==0)    /* crossover to left of cutpoint position*/
      for(j=0;j<=cut-1;++j)
      {
          t=newpop[2*i].chromosome[j];
          newpop[2*i].chromosome[j]= newpop[(2*i+1)].
          chromosome[j];
          newpop[(2*i+1)].chromosome[j]= t;
      }
   else   /* crossover to the right of cutpoint position*/
      for(j=cut+1;j<=4;++j)
      {
          t=newpop[2*i].chromosome[j];
          newpop[2*i].chromosome[j]= newpop[(2*i+1)].
          chromosome[j];
          newpop[(2*i+1)].chromosome[j]= t;
      }
      for(j=4; j>=0; -j)
      printf("%1d", newpop[2*i].chromosome[j]);
      printf("\n");
      for(j=4; j>=0; -j)
      printf("%1d", newpop[2*i+1].chromosome[j]);
      printf("\n");
}
/* copy newpopulation to initial population*/
for(i=0; i<10; ++i)
{
      for(j=0; j<5;++j)
          ipop[i].chromosome[j] = newpop[i].chromosome[j];
  }
  gen=gen+1;
 }
 l1:
   printf("end\n");
}
int f(int x)
{
 return ( x*x + 5*x + 6);
}
```

Output

Generation- 1
Chromosome	decimalvalue	Fittnessvalue
11010	-6	12
00100	4	42
01010	10	156
01000	8	110
00110	6	72
00001	1	12
10001	-15	156
00101	5	56
11100	-4	2
00100	4	42

Generation- 2
Chromosome	decimalvalue	Fittnessvalue
11100	-4	2
00100	4	42
00001	1	12
11100	-4	2
00101	5	56
00100	4	42
00101	5	56
00000	0	6
00110	6	72
11001	-7	20

Generation- 3
Chromosome	decimalvalue	Fittnessvalue
00000	0	6
11100	-4	2
11100	-4	2
11100	-4	2
00100	4	42
11001	-7	20
00001	1	12
00110	6	72
00100	4	42
00001	1	12

Generation- 4
Chromosome	decimalvalue	Fittnessvalue
11100	-4	2
11100	-4	2
00001	1	12
11001	-7	20
00001	1	12

```
00000  0   6
11000 -8  30
00001  1  12
00100  4  42
00001  1  12
```

Generation- 5

Chromosome	decimalvalue	Fittnessvalue
00001	1	12
11100	-4	2
11100	-4	2
00000	0	6
00001	1	12
00001	1	12
10001	-15	156
01100	12	210
11101	-3	0
00000	0	6

stop generations
result = -3

9.9 Summary

Thus in this chapter the implementation of genetic algorithm concept using C/C++ has been dealt. The various problems of maximizing and minimizing the functions, finding the roots of a quadratic equation, traveling salesman problem, word-matching problem has been included. C being a universal language helps in evolving the genetic process and since it is portable, GA programs written in C for one computer can be run on another with little or no modification. With the availability of large number of functions, the programming task becomes simple. C++, an evolution of C, has helped genetic algorithm to run in an object oriented programming environment. As a result, this can further be extended to implement parallel genetic algorithms using C/C++.

9.9.1 Projects

1. Implement the optimization of Ackley's function using a C program
2. Maximize Rosenbrock's function using a C++ program
3. Minimize Rastrigin's function using structure oriented programming language
4. Choose a vectorized objective function of your own and try to find a solution to the function using object oriented programming language
5. Given a polynomial equation of the form $f(x) = x^4 + x^3 + x^2 + x + 1$. Find the roots of this polynomial using GA approach

6. Consider a hyperbolic sine function. Maximize it within the range 0<x<22/7 using a C program. Apply two-point crossover and tournament selection process.
7. Implement a Hybrid Genetic Algorithm for an application of your own using C++ approach
8. For the Traveling sales man problem in Sect. 9.2, use two-point crossover and obtain optimized solution.
9. Find the roots of the quadratic equation using genetic algorithm The quadratic equation is $f(x) = x^2 + 3x + 2$.
10. Find the solution of the function $f(x) = \sin(3\pi x) + 10$ with the constraint $-3 < x < 3$ by using genetic algorithm.

Chapter 10
Applications of Genetic Algorithms

10.1 Introduction

Genetic algorithms have been applied in science, engineering, business and social sciences. Number of scientists has already solved many engineering problems using genetic algorithms. GA concepts can be applied to the engineering problem such as optimization of gas pipeline systems. Another important current area is structure optimization. The main objective in this problem is to minimize the weight of the structure subjected to maximum and minimum stress constrains on each member. GA is also used in medical imaging system. The GA is used to perform image registration as a part of larger digital subtraction angiographies. It can be found that Genetic Algorithm can be used over a wide range of applications. In this chapter a few topics of its application are being covered. This includes the application of Genetic Algorithm in to main engineering applications, data mining and in various other image processing applications. Hope the chapter would give the reader a brief idea of how the genetic algorithm can be applied to any practical problems.

10.2 Mechanical Sector

10.2.1 Optimizing Cyclic-Steam Oil Production with Genetic Algorithms

The Antelope reservoir in the Cymric field, in the San Joaquin Valley, is a siliceous shale reservoir containing 12 to 13°API heavy oil. The reservoir consists primarily of diatomite, characterized by its high porosity, high oil saturation, and very low permeability. Approximately 430 wells are producing from this reservoir, with an average daily production of 23,000 bbl. The oil from the field is recovered using a Chevron-patented cyclic-steam process. A fixed amount of saturated steam is injected into the reservoir during a 3- to 4-day period. The high-pressure steam fractures the rock, and the heat from the steam reduces oil viscosity. The well is shut in during the next couple of days, known as the soak period. Condensed steam is absorbed by the diatomite, and oil is displaced into the fractures and wellbore.

After the soak period, the well is returned to production. The flashing of hot water into steam at the prevailing pressure provides the energy to lift the fluids to the surface. The well flows for approximately 20 to 25 days. After the well dies, the same cycle is repeated. Cycle length is 26 to 30 days.

Because there is no oil production during the steaming and soaking period, there is an incentive to minimize the steaming frequency and increase the length of the cycle. But because well production is highest immediately after returning to production and declines quickly thereafter, a case can be made for increasing the steaming frequency and reducing the length of the cycle. This suggests that there is an optimum cycle length for every well that results in maximum productivity during the cycle. Because there are more than 400 wells in the field, and there are constraints of steam availability and distribution system, as well as facility constraints, the result is a formidable scheduling problem.

10.2.1.1 Genetic Algorithms

Genetic algorithms (GAs) are global optimization techniques developed by John Holland in 1975. They are one of several techniques in the family of evolutionary algorithms—algorithms that search for solutions to optimization problems by "evolving" better and better solutions. A genetic algorithm begins with a "population" of solutions and then chooses "parents" to reproduce. During reproduction, each parent is copied, and then parents may combine in an analog to natural crossbreeding, or the copies may be modified, in an analog to genetic mutation. The new solutions are evaluated and added to the population, and low-quality solutions are deleted from the population to make room for new solutions. As this process of parent selection, copying, crossbreeding, and mutation is repeated, the members of the population tend to get better. When the algorithm is halted, the best member of the current population is taken as the solution to the problem posed.

One critical feature of a GA is its procedure for selecting population members to reproduce. Selection is a random process, but a solution member's quality biases its probability of being chosen. Because GAs promote the reproduction of high-quality solutions, they explore neighboring solutions in high-quality parts of the solution search space. Because the process is randomized, a GA also explores parts of the search space that may be far from the best individuals currently in the population. In the last 20 years, GAs have been used to solve a wide range of optimization problems. There are many examples of optimization problems in the petroleum industry for which GAs are well suited. At ChevronTexaco, in addition to the cyclicalsteam scheduling problem, well placement, rig scheduling, portfolio optimization, and facilities design have been addressed with GAs. At NuTech Solutions, GAs have been used in planning rig workover projects so that overall workover time is reduced, planning production across multiple plants to reduce costs, planning distribution from multiple plants to a large number of customers to reduce costs, and controlling pipeline operations to reduce costs while satisfying pipeline constraints.

10.2.1.2 Problem Formulation

The cyclic-steam scheduling problem is formulated as a GA optimization problem in which the objective is to maximize cumulative production over a 2-month period. The fitness function is calculated as the cumulative production minus the penalties for violating the soft constraints. The problem has many constraints. The field-level constraints include steam availability and the maximum number of wells steaming. Gauge-station constraints include minimum amount of steam used and maximum number of wells on production. The header-level constraints include maximum number of wells steaming, and the individual-well constraints include maximum/minimum number of production days.

Additionally, there are operational constraints such as communication where multiple wells must be steamed together and wells blocked because of rig activities. Although all of the field constraints could have been incorporated in the problem formulation as hard constraints, constraints that absolutely cannot be violated, the decision was made to make many of the constraints soft constraints, constraints that can be violated but with an associated penalty.

An example of a hard constraint is the total steam available on a given day for the whole field, whereas some soft constraints are maximum amount of steam used by a well group and minimum number of wells steaming in a header. The optimization is stopped when one of the following criteria is satisfied.

- A specified number of generations have been created.
- A specified amount of time has elapsed.
- The fitness function has not improved over a specified number of generations.

The GA used multiple heuristics to enhance its performance and speed up its search for high-quality solutions. To begin with, when it created the initial population of solutions, "the seed," it used heuristics based on those that the well operators and steam operators used at the oil field. It also used some heuristics developed for the project to find good initial schedules. An example of such a heuristic is "attempt to steam high-production wells at their optimal cycle length—the length of time between steaming at which a well's average daily production is maximized." The constraints of the problem made it impossible to steam all wells at their optimal cycle length, but inserting schedules based on this as a goal into the initial population gave the algorithm some high-quality solutions that could be mutated and crossbred with other types of solutions to find even better solutions.

The technique used for representing solutions was not the approach commonly found in GA textbooks. An indirect encoding approach was used in which each solution was a permutation list of wells, with multiple entries allowed for the same well. Then a decoding procedure was used that simulated the effects of various schedules to translate the permutation list into an actual schedule. The schedule builder looks at the first well on the list and simulates steaming it on Day 1. If this process violates no hard constraints, then the well is scheduled for steaming on Day 1. The schedule builder then looks at the second well on the list. It simulates the effects of steaming that well together with the first well on Day 1. If no hard constraints are violated, this well is added to the schedule for Day 1. If hard constraints are violated, the well is not added to the schedule. The process continues, considering each well for steaming on Day 1, and adding each well, in order, that can be steamed without violating a hard constraint. Then the process continues with Day 2, considering each well, in order, that was not already steamed on Day 1. The process is repeated for Days 3 and 4. The critical point is that the schedule-building process will not build a schedule that violates a hard constraint. Also, this schedule-building process uses some clever heuristics and a simulator to transform a list of wells into a feasible steaming schedule. Once a schedule is built, it can be evaluated, and its "score" is returned to the GA as the evaluation of the original solution, the list of wells.

The optimization process uses heuristics to initialize the population, as well as randomly generated solutions to fill out the initial population. The process includes intelligent heuristics in the procedures used to modify new solutions. Also used are crossbreeding procedures appropriate to combining different permutations to combine two parents to produce a child. The process includes a good deal of domain knowledge in the schedule builder to produce feasible schedules. A post-processor is included that checks to find simple changes that could be made to the best solution found to improve its quality. The interface to the optimizer gives the well operators and steam operators at the field a great deal of power and flexibility in their interactions with the system. The operators can edit the well data that are entered into the optimizer. They can select optimization heuristics and procedures used in a run. They can parameterize the objective function that specifies the goals of the run. They can activate, deactivate, and parameterize the hard and soft constraints. They also can edit the solutions found by the optimizer in cases in which there is a constraint known to the operators that is not reflected in the databases available to the optimizer.

10.2.2 Genetic Programming and Genetic Algorithms for Auto-tuning Mobile Robot Motion Control

Robotic soccer is an entertaining but very complex problem domain. A subsection of which is path planning and motion control, itself a complex and challenging field. An alternate technique to hand-coding control strategies is highly desirable. Both genetic programming (GP) and genetic algorithms (GA) are such techniques; they have the potential to remove the burden of programming from humans. This section will compare the two techniques and discuss their use.

Mobile robot path planning and motion control has tended to be treated independently in the literature. Path planning is often a slow process that assumes that the current state of the world is static and that the time taken to create the plan will not have significant effects on the performance of the robot. Motion control, in contrast, assumes that a path plan exists and that the motion controller has to follow the plan as closely as possible. This is, however, difficult to implement in a real system, as combining path planning with motion control in a fast dynamic environment tends to be unsuccessful, as the path planned does not reflect the current state of the world.

10.2 Mechanical Sector

Once a new path planning and motion control technique is developed it is applied to a real robot system. However, tuning the system tends to be a tedious process. Auto-tuning techniques are essential for quick development and deployment. GP and GAs have been applied to various robotic control applications.

10.2.2.1 Genetic Programming

Two independently driven wheels control the robot design that was simulated. Thus: if the two wheels are made to go at the same speed the robot will move straight ahead; if the wheel speeds are slightly different, the robot will turn in an arc to the side of the slower moving wheel, eventually coming around full circle; and if the wheel speeds are the same but their direction different, the robot will pivot on the spot. Equation 10.1 describes the robot behaviour formally.

$$\begin{bmatrix} 1 & 0 & 0 \\ 0 & 1 & 0 \\ 0 & 0 & \frac{d}{2\pi} \end{bmatrix} \begin{bmatrix} \frac{dx}{dt} \\ \frac{dy}{dt} \\ \frac{d\theta}{dt} \end{bmatrix} = \begin{bmatrix} \frac{1}{2}\cos\theta & \frac{1}{2}\cos\theta \\ \frac{1}{2}\sin\theta & \frac{1}{2}\sin\theta \\ \frac{-1}{2} & \frac{1}{2} \end{bmatrix} \begin{bmatrix} V_L \\ V_R \end{bmatrix} \quad (10.1)$$

A natural design for a low-level control strategy is to produce velocity setpoints for each wheel. Genetic programming allows for such a design. Canonical GP evolves individuals that are a single program tree, but if each individual was made up of two trees (where each tree produced a wheel setpoint) it would naturally produce robot-controlling code.

The genetic operators can be easily extended. The crossover operation requires two subtrees from two individuals. After normal selection of the individuals, one of the two trees is randomly selected from each of the two individuals. Normal selection of subtrees occurs from the selected trees. The mutation operation requires just one individual's tree. Having selected the individual, the tree-to-mutate is randomly chosen from the two available trees. Other operations can be extended similarly. With the appropriate structure in place, the focus now moves to gene selection. It was desired that the least amount of human assistance be given and in this vain, only very basic genes were chosen:

- the coordinates of the robot (robot x and robot y),
- the coordinates of the ball (ball x and ball y),
- a number of random constants from $< 001 >1$ to 1,
- arithmetic operations (add, subtract, etc.),
- trigonometric operations (sine, cosine), and
- an error-protected logarithmic function (log).

The control-strategy problem, based on a robot soccer simulator, was to learn to follow a ball. The success of an individual was initially measured by allowing the simulator to run for a short amount of time and taking the distance from the final ball position to the final robot position. The smaller the distance the greater the performance on behalf of the robot. The genetic programming kernel was left to work through the generations, and when finished we studied the best individual of each generation. This process generated individuals that clearly improved their ball-following ability, however their performance was unimpressive. In light of Koza's work and to improve performance, two more functions were added to the gene pool:

- the difference between the ball's and robot's x-coordinates (delta x), and
- the difference between the ball's and robot's y-coordinates (delta y).

Along with the addition of the new genes, the fitness test was altered. Rather than a single-time distance measurement (i.e. at the end of the simulation the distance between the ball and the robot was measured) we introduced a multi-time measurement. For each frame of the simulation (which ran at 33 frames a second) the distance from ball to robot was measured; these distances were summed to give a final fitness scored. Again, the smaller the value the better the performance, however it was impossible to achieve a zero-value fitness score. These changes had a positive effect on the individuals' ball-following ability. Figure 10.1 shows an example of the best-of-run individual executing its evolved code to follow the ball. (The run went to 50 generations with a population of 500.) Although it is far from an efficient ball-follower, it does demonstrate an interesting trait: when it comes very close to the ball it describes very tight circles—a behavior that optimizes the performance measure.

10.2.2.2 Genetic Algorithms

Genetic algorithms are quite different to genetic programming. The use of the process was not as intuitive as GP because rather than immediately producing velocity setpoints, the GA needed an extra layer of human intervention, a gain-scheduled controller. The GA system produced individuals that generated the required velocity set-points via the control formulae given in (10.2).

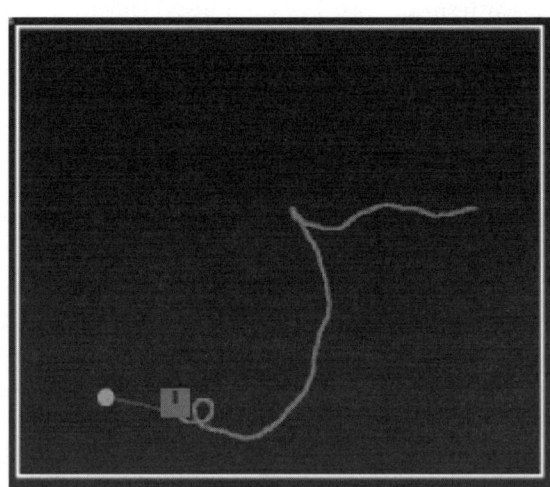

Fig. 10.1 The best-of-run individual found using genetic programming

10.2 Mechanical Sector

$$V_L = K_d \times distance + \frac{K_A}{100} \times angleError$$
$$V_R = K_d \times distance - \frac{K_A}{100} \times angleError \tag{10.2}$$

Where *distance* is the distance of the robot to the target, while *angleError* is the angular difference between the direction from the robot to the target and the heading of the robot. These formulae represent considerable domain knowledge given to the GA system in comparison to that given to GP. For the genetic algorithm system the problem environment was made simpler. From an arbitrary position, the robot was given a target (the origin) and a small amount of time with which to reach it. The success of the robot was measured by the robot's final distance from the target position. Note that this is a significantly easier problem and that is effectively the same as that posed for the GP system, except that the ball is "frozen", unable to be knocked away by the robot.

The population size used was 40 while the number of generations was 20. Each individual consisted of 11 K_A gains and 7 K_d gains, (18 in total). Each gain was coded as a 20 bit precision gene. Each individual was evaluated from 20 random positions around the goal position. The sum of the distances from the target position was used as the fitness function. There was rapid convergence within 10 generations to an almost perfect solution. Figure 10.2 shows the performance of the best of run

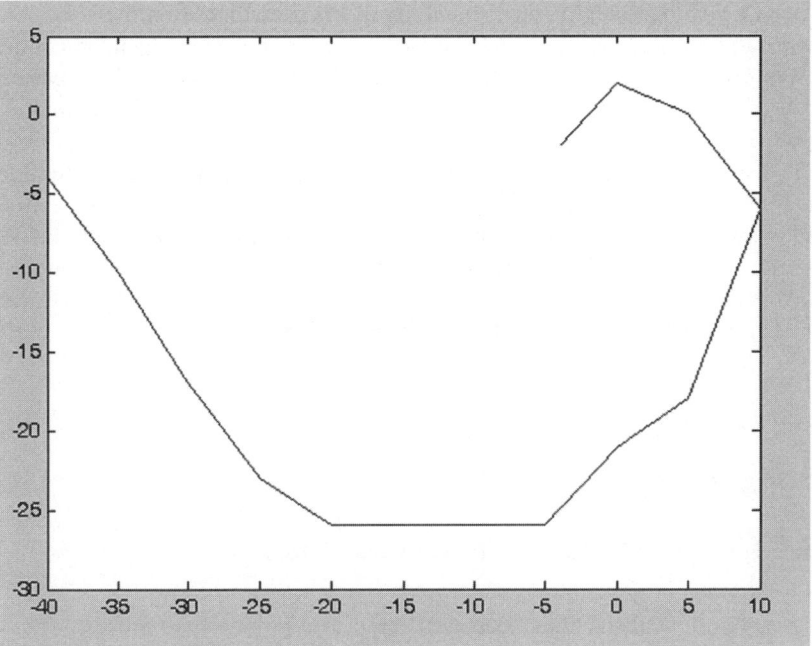

Fig. 10.2 A plot of performance of the best-of-run individual found using genetic algorithms

individual after the 20 generations of learning. The performance is good; the robot reaches the target position within the simulation time, however it is not optimal.

10.2.2.3 Comparison

The results of the GP and GA evaluations showed that both techniques have merits in developing and auto-tuning mobile robot controllers. Superficially, the GA solution seems to be better than the GP solution, but this is due to the different problem scenarios and evaluation functions used. The GA is evaluated on the distance to the target position at the end of the simulation time, while the GP is evaluated on the total distance from the ball. The target position is fixed while the ball can be hit away, so the GP solution is trying to get near the ball while actually avoiding hitting it further away.

It can be seen that the two approaches vary fundamentally in the amount and type of information required to setup the infrastructure for applying the techniques. The GP technique required only the information that two output values need to be provided (the wheel velocity set-points) and the processing that follows is automatic. Beyond ensuring there are suitable atoms available (such as delta x and delta y), the GP technique naturally searches the space of possible solutions. If the search space is too large (for instance, neglecting to provide the atoms delta x and delta y but rather only the absolute robot and target positions) the increase in problem complexity means that finding the solution can become intractable.

The GA technique on the other hand requires a transformation from the genetic representation to a computational execution system. The gain-scheduled controller that was used contains a controller that can solve the problem if a correct set of gains is given. The complexity of the search space is much smaller than that of the GP problem.

The evaluations presented in this section show that GAs are a more natural technique for including knowledge that already exist in a domain as this can be directly coded in the transformation of the genetic code to the implementation. With GP, the only natural position for including prior knowledge is in the atoms that are available; while it is difficult to force the GP to progress in a direction that is better structured even when more domain knowledge exists.

10.3 Electrical Engineering

10.3.1 Genetic Algorithms in Network Synthesis

Digital network synthesis has been developed to a point where most designs can be generated automatically on a computer. A specification written in a hardware description language such as VHDL or Verilog can be compiled into a form suitable

10.3 Electrical Engineering

for programming a FPGA or fabricating a full custom integrated circuit, with little or no human intervention.

The situation is very different for analogue network synthesis, however. With the exception of the limited number of problems for which formal design solutions exist, there are no automatic design tools available for analogue networks. As a consequence most analogue network design is still performed manually by skilled engineers. Although the analogue part of most electronic systems is often a small part of the overall system, it is usually the most expensive part in terms of design effort. Recently there have been attempts to automate the analogue network design process by the use of Genetic Algorithms (GAs) in conjunction with high-level statements of the desired response.

There is no reason in principle why the network topology and the component values should not be chosen using entirely different methods. Clearly there is a natural hierarchy: the network topology must be determined before the component values can be selected. Nevertheless, these two operations can be performed separately, and different optimization techniques can be used. Specifically, the network topology can be optimized using GAs; for each network topology generated the component values can then be determined by numerical optimization. The performance of the numerically optimized network is returned as the fitness function to the GA.

For optimization problems involving well-behaved objective functions dependent on the values of fixed number of variables, it is well established that numerical optimization methods converge much faster and involve fewer objective function evaluations than GAs. It seems likely (although it cannot be guaranteed) that a hybrid approach will be more efficient than a GA generating both structure and values.

Of course no optimization method guarantees to find the global optimum, but if this hybrid approach is to be successful the numerical optimization of the component values should achieve a high proportion of results close to the global optimum. This section aims to show that, at least for one network topology, this is in fact the case. The performance of the hybrid GA will be illustrated by applying it to a filter design problem applying it to an established filter design problem.

10.3.1.1 Component Value Selection by Numerical Optimization

The problem of designing a normalised sharp cut-off lowpass filter with the following specification will be considered:

Pass-band edge: 1.0 rad/s
Stop-band edge: 1.5 rad/s
Maximum pass-band gain: -6 dB
Minimum pass-band gain: -7 dB
Maximum stop-band gain: -52 dB

Fig. 10.3 Equally terminated ladder filter

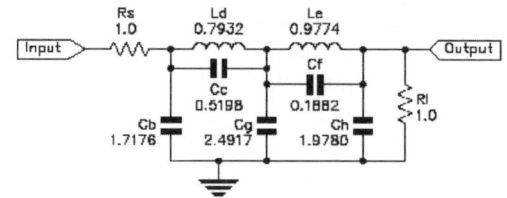

A formal design procedure exists for this problem: a 5th order normalized low-pass elliptic transfer function is implemented as an equally terminated ladder filter as shown in Fig. 10.3. Figure 10.4 shows the corresponding frequency response.

The network topology shown in Fig. 10.3 can be used to test the effectiveness of numerically optimizing the component values. First one component (*Ra*) is selected and given a value of 1.0, and the other components are assigned random values in the range 0.0 to 1.0. Then the component values (except *Ra*) are optimized numerically against an objective function incorporating the specification using a quasi-Newton optimization algorithm based on the Davidon-Fletcher-Powell (DFP) method. In 1000 trials with random initial sets of component values the results were fully compliant with the specification in 84% of cases.

DFP therefore failed to produce fully compliant designs in only a small fraction of attempts for this network topology. Using instead a GA to select the component values could not achieve a much higher success rate, and would probably be much less efficient in terms of computational effort. Of course, setting the pass-band edge to 1.0 rad/s and choosing *Ra* = 1.0 leads to the component values clustering around 1.0, and this makes it more likely that the numerical optimization will succeed.

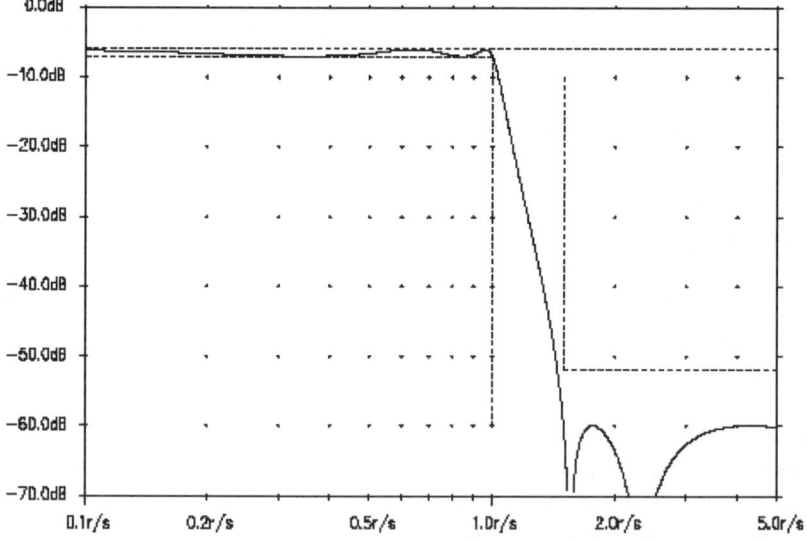

Fig. 10.4 Frequency response corresponding to equally terminated ladder filter

10.3 Electrical Engineering

Fortunately any frequency domain design problem can be transformed to a response centered 1.0 rad/s, and impedance levels can be scaled appropriately.

10.3.1.2 Network Synthesis Using the Hybrid Genetic Algorithm

It has been established, at least for a simple filter synthesis, that only a small proportion of potentially successful network topologies are likely to be rejected because of the failure of numerical optimization to find a near-global minimum. Consequently a network synthesis method based on the use of a GA to select topologies, followed by numerical optimization to determine component values, appears to be an attractive option.

To test the effectiveness of the hybrid GA network synthesis program, it was used to design a passive LCR filter to the specification given above. The basic network synthesis program incorporates no design rules and simply works towards satisfying the specified design goals; it is therefore the complete antithesis of an "expert system". Applied to this filter design problem it automatically generates a fully-compliant LCR network, but one which may be sub-optimal with respect to performance factors (such as component value sensitivity) that are not included in the design goals. Of course the design goals can be modified to include these factors, but this would result in a significantly increased computational effort.

In the case of frequency-domain filters it is well known that an equally terminated network provides low component value sensitivity. The synthesis program was therefore constrained to generate only LC networks between equal value termination resistances. A population size of 80 networks was used, and the program was run for 100 generations. This took around 4 hours on a PC (300 MHz Pentium II) and the synthesized network is shown in Fig. 10.5. The corresponding frequency response is shown in Fig. 10.6.

Significantly the synthesized network is fully compliant with the specification, while using fewer components than the filter resulting from the traditional formal design process based on an elliptic response. The synthesized filter has a 5th-order response with one pair of imaginary zeros; by contrast the elliptic filter has a 5th-order response, but two pairs of imaginary zeros. The number of imaginary zeros in an elliptic response is determined by the filter order, but in networks synthesized us-

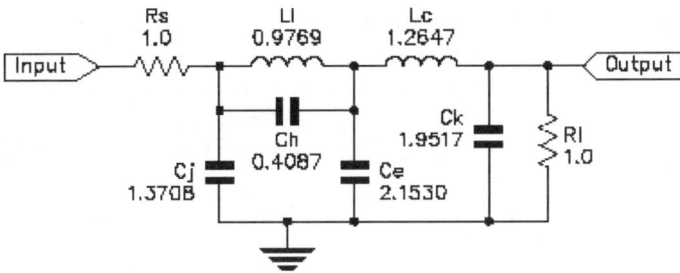

Fig. 10.5 Synthesized network

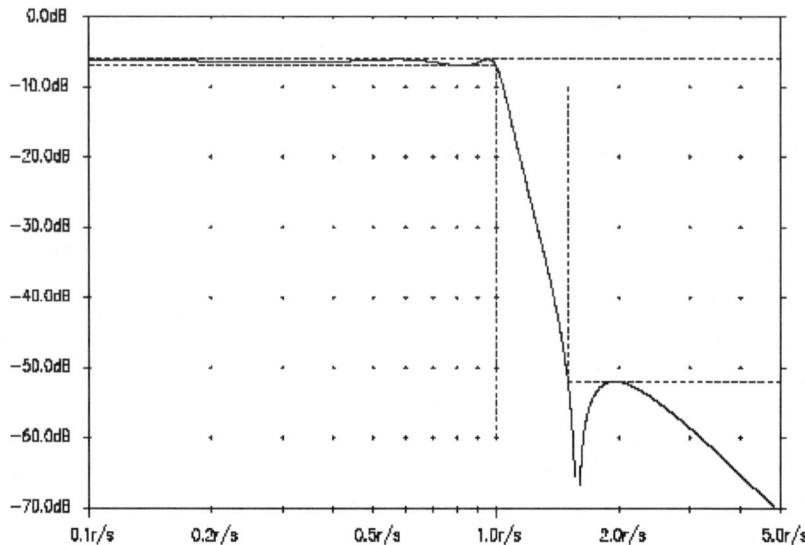

Fig. 10.6 Frequency response corresponding to synthesized network

ing GAs there is no such constraint. It is this flexibility which allows the component count to be reduced below that obtained using the traditional design process. Numerical optimization can be used to determine the component values for a network configuration created by a GA. This hybrid GA approach to network synthesis has successfully been applied to an existing filter design problem and has been shown to have significant advantages over a pure GA approach. Networks synthesized for this design problem are more economical than a filter designed by hand.

In principle the hybrid GA approach can be applied to any network design problem for which a means of evaluating potential networks against the design goals is available. Computational effort is the only limiting factor.

10.3.2 Genetic Algorithm Tools for Control Systems Engineering

There has been widespread interest from the control community in applying the Genetic Algorithm (GA) to problems in control systems engineering. Compared to traditional search and optimization procedures, such as calculus-based and enumerative strategies, the GA is robust, global and generally more straightforward to apply in situations where there is little or no a priori knowledge about the process to be controlled. As the GA does not require derivative information or a formal initial estimate of the solution region and because of the stochastic nature of the search mechanism, the GA is capable of searching the entire solution space with more likelihood of finding the global optimum.

GAs have been shown to be an effective strategy in the offline design of control systems by a number of practitioners. MATLAB has become a de-facto standard

10.3 Electrical Engineering

in Computer Aided Control System Design (CACSD) for the control engineer. The complete design cycle from modeling and simulation through controller design is addressed with a wide range of toolboxes, notably the Control System and Optimization Toolboxes, and the SIMULINK nonlinear simulation package along with extensive visualization and analysis tools. In addition, MATLAB has an open and extensible architecture allowing individual users to develop further routines for their own applications. These qualities provide a uniform and familiar environment on which to build genetic algorithm tools for the control engineer. This section describes the development and implementation of a Genetic Algorithm Toolbox for the MATLAB package and provides examples of a number of application areas in control systems engineering.

Whilst there exist many good public-domain genetic algorithm packages, such as GENESYS and GENITOR, none of these provide an environment that is immediately compatible with existing tools in the control domain. The MATLAB Genetic Algorithm Toolbox aims to make GAs accessible to the control engineer within the framework of an existing CACSD package. This allows the retention of existing modelling and simulation tools for building objective functions and allows the user to make direct comparisons between genetic methods and traditional procedures.

10.3.2.1 Data Structures

MATLAB essentially supports only one data type, a rectangular matrix of real or complex numeric elements. The main data structures in the GA Toolbox are chromosomes, phenotypes, objective function values and fitness values. The chromosome structure stores an entire population in a single matrix of size $Nind \times Lind$, where $Nind$ is the number of individuals and $Lind$ is the length of the chromosome structure. Phenotypes are stored in a matrix of dimensions $Nind \times Nvar$ where $Nvar$ is the number of decision variables. An $Nind \times Nobj$ matrix stores the objective function values, where $Nobj$ is the number of objectives. Finally, the fitness values are stored in a vector of length $Nind$. In all of these data structures, each row corresponds to a particular individual.

10.3.2.2 Toolbox Structure

The GA Toolbox uses MATLAB matrix functions to build a set of versatile routines for implementing a wide range of genetic algorithm methods. In this section we outline the major procedures of the GA Toolbox.

Population Representation and Initialisation: crtbase, crtbp, crtrp

The GA Toolbox supports binary, integer and floatingpoint chromosome representations. Binary and integer populations may be initialised using the Toolbox function to create binary populations, crtbp. An additional function, crtbase, is provided that builds a vector describing the integer representation used. Real-valued populations may be initialised using crtrp. Conversion between binary and real-values is provided by the routine bs2rv that also supports the use of Gray codes and logarithmic scaling.

Fitness Assignment: ranking, scaling

The fitness function transforms the raw objective function values into non-negative figures of merit for each individual. The Toolbox supports the offsetting and scaling method of Goldberg and the linear-ranking algorithm. In addition, non-linear ranking is also supported in the routine ranking.

Selection Functions: reins, rws, select, sus

These functions select a given number of individuals from the current population, according to their fitness, and return a column vector to their indices. Currently available routines are roulette wheel selection, rws, and stochastic universal sampling, sus. A high-level entry function, select, is also provided as a convenient interface to the selection routines, particularly where multiple populations are used. In cases where a generation gap is required, i.e. where the entire population is not reproduced in each generation, reins can be used to effect uniform random or fitness-based reinsertion.

Crossover Operators: recdis, recint, reclin, recmut, recombin, xovdp, xovdprs, xovmp, xovsh, xovshrs, xovsp, xovsprs

The crossover routines recombine pairs of individuals with given probability to produce offspring. Single-point, double-point and shuffle crossover are implemented in the routines xovsp, xovdp and xovsh respectively. Reduced surrogate crossover is supported with both single-, xovsprs, and double-point, xovdprs, crossover and with shuffle, xovshrs. A general multi-point crossover routine, xovmp, that supports uniform crossover is also provided. To support real-valued chromosome representations, discrete, intermediate and line recombination are supplied in the routines, recdis, recint and reclin respectively. The routine recmut performs line recombination with mutation features. A high-level entry function to all the crossover operators supporting multiple subpopulations is provided by the function recombin.

Mutation Operators: mut, mutate, mutbga

Binary and integer mutation are performed by the routine mut. Real-value mutation is available using the breeder GA mutation function, mutbga. Again, a high-level entry function, mutate, to the mutation operators is provided.

10.3 Electrical Engineering

In the forth coming section, an example is used to illustrate a number of features of the GA that make it potentially attractive to the control engineer. The example deals with the design of an aerospace control system demonstrates how GAs can be used to search a range of controller structures to satisfy a number of competing design criteria.

MIMO Controller Design

This design example demonstrates how GAs may be used to select the controller structure and suitable parameter sets for a multivariable system. The system is a propulsion unit for an Advanced Short Take-Off, Vertical Landing (ASTOVL) aero-engine, shown in Fig. 10.7. It is required that the pilot have control of the fore-aft differential thrust (XDIFF) and the total engine thrust (XTOT). The inputs to the system are XTOTD and XDIFFD. The design problem is to find a set of pre-compensators that satisfy a number of time response specifications whilst minimizing the interaction between the loops.

The time domain performance requirements, in response to a step in demand at one of the inputs, are

i. Maximum overshoot $\leq 10\%$
ii. 70% rise time ≤ 0.35 seconds
iii. 10% settling time ≤ 0.5 seconds at the associated output.

The amount of interaction, or cross-coupling, between modes is measured as

$$\int_0^\infty (XTOT)^2 dt \qquad (10.3)$$

when exited by a step input to XDIFFD, and vice-versa, and should be less than 0.05 for this design example. Thus, a total of eight design objectives must be satisfied.

The ASTOVL propulsion unit is modelled directly in the SIMULINK package as shown in Fig. 10.7. To simplify the problem, pre-compensators are selected to be either first or second order or simple gains and pre-compensator parameters are represented using real values. Using a structured chromosome representation, it is possible to allow the free parameters for each possible recompensator to reside in all individuals, although only certain parameter sets are active in any given individual at any time. The active parts of a chromosome are controlled by high-level genes in an individuals representation. Thus, an individual may contain a number of possibly good representations at any time.

In order to solve this problem using a simple GA, the design objectives are reformulated as a single minimax function thus:

$$f = \max_i \left(\frac{F_i - Goal_i}{Goal_i} \right) \qquad (10.4)$$

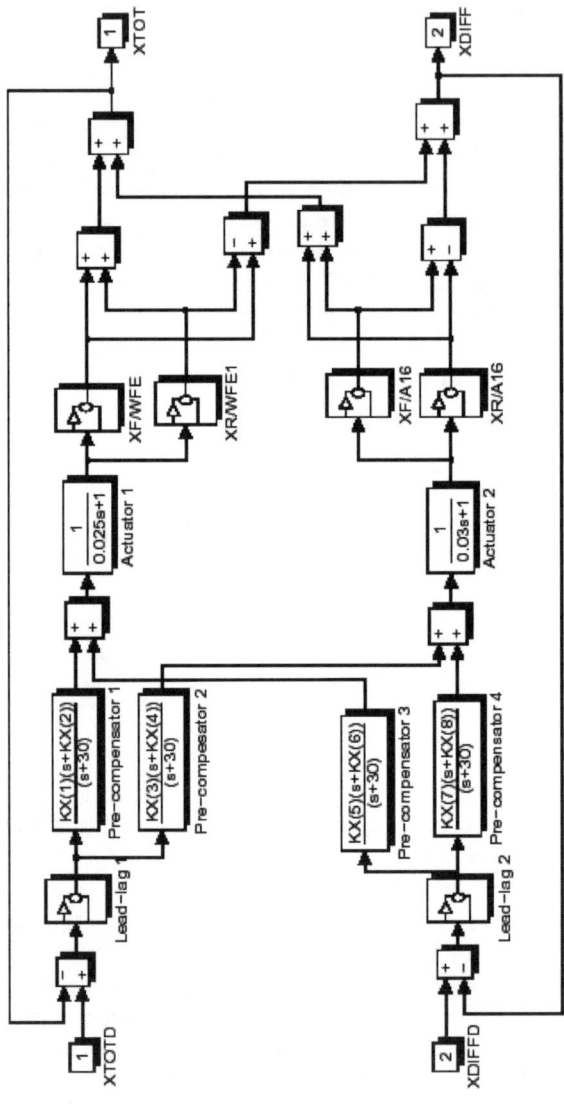

Fig. 10.7 SIMULINK model of the ASTOVL propulsion unit

10.3 Electrical Engineering

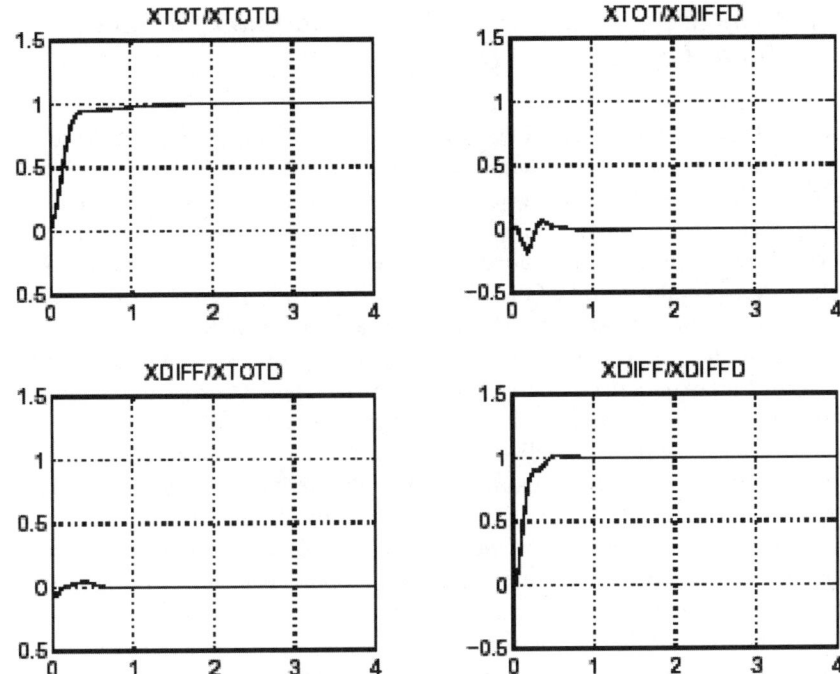

Fig. 10.8 Response of best ASTOVL controller found

where *Goali* are the design goals and F_i are the individual objective functions associated with each design criteria.

Using a population size of 40, the GA was run for 100 generations. A list of the best 50 individuals was continually maintained during the execution of the GA allowing the final selection of controller to be made from the best structures found by the GA over all generations.

Figure 10.8 shows the response of the best controller found by the GA. In this case, all of the pre-compensator structures were of first order complexity and the design objectives where over obtained by a factor of 38.18%.

The GA approach has the clear advantage over conventional optimization approaches in that it allows a number of controller structures to be examined in a single design cycle. The final choice of controller being made from a selection of different structures and parametric values. A minimax approach to this multiobjective problem has been described here that is simple to formulate and requires no special fitness assignment or selection methods.

10.3.3 Genetic Algorithm Based Fuzzy Controller for Speed Control of Brushless DC Motor

Brushless DC motors are reliable, easy control, and inexpensive. Due to their favorable electrical and mechanical properties, high starting torque and high efficiency, the BLDCM are widely used in most servo applications such as actuation, robotics, machine tools, and so on. The design of the BLDCM servo system usually requires time consuming trial and error process, and fail to optimize the performance. In practice, the design of the BLDCM drive involves a complex process such as model, devise of control scheme, simulation and parameters tuning. Usually, the parameters tuning for a servo system involves a sophisticated and tedious process and requires an experienced engineer in doing so. Application of intelligent optimization technique in tuning critical servo parameters remains an interesting and important issue to be further studied. Many sections have presented different design approaches and control structures in designing the digital servo controller. The PI controller can be suitable for the linear motor control. However, in practice, many non-linear factors are imposed by the driver and load, the PI controller cannot be suitable for non-linear system. Fuzzy control is a versatile and effective approach to deal with the non-linear and uncertain system. Even if a fuzzy controller (FLC) can produce arbitrary non-linear control law, the lack of systematic procedure for the configuration of its parameters remains the main obstacle in practical applications. In FLC for BLDCM, the parameters of the FLC cannot be auto-tuning and not be suitable for difference conditions.

Recently, the design of FLC has also been tackled with genetic algorithm (GA). These are optimization algorithm performing a stochastic search by iteratively processing "populations" of solutions according to fitness. In control applications, the fitness is usually related to performance measures as integral error, setting time, etc. GA based FLC have been used in induction motor control system design successful, but the application in BLDCM servo system is few. The GA based FLC has been applied to the control system of BLDCM by using digital signal processor (DSP) TMS320LF2407A and controller improves the performance and the robustness of the BLDCM servo system.

10.3.3.1 BLDCM Servo System

Figure 10.9 shows the block diagram of the configuration of fuzzy model control system for BLDCM. The inner loop of Fig. 10.9 limits the ultimate current and ensures the stability of the servo system. The outer loop is designed to improve the static and dynamic characteristics of the BLDCM servo system. In this section, a fuzzy control is used to make the outer loop more stable. To make the fuzzy controller more robust, this section presents the genetic algorithm to optimize the fuzzy rules, and auto-tuning the coefficient of the controller. Figure 10.10 is the control configuration of the BLDCM servo system. The TMS320LF2407A DSP is used to generate the PWM and an IR2130 is used to drive the MOSFET. The A/D Unit is

10.3 Electrical Engineering

used to sample the current of the motor. The position signal of the rotor is gained by the Capture Unit of the DSP, and the speed value is calculated from the position information.

Fuzzy Control

Fuzzy logic provides an approximate effective mean of describing the behavior of some complex system. Unlike traditional logic type, fuzzy logic aims to model the imprecise modes of human reasoning and decision making, which are essential to our ability to make rational decisions in situations of uncertainty and imprecision. Figure 10.9 shows the block diagram of speed control system using a fuzzy logic controller.

The most significant variables entering the fuzzy logic speed controller have been selected as the speed error and its time derivative. The output this controller is U. The two input variables e (speed error) and e_c (change in error) are calculated at each sampling time as

$$e(k) = n^*(k) - n(k)$$
$$ec(k) = e(k) - e(k-1) \quad (10.5)$$

where $n^*(k)$ is the reference speed that time, and $n(k)$ is the actual rotor speed at that sampling.

The FLC consists of three stages: fuzzy, rule execution and de-fuzzy operations.

Fuzzy Operation

In this stage, the crisp variables are converted into fuzzy variables as

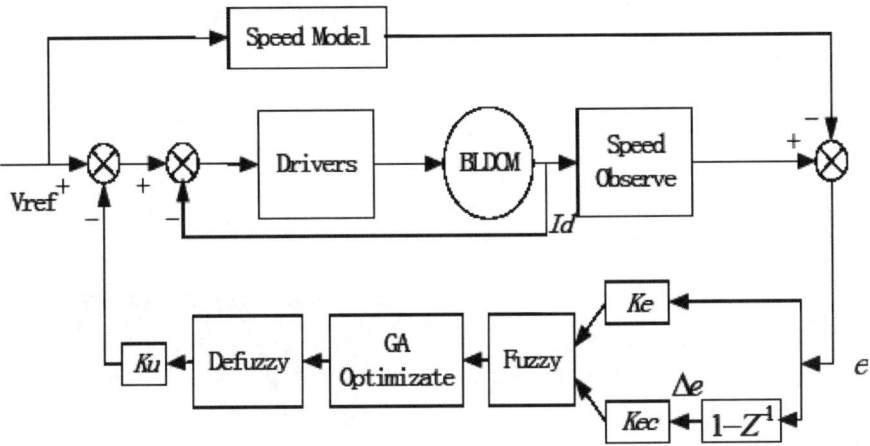

Fig. 10.9 Configuration of fuzzy model control system for BLDCM

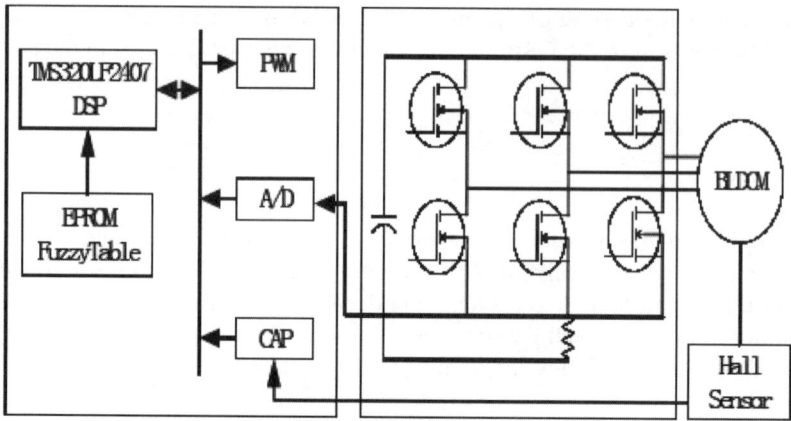

Fig. 10.10 Control configuration of the BLDCM

$$E = K_e e$$
$$EC = K_{ec} ec$$
$$U = u/K_u \qquad (10.6)$$

In (10.6), *Ke* and *Kec* are the proportion coefficients. They transform the inputs to universe of fuzzy sets. And use *Ku* to transform the output of the fuzzy control to actual control value. These transformations are closely according to the prescribed membership functions associate with the control variables, the membership functions have been chosen with triangular shapes as shown in Fig. 10.11.

The universe of discourse of input variables and *ec* and output U are divided from −6 to +6. Each universe of discourse is divided into seven fuzzy sets: NB, NM, NS, Z, PS, PM and PB. Each fuzzy variable is a member of the subsets with a degree of between 0 (non member) and 1 (full member) as

$$\mu_A(x) = \begin{cases} 1 & if \ \mu_A \in A \\ 0 & if \ \mu_A \notin A \end{cases} \qquad (10.7)$$

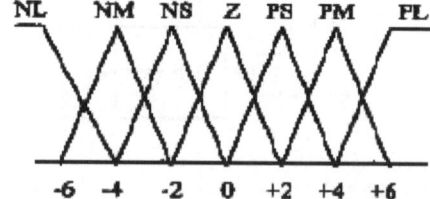

Fig. 10.11 The shape of membership function for fuzzy logic

10.3 Electrical Engineering

Rule Execution

The fuzzy rules are actually experience rules based on expertise or operators' long-time experiences. Table 10.1 shows the fuzzy rules. The variables are processed by an inference engine executes 49 rules (7*7). Each rule is expressed in the form as,

If e is NB and *ec* is PM then U is PM
If e is PM and *ec* is NB then U is PS and so on.

Table 10.1 The fuzzy linguistic rule table

e	e_c						
	NB	NM	NS	ZE	PS	PM	PB
NB	PB	PB	PM	PM	PS	PS	NM
NM	PB	PB	PM	PS	PS	ZE	NM
NS	PB	PM	PS	PS	ZE	NS	NB
ZE	PB	PS	PS	ZE	NS	NS	NB
PS	PB	PS	ZE	NS	NS	NM	NB
PM	PM	ZE	NS	NS	NS	NM	NB
PB	PM	NS	NS	NM	NM	NB	NB

De-fuzzy Operation

In this stage, a crisp value of the output variable U is obtained by using the de-fuzzy method, in which the centroid of each output membership function for each rule is first evaluated. The final output is then calculated as the average of the individual Centroid.

10.3.3.2 Genetic Algorithms

GA is a stochastic optimization algorithm is originally motivated by the mechanisms of natural selection and evolutionary genetics. The GA serves as a computing mechanism to solve the constrained optimization problem resulting from the motor control design where the genetic structure encodes some sort of automation. The basic element processed by a GA is a string formed by concatenating sub-strings, each of which is a binary coding (if binary GA was adopted) of a parameter. Each string represents a point in the search space. The Selection, Crossover and Mutation are the main operations of GA. Selection direct the search of Gas toward the best individual. In the process, strings with high fitness receive multiple copies in the

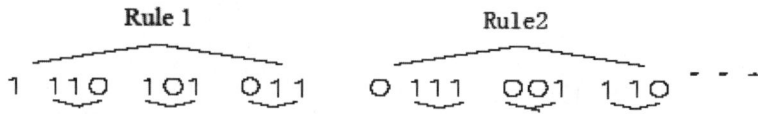

Fig. 10.12 Coding method for GA

next generation while strings with low fitness receive fewer copies or even none at all.

Crossover can cause to exchange the property of any two chromosomes via random decision in the mating pool and provide a mechanism to product and match the desirable qualities through the crossover.

Although selection and crossover provide the most of the power skills, but the area of the solution will be limited. Mutation is a random alternation of a bit in the string assists in keeping delivery in the population.

The optimization step of GA is follow:

A. Code the parameter
B. The initialization of the population
C. Evaluate the fitness of each member
D. Selection
E. Crossover
F. Mutation
G. Go to step B until find the optimum solution.

GA based Fuzzy Controller

Since the fuzzy inference is time-consuming, and the DSP used in motor control is speed-limited, so real-time inference method cannot be chosen. Here by using the synthetic fuzzy inference algorithm, the computer makes a query table off-line in advance and stores it in the memory of DSP. In a practical control, the control value can be obtained according to the query table, and tuning the *Ke, Kec* and *KU* on-line.

The design of the fuzzy controller is base on the genetic algorithm. Figure 10.12 shows the coding formulation when using GA to optimize the fuzzy controller. Here using 10 bits binary code to denote one fuzzy inference rule. The first binary code is the flag whether the rule is used. The 2~4, 5~7 and 8~10 refer to the error, change in error and the output variable. And001,010,011,100,101,110 and 111 refer to NB, NM, NS, ZE, PS, PM and PB respectively.

For example, the first rule of Fig. 10.12 shows that if e is PB and ec is NB then

Table 10.2 The parameters of genetic algorithm

Crossover Possibility	0.85
Mutation Possibility	0.002
Generation Numbers	30

U is PM, and the first bit binary code "0" indicate that this rule will be eliminated through optimization. In order to improve the speed of the optimization, this section chooses 30 candidates as the initialization population, and these candidates are proved to be able to make the motor run steadily. Table 10.2 shows the parameter of GA used in this section, and Table 10.3 shows the fuzzy rules intimidated through GA method. Through the optimization, 6 rules are eliminated and 4 rules are optimized.

On-Line Tuning

In order to improve the dynamic performance of the BLDCM servo system, the elements of the query table need to be adjusted according to the input variables. To do this, this section adjusts the coefficients (K_e, K_{ec} and K_U) to tuning the control system on-line. The basic principle is the "rough adjustment" and "accurate adjustment", namely, constantly adjusting the coefficients according to actual e, e_c. If the e and e_c are large, K_e and K_{ec} should be reduced while K_U should be increased because the main objective is diminishing the errors. When e and ec are small, because the main aim is to diminish the overshoot and steady-state error, K_e and K_{ec} should be increased to increase the resolution of e and e_c while K_U should be reduced to obtain small control value to reduce the overshoot and steady-state error. The adjust function as follow

$$K_e = \begin{cases} K_{e0} + K_1 \times e, & |e| \leq \frac{e_{max}}{2} \\ K_{e0} + K_1 \times \frac{e_{max}}{2}, & |e| > \frac{e_{max}}{2} \end{cases}$$

$$K_{ec} = \begin{cases} K_{ec0} + K_1 \times e, & |e| \leq \frac{e_{max}}{2} \\ K_{ec0} + K_1 \times \frac{e_{max}}{2}, & |e| > \frac{e_{max}}{2} \end{cases}$$

$$K_u = \begin{cases} K_{u0} + K_1 \times e, & |e| \leq \frac{e_{max}}{2} \\ K_{u0} + K_1 \times \frac{e_{max}}{2}, & |e| > \frac{e_{max}}{2} \end{cases} \quad (10.8)$$

In order to verify the validity of the proposed controller, conventional fuzzy controller is compared with GA based fuzzy controller. In the case of changing motor, all the system parameters are varied. Thus, GA fuzzy controller will be adaptable to uncertain control parameters.

A simulation program is designed to compare the stable and dynamic performances. Figure 10.13 shows the speed curve when the motor speed is 2100r/m for GA based fuzzy controller.

Figure 10.13 shows that GA based fuzzy controller has less overshoot and more stable performance.

This section simulates the situation when the load is change, Fig. 10.14 shows the simulation result when using GA based fuzzy controller. GA based fuzzy controller when the load change shows that fuzzy controller has good dynamic performance.

Table 10.3 The linguistic rule table after optimization

e	e_c						
	NB	NM	NS	ZE	PS	PM	PB
NB	PB	PB	PM		PS	PS	NM
NM		PM	PS	PS	PS	ZE	NM
NS	PB	PM	PS	PS	ZE	NS	NB
ZE	PB	PS	PS	ZE	NS	NS	NB
PS		PS	ZE	NS	NS	NM	
PM	PM	ZE	ZE	NS	NS	NB	NB
PB	PM		NS		NM	NB	NB

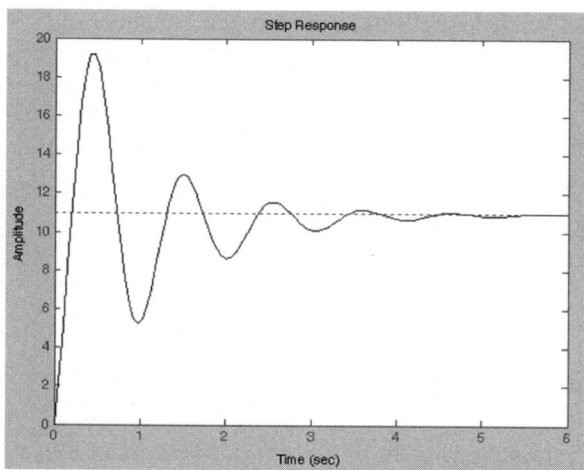

Fig. 10.13 Rotate speed simulation curve when adopting fuzzy controller based on GA

This section uses the GA based fuzzy controller as the speed controller of the BLDCM servo system. This method is more robust and can improve dynamic performance of the system. The off-line adjust optimize the fuzzy rules, and the on-line tuning of the parameters of the fuzzy controller make the controller has good dynamic and robust performance. Table 10.4 gives the specifications of BLDCM.

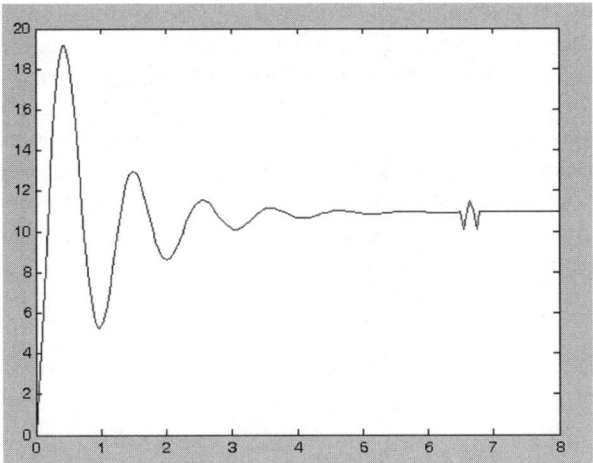

Fig. 10.14 Rotate speed simulation result using Fuzzy controller based on GA when the load

10.4 Machine Learning

10.4.1 Feature Selection in Machine learning using GA

In recent years there has been a significant increase in research on automatic image recognition in more realistic contexts involving noise, changing lighting conditions, and shifting viewpoints. The corresponding increase in difficulty in designing effective classification procedures for the important components of these more complex recognition problems has led to an interest in machine techniques as a possible strategy for automatically producing classification rules. This section describes part of a larger effort to apply machine learning techniques to such problems in an attempt to generate and improve the classification rules required for various recognition tasks.

The immediate problem attacked is that of texture recognition in the context of noise and changing lighting conditions. In this context standard rule induction systems like AQ15 produce sets of classification rules which are sub-optimal in two respects. First, there is a need to minimize the number of features actually used for classification, since each feature used adds to the design and manufacturing costs as well as the running time of a recognition system. At the same time there is a need

Table 10.4 Parameters of motor

Type	70BL1030-12
Voltage	12VDC
Power	120W
Rotate Speed	3000rpm

to achieve high recognition rates in the presence of noise and changing environmental conditions. This section describes an approach being explored to improve the usefulness of machine learning techniques for such problems. The approach described here involves the use of genetic algorithms as a "front end" to traditional rule induction systems in order to identify and select the best subset of features to be used by the rule induction system. The results presented suggest that genetic algorithms are a useful tool for solving difficult feature selection problems in which both the size of the feature set and the performance of the underlying system are important design considerations.

10.4.1.1 Feature Selection

Since each feature used as part of a classification procedure can increase the cost and running time of a recognition system, there is strong motivation within the image processing community to design and implement systems with small feature sets. At the same time there is a potentially opposing need to include a sufficient set of features to achieve high recognition rates under difficult conditions. This has led to the development of a variety of techniques within the image processing community for finding an "optimal" subset of features from a larger set of possible features. These feature selection strategies fall into two main categories.

The first approach selects features independent of their effect on classification performance. The difficulty here is in identifying an appropriate set of transformations so that the smaller set of features preserves most of the information provided by the original data and are more reliable because of the removal of redundant and noisy features. The second approach directly selects a subset "d" of the available "m" features in such a way as to not significantly degrading the performance of the classifier system. The main issue for this approach is how to account for dependencies between features when ordering them initially and selecting an effective subset in a later step. The machine learning community has only attacked the problem of "optimal" feature selection indirectly in that the traditional biases for simple classification rules (trees) leads to efficient induction procedures for producing individual rules (trees) containing only a few features to be evaluated. However, each rule (tree) can and frequently does use a different set of features, resulting in much larger cumulative features sets than those typically acceptable for image classification problems. This problem is magnified by the tendency of traditional machine learning algorithms to overfit the training data, particularly in the context of noisy data, resulting in the need for a variety of ad hoc truncating (pruning) procedures for simplifying the induced rules (trees).

The conclusion of these observations is that there is a significant opportunity for improving the usefulness of traditional machine learning techniques for automatically

10.4 Machine Learning

10.4.1.2 Feature Selection Using GAs

Genetic algorithms (GAs) are best known for their ability to efficiently search large spaces about which little is known a priori. Since genetic algorithms are relatively insensitive to noise, they seem to be an excellent choice for the basis of a more robust feature selection strategy for improving the performance of our texture classification system.

10.4.1.3 Genetic Algorithms

Genetic algorithms (GAs), a form of inductive learning strategy, are adaptive search techniques which have demonstrated substantial improvement over a variety of random and local search methods. This is accomplished by their ability to exploit accumulating information about an initially unknown search space in order to bias subsequent search into promising subspaces. Since GAs are basically a domain independent search technique, they are ideal for applications where domain knowledge and theory is difficult or impossible to provide.

The main issues in applying GAs to any problem are selecting an appropriate representation and an adequate evaluation function. In the feature selection problem the main interest is in representing the space of all possible subsets of the given feature set. Then, the simplest form of representation is binary representation where, each feature in the candidate feature set is considered as a binary gene and each individual consists of fixed-length binary string representing some subset of the given feature set. An individual of length l corresponds to a l-dimensional binary feature vector X, where each bit represents the elimination or inclusion of the associated feature. Then, $x_i = 0$ represents elimination and $x_i = 1$ indicates inclusion of the ith feature.

10.4.1.4 Evaluation function

Choosing an appropriate evaluation function is an essential step for successful application of GAs to any problem domain. The process of evaluation is similar to the regular process. The only variation was to implement a more performance-oriented fitness function that is better suited for genetic algorithms. In order to use genetic algorithms as the search procedure, it is necessary to define a fitness function which properly assesses the decision rules generated by the AQ algorithm. Each testing example is classified using the AQ generated rules as described before. If this is the appropriate classification, then the testing example has been recognized correctly. After all the testing examples have been classified, the overall fitness function will be evaluated by adding the weighted sum of the match score of all of the correct recognitions and subtracting the weighted sum of the match score of all of the incorrect recognitions, i.e.

Fig. 10.15 The improvement in feature set fitness over time

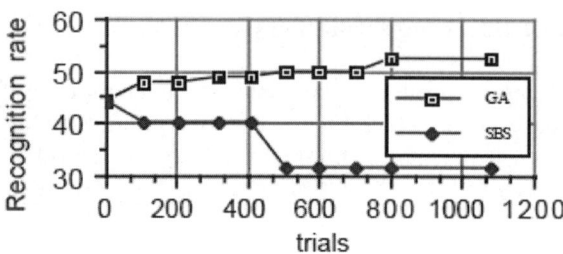

$$F = \sum_{i=1}^{n} S_i * W_i - \sum_{j=n+1}^{m} S_j * W_j \qquad (10.9)$$

The range of the value of **F** is dependent on the number of testing events and their weights. In order to normalize and scale the fitness function **F** to a value acceptable for GAs, the following operations were performed:

$$\text{Fitness} = 100 - [(F/TW)*100] \qquad (10.10)$$

where:

$$TW = \text{total weighted testing examples} = \sum_{i=1}^{m} W_i$$

As indicated in the above equations, after the value of **F** was normalized to the range [-100, 100], the subtraction ensures that the final evaluation is always positive (the most convenient form of fitness for GAs), with lower values representing better classification performance.

10.4.1.5 Performance Evaluation

In performing the evaluations reported here, the same AQ15 system was used for rule induction. In addition, GENESIS, a general purpose genetic algorithm program, was used as the search procedure (replacing Sequential Backward Selection (SBS)).

In the GA-based approach presented here, equal recognition weights (i.e., W=1) were assigned to all the classes in order to perform a fair comparison between the two presented approaches. The evaluations were performed on the texture images. The results are summarized in Figs. 10.15 and 10.16 and provide encouraging support for the presented GA approach.

Figure 10.15 shows the steady improvement in the fitness of the feature subsets being evaluated as a function of the number of trails of the genetic algorithm. This indicates very clearly that the performance of rule induction systems (as measured by recognition rates) can be improved in these domains by appropriate feature subset selection. Figure 10.16 shows that the number of features in the best feature set decreased for both approaches. However, the feature subset found

10.5 Civil Engineering

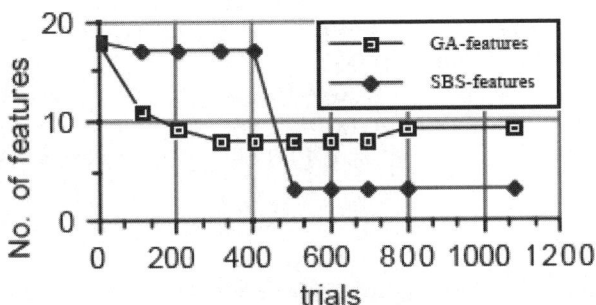

Fig. 10.16 The number of features used by the best individual

by statistical measures was substantially smaller than that found by the GA-based system. Figure 10.15 indicates that this was achieved at the cost of poorer performance. The advantage of the GA approach is to simultaneously improve both figures of merit.

10.5 Civil Engineering

10.5.1 Genetic Algorithm as Automatic Structural Design Tool

The simple GA while powerful, is perhaps too general to be efficient and robust for structural design problems. First, function (or, fitness) evaluations are computationally expensive since they typically involve finite element analysis. Second, the (feasible) design space is at times disjointed with multiple local minima. Third, the design space can be a function of boolean, discrete and continuous design variables. The use of GA to find the optimal solution(s) of engineering design problems is still an open research area. Experience with GA has indicated that more often than not, tuning the GA strategy and parameters can lead to more efficient solution process for a class of problems. Researchers have proposed modifications, such as parameters-pace size adjustment and adaptive mutation for continuous problems, which focus on refining the searching space adaptively, niching genetic algorithms that emphasizing on repeating the fitter individuals and special modification for construction time-cost optimization problems. Research has also made it possible to combine genetic algorithms and gradient-based techniques for handling constraints for aerodynamic shape optimization problems. In this section, let's discuss how GA has been used as an automatic structural design tool.

10.5.1.1 Formulation of the Design Problem

The design of three-dimensional frames can be stated as follows.

10 Applications of Genetic Algorithms

$$
\begin{aligned}
&\text{Find} && \mathbf{x} = \lfloor {}^b x_1, \ldots, {}^b x_{nb}, {}^1 x_1, \ldots, {}^1 x_{nd}, {}^2 x_1, \ldots, {}^3 x_{ni} \rfloor \\
&\text{to minimize} && f(\mathbf{x}) \\
&\text{subject to} && g_i(\mathbf{x}) \le 0 && i = 1, \ldots, ni \\
& && h_j(\mathbf{x}) = 0 && j = 1, \ldots, ne \\
& && {}^b x_p \in \{0, 1\} && p = 1, \ldots, nb \\
& && {}^1 x_q \in \{x_q^1, x_q^2, \ldots, x_q^{nq}\} && q = 1, \ldots, nd \\
& && {}^2 x_r^l \le {}' x_r \le {}^2 x_r^U && r = 1, \ldots, ns
\end{aligned}
\qquad (10.11)
$$

where $\mathbf{x}$ is the design variable vector, $f(x)$ is the objective function, ni is the number of inequality constraints, ne is the number of inequality constraints, nb is the number of Boolean design variables, nd is the number of discrete design variables selected from a list of nq values, and ns is the number of continuous design variables. All structural design problems do not lend themselves to a simultaneous consideration of all of the above-mentioned constraints and design variables. Design problems are usually categorized as sizing, shape or topology design problems or combinations thereof.

This section deals with the solution to the above-mentioned problem.

10.5.1.2 Genetic Algorithm as a Design Automation Tool

GAs were developed to solve unconstrained optimization problems. However, engineering design problems are usually constrained. They are solved by transforming the problem to an unconstrained problem. The transformation is not unique and one possibility is to use the following strategy.

$$
\begin{aligned}
&\text{find}: && \mathbf{x} = \lfloor {}^b x_1, \ldots, {}^b x_{NBDV}; {}^i x_1, \ldots, {}^i x_{NIDV}; {}^s x_1, \ldots, {}^s x_{NSDV} \rfloor \\
&\text{minimize}: && f(\mathbf{x}) + \sum_i c_i \cdot \max(0, g_i) + \sum_j c_j |h_j|
\end{aligned}
\qquad (10.12)
$$

where c_i and c_j are penalty parameters used with inequality and equality constraints. Determining the appropriate penalty weights c_i and c_j is always problematic. We propose an algorithm here where the penalty weight is computed automatically and adjusted in an adaptive manner. First the objective function is modified as

$$
f(\mathbf{x}) + c_a \left(\sum_i \max(0, g_i) + \sum_j |h_j| \right) \qquad (10.13)
$$

The following rules are used to select c_a.

(1) If there are feasible designs in the current generation, c_a is set as the minimum f among all feasible designs in the current generation. The rationale is that for the design with minor violations and smaller objective value, the probability of survival is not eliminated. If, on the other hand, the maximum f among all feasible

10.5 Civil Engineering

designs is used, infeasible designs will have a smaller probability to survive even if the constraint violations are small.

(2) If there is no feasible design, c_a is set as the f that has the least constraint violation. The motivation idea has the effect of both pushing the design into feasible domain as well as preserving the design with the smallest fitness.

In this case, the one-point crossover is preferred for continuous domains, and the uniform crossover for discrete domains. However, schema representation still plays a pivotal role in the efficiency of the GA. If one uses a one-point crossover then it is obvious that the ordering of the design variables is an important issue.

The selection schemes (for generating the mating pool) together with the penalty function dictate the probability of survival of each string. While it is very important to preserve the diversity in each generation, researchers have also found that sometimes it may be profitable to bias certain schema. However, results from most of the selection rules, like roulette wheel; depend heavily on the mapping of fitness function. Here the tournament selection is used. There are at least two reasons for this choice. First, tournament selection increases the probability of survival of better strings. Second, only the relative fitness values are relevant when comparing two strings. In other words, the selection depends on individual fitness rather than ratio of fitness values.

It is found that, during the evolutionary process, the same chromosomes at times are repeatedly generated. Since the fitness evaluation in structural design involves finite element analysis, a computationally expensive step, all generated chromosome and the associated fitness information are saved in memory. In this way, if a chromosome is repeated, a finite element analysis is not necessary. Saved chromosomes may also be helpful for further processing.

The initial population should contain uniformly distributed alleles. By this, it is meant that no chromosome pattern should be missed. Each chromosome is represented by n bits with each bit being either 1 or 0. If the distribution of 1's in each bit location is to be uniform, the initial population size should be at least n. During the evolution, it is expected that that the chromosome converges to some special pattern with the (0–1) choice decided for n locations.

Assume that the choice of each bit is independent of all the other bits. Since the population size is n in each generation, after every generation from the statistical viewpoint we can expect to learn about at least one bit. Ideally then after n generations, one can expect to learn about all the n bits forming the chromosome. However, since each bit is not independent of the others, more than n generations are perhaps necessary to obtain a good solution. This suggests that the population size and the number of generations should be *at least* n.

10.5.1.3 The Improved GA Optimizer

As mentioned before selective improvement can be made to obtain a more robust solution methodology for a class of problems. The primary focus in this research is to make the GA a powerful and reliable optimizer for structural optimization. Table 10.5 shows the proposed improvements.

10.5.1.4 Sizing, Shape and Topology Optimization of Space Frames

The structural optimization problem involving sizing, topology and shape parameters has always been a difficult problem to handle. Some of the design variables are discrete, the design space is disjoint and traditional gradient-based methods cannot be employed. The design problem of a three-dimensional frame can be stated as shown in (10.11).

Researchers working in this area have divided the existing algorithms for discrete variables into three types—branch and bound, approximation, and ad-hoc methods. The solution techniques such as approximation methods, branch and bound methods, and ad hoc strategies of adapting continuous design variables in NLP techniques suffer from several drawbacks. These methods either are inefficient, or do not really converge to the optimal solution or can be used under very restrictive conditions. For example, the approximation method allows the candidate solution to be discrete, but still require the whole design domain to be differentiable and continuous.

In the case of topology optimization, approximation methods and branch and bound techniques cannot be applied since the methods cannot handle the presence or absence of members as design variables. The design problem can be solved more easily using GAs since they can be adapted to work with discrete and boolean design variables.

10.5.1.5 Design Variable Linking

As shown in Table (10.6), GAs essentially can handle three types of design variables—discrete or integer, real, and boolean. These design variables capture all the possible structural design parameters. The sizing design variables considered may be either cross-sectional dimensions or available cross-section. The former can be described using continuous design variables since these dimensions can vary continuously. The latter is described in terms of integers (an integer index that points to a row

Table 10.5 Differences between traditional and proposed GA

	Traditional GA	Proposed GA
Penalty Function	ad hoc	Automatic
Schema	ad hoc	Ordered
Cross-over Probability	ad hoc	Adaptive
Population/Max Generation Size	ad hoc	Suggested as $2n$

10.5 Civil Engineering

in a table of available cross-sections). The table search is carried out by using a table of ordered available cross-sections with the lower and upper bound candidate cross-sections specified by the user. The shape design variables are the nodal locations. These are real design variables. The topology (boolean) design variables can be structural parameters such as the presence or absence of members, and presence or absence of fixity conditions at supports or connections.

10.5.1.6 Special Considerations

When topology design is considered, several problems should be handled very carefully.

(i) There may be elements not connected to the structure during design, if topology design is performed. This can be detected by examining the singularity of the stiffness matrix.
(ii) There may be "null" nodes during the design. A null node is one to which no element is attached. Such nodes need to be suppressed (from the finite element analysis) in order to find the response of the remaining structure.
(iii) Sometimes, crisscrossing members are not allowed in frame structures. This situation is detected by testing the possible intersection of a member with all other members. It should be noted that handling such a constraint by traditional (gradient-based) optimization approach can be very challenging.

Thus the above discussed GA based concept can be applied to Roof frame design, Ten story frames and so on.

Table 10.6 Linking of design variables and the physical meaning

Optimization	Physical Meaning	Design Variable Type in GA	Note
Topology	Element Existence	Boolean	
Sizing	Cross-sectional selection	Integer	Search through a given table
Shape	Nodal Coordinates	Real	Varies between upper and lower bounds

10.5.2 Genetic Algorithm for Solving Site Layout Problem

Construction site layout involves coordinating the use of limited site space to accommodate temporary facilities (such as fabrication shops, trailers, materials, or equipment) so that they can function efficiently on site. The layout problem is generally defined as the problem of

(1) identifying the shape and size of the facilities to be laid out;
(2) identifying constraints between facilities; and
(3) determining the relative positions of these facilities that satisfy the constraints between them and allow them to function efficiently.

There are different classes of layout problems. The variations stem from the assumptions made on the shape and size of facilities and on the constraints between them. Facilities may have a defined shape and size or a loose shape, in which case they will assume the shape of the site to which they have been assigned (f or example, bulk construction material). The constraints can vary from simple non-overlap constraints to other geometric constraints that describe orientation or distance constraints between facilities. In the layout problem addressed here, the shape and size of facilities are fixed. Facilities can have 2D geometric constraints on their relative positions along with proximity weights describing the level of interaction or flow between them.

The layout problem is an NP-complete combinatorial optimization problem, that is, optimal solutions can be computed only for small or greatly restricted problems. Hence, layout planners often resort to using heuristics to reduce their search for acceptable solutions.

The application of genetic algorithms to solving layout problems is relatively recent. GAs work with a family of solutions, known as the "current population," from which we obtain the "next generation" of solutions. When the algorithm is designed properly, we obtain progressively better solutions from one generation to the next. The main advantage of using GAs is in the fact that it only needs an objective function with no specific knowledge about the problem space. The challenge, however, remains in finding an appropriate problem representation that result in an efficient and successful implementation of the algorithm

10.5.2.1 Constrained Site Layout Problem

The layout problem as modeled in this section is characterized by rectangular layout objects with fixed dimensions representing the facilities to be positioned on site. Facilities can be positioned in one of two orientations only: a 0 or 90° orientation. In addition, facilities can have 2D constraints on their relative positions: namely, minimum and maximum distance, orientation, and nonoverlap constraints. Minimum and maximum distance constraints limit the distance between the facing sides of two facilities in the x- or y-direction to be greater than or less than a predefined value, respectively. Distance constraints can be used to model equipment reach or general clearance requirements. Orientation constraints limit a facility's position to be to the north, south, east, or west of another reference facility. These constraints can be used to locate access roads or gates with respect to the main facility. Nonoverlap constraints are default constraints that restrict the positions of any two facilities from overlapping. The geometric constraints are considered hard constraints that should be satisfied for the layout to be feasible.

The objective is to find a feasible arrangement for all layout objects within the site space that minimizes the sum of the weighted distances separating the layout objects (Z):

$$Z = \Sigma \Sigma_{j<i}(w_{ij} \times d_{ij}) \tag{10.14}$$

where w_{ij} is the affinity weight between objects i and j that could be used to represent the flow or the unit transportation cost between i and j, and d_{ij} is the rectilinear distance separating objects i and j. A feasible arrangement is obtained by finding positions for all layout objects that satisfy the 2D constraints between them.

10.5.2.2 Genetic Algorithm approach

The basic notion of evolutionary computation is to mimic some principles of natural evolution in order to solve optimization problems of high complexity. A group of randomly initialized points of the search space (*individuals*) is used to search the problem space. Each individual encodes all necessary problem parameters (*genes*) as bit strings, vectors, or graphs. The iterative process of selection and combination of "good" individuals should yield even better ones, until a solution is found or a certain stop criterion is met.

A population is a collection of chromosomes where each chromosome represents a layout solution. Every chromosome is coded as a vector whose length is equal to the number of facilities that exist on site n. Each facility i is represented by the coordinates of its position on site: Xi, Yi, by its dimensions: Li, Wi; and by a series of pointers pointing at the facilities that surround it in the four directions: north, south, east, and west. These pointers will be used to facilitate the check for overlap. The fitness function used is Z.

A number of genetic operators are used to evolve an initial population to the optimal solution. The flowchart for the proposed GA approach is as follows:

Following the generation of the initial population, the genetic operators are applied to evolve the initial population into better ones as depicted in the flow chart of Fig. 10.17. Thus the proposed GA approach can be applied to the following cases and tested: equal size with equal weight objects, unequal size with unequal weight objects, unequal size with unequal weight objects, and 2D constraints between objects.

10.6 Image Processing

10.6.1 Designing Texture Filters with Genetic Algorithms

Several techniques have been employed for texture based segmentation. Most of them derive categories of texture descriptors and then, during a training phase, cluster these descriptors to achieve discrimination. Traditional methods of texture feature extraction are based either on statistical or structural models. In the statistical model texture is defined by a characteristic set of relationships between image

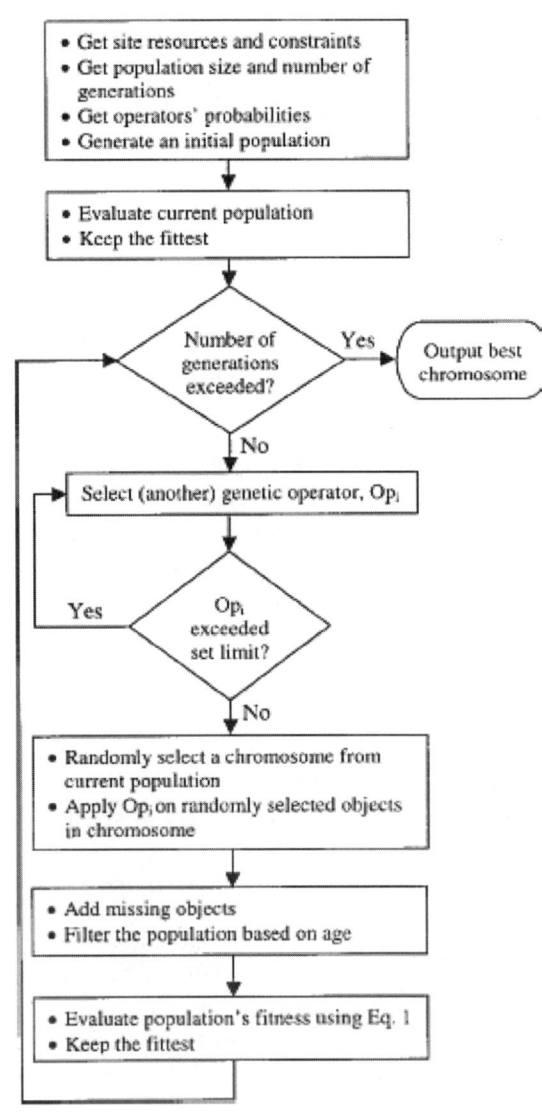

Fig. 10.17 Flow chart of proposed genetic algorithm

10.6 Image Processing

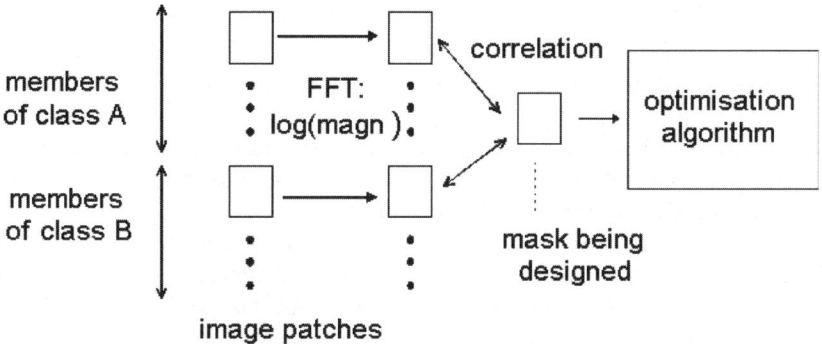

Fig. 10.18 Correlation process

elements, and for most practical purposes these are determined from tonal values. We will be using one or more of these as comparators, namely

- Grey level spatial dependency (GLSD) matrices, or co-occurrence matrices and the simplified approach from user based on sum and difference histograms.
- Texture energy in the spatial domain derived by convolution as described by Laws
- Methods based on the use of fractals

10.6.1.1 Texture Discrimination Using Genetic Algorithms

In this section, it is to design a mask which, when correlated with the Fourier spectrum of each of the given patterns, will produce a response such that the inter-class difference will be maximized and the intra-class differences will be minimized. Now let's use GA to solve the optimization over all possible masks, by minimizing (or maximizing) an objective function based on the correlation.

Correlation Based Optimization

The steps of the algorithm (Fig. 10.18) can be summarised as:

1. Rectangular patches are selected from a given image as members (training templates) representing each class of texture to be detected.
2. A Fourier Transform (FT) is performed on each of the patches and the resulting spectra are logarithmically transformed to reduce the range of the values appearing in the patch spectra, simplifying the encoding into binary needed by the GAs.
3. An objective function involving the responses of the correlation of the mask with the results of step 2 is evaluated over the set of all possible masks.
4. An optimisation algorithm (in this case a Genetic Algorithm) is applied to the objective function over the set of all possible masks.

Choices for Correlation

The application of the Fourier transform to a texture image leads to a choice of whether to examine the magnitude or phase components of the transform. Although a mathematical background to the spectral properties of texture has suggested the use of transform magnitude, phase is an important feature of signals and in computation involving 2D transforms is often overlooked.

To discover which spectral quantity carries most textural information for our medical images, we applied a forward 2D FFT on a 128×128 pixel central section of the image and formed a reconstruction with an inverse FFT after encoding the magnitude at different resolutions (from 2 to 8 bits), whilst keeping the phase component unchanged and vice-versa. Table 10.7 shows the Mean Average Error (MAE) between the original and reconstructed image. Altering the encoding resolution for the magnitude results in slower rate of increase in MAE than for phase. MAE is therefore more sensitive to phase encoding resolution, justifying use of the magnitude of the spectrum for texture matching at a given resolution.

10.6.1.2 The Objective Function

The aim is to design a filter which when correlated with all the members of class A will produce a high response and when correlated with the members of class B will produce a low response. As a starting point the objective function should incorporate the terms:

$$F_1 = \frac{(n_a + n_b)}{2} \frac{1}{n_a} \sum_i c_i^a \qquad (10.15)$$

where $0 < i < n_a, n_b$ and n_a and n_b are the number of members in class A and class B respectively. c_i^a denotes the correlation coefficient of (10.15), between the designed mask and the logarithm of the spectrum of the ith member of class A,

Table 10.7 Relative importance of magnitude and phase

Encoding Bits	Mean Average Error (MAE)	
	Magnitude	Phase
2	18.65	156.66
3	12.96	62.44
4	8.00	31.78
5	4.83	15.96
6	2.53	7.98
7	1.36	3.97
8	0.68	1.97

10.6 Image Processing

arranged such that $0 \leq c_i^a \leq 1$. This function maximises the responses with all members of class A. Similarly:

$$F_2 = \frac{(n_a + n_b)}{2} \frac{1}{n_b} \sum_i (1 - c_i^b) \tag{10.16}$$

minimizes the responses with all members of class B, and:

$$F_3 = \frac{\sum_{i=1}^{n_a} \sum_{j>i} |c_i^a - c_j^a|}{n_a(n_a - 1)} + \frac{\sum_{i=1}^{n_b} \sum_{j>i} |c_i^b - c_j^b|}{n_b(n_b - 1)} \tag{10.17}$$

will minimize all the intra-class distances, achieving uniformity of the response of the correlation within each of the classes. Since discrimination consists of three components: *acceptance*, *rejection* and *uniformity of response*, the objective function adopted is the sum of the three terms. The weighting factor $(n_a + n_b)/2$ is applied to the first two terms so that their contribution balances that of the final term.

10.6.1.3 Filter Realisation Using Genetic Algorithms

For a GA to be used the problem has to be encoded for genetic search, i.e. the parameters have to be mapped to a finite length symbol string, using an appropriate conversion alphabet.

The Chromosome

The main new element in the design of the chromosome is that it is two-dimensional. This gives it physical significance and makes the implementation of the genetic operators more meaningful. Being the output of a 2D FFT, the chromosome is a two dimensional matrix with n rows and n columns. Since Fourier spectra normally exhibit a considerable dynamic range in amplitude between basic and higher harmonics the most convenient way to proceed is via a logarithmic transform, quantized to an minimal number of bits. Each gene, representing logarithmic spectral magnitude is encoded in binary.

A significant factor affecting the GA's performance is the number of harmonics used to evaluate the correlation coefficient. Too few result in non-robust filters whilst too many produce over-sensitive filters with a sharp response to a specific image feature rather than mean regional texture. Eight harmonics were used in this work.

The Genetic Operators

Selection is performed in the usual way enhanced by pre-computing for each generation the number of offspring each chromosome is allowed to have according to its fitness. The nature of the problem and the design of the chromosome encouraged the implementation of crossover by exchanging rectangular segments of chromosome, with the breakpoints aligned on gene boundaries. Each bit of each segment is copied perfectly otherwise a mutation occurs which acts as a logical NOT on the value of the bit.

Choice of GA Parameters

The GA population was limited to 100 and 100 generations of evolution allowed. Crossover probability was typically 0.8 and 2 breakpoints allowed in each dimension. Bit mutation was 0.5%. The quantities correlated were the logarithm of the real part of the spectral magnitude, encoded to 8 bits. Experimental variation of the genetic parameters usually failed to alter the final convergence, although varying the crossover strategy from 1-breakpoint to n-breakpoints resulted in a halving of the number of generations needed to achieve a given convergence.

Implementation of the System

- The starting point is an image with regions classified into two different classes. One or more training patches (32 32 pixels) are selected as members of each class.
- The first generation of the GAs is initialized with random values. The design of a discriminating filter now proceeds by GA optimization
- This filter can be used either to find additional regions of these classes in the same or different image.

Post-Processing the Results of Texture-Based Segmentation

The result of the application of the GA-designed filter to an image is likely to be an incompletely segmented image, where regions containing texture which belongs to class A have high values and regions similar to members of class B have low values. To improve the segmentation and apply the method to more than two textures, a maximum likelihood decision rule that minimizes the probability of false classification was used and a 5×5 median filter applied to the segmented image.

Derivation of an Enclosing Contour

The candidate boundary that has been obtained so far may not accurately enclose the region of contrasting texture because of the filter's inherent spatial resolution, but it is likely that it will follow its boundary and have a similar shape. It can be further refined if there is strong edge information to be exploited.

Output

The output of the GA-based texture classification is given here. Figure 10.19 shows the result of the GA based texture filter, using 32×32 pixel training segments.

A system for texture discrimination, based on the spectral frequency properties is described and results produced using images containing standard textures has been dealt in this section. The system exploits well-established Fourier spectral properties.

10.6.2 Genetic Algorithm Based Knowledge Acquisition on Image Processing

Easy and immediate acquisition of large numbers of digital color images, for example, of the daily growth of plants in remote fields, has been made possible via the Internet nowadays. From such images, we can expect that detailed information concerning the shape, growth rate and leaf colors of plants will be obtained. Vast quantities of image data, however, increase the time spent extracting such information from the data. This is because the extraction procedure needs human aid—empirical knowledge of image processing and the features of target objects. Thus, image analysis, segmenting images of objects and deriving their outlines or areas, commonly invokes procedures based not only on routine, but also on trial and error performed by hand.

Automated image processing systems, such as expert systems, have been studied in various areas of engineering. In this section, certain procedures are discussed for selecting filtering algorithms and for adjusting their parameters to segment target components in images. Genetic algorithms (GAs) are suitable for this purpose because the algorithms involve optimization and automation by trial and error. For instance, researchers have applied GAs for obtaining optimal image processing transformations mapping the original image into the target. From the viewpoint of segmenting images of plants, we present application software based on GAs, not only for segmenting images, but also for acquiring knowledge on the operations.

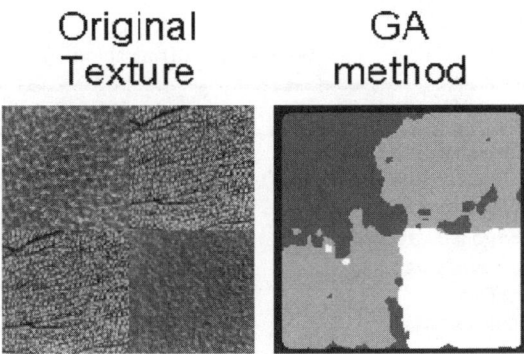

Fig. 10.19 GA Texture Analysis

10.6.2.1 Image Segmentation Strategy

Many kinds of efficient filtering algorithms for image segmentation, such as noise elimination and edge enhancement, have been contrived. Implementing all of them into our algorithms, however, is unrealistic because the increase in operations invokes a proportional increase in processing time. Based on our empirical knowledge of the segmentation of plant images, several filtering algorithms commonly used are selected and implemented in the developed algorithm as shown in Table 10.8. The thresholding and reversion algorithms are performed on a focused pixel of the image processed in serial order, and others have spatial mask operators. Figure 10.20 shows the common procedures to segment targets in color images. The procedures are explained as follows:

1) Color of component areas in the images is averaged using smoothing (SM).
2) Target components are enhanced using thresholding on hue (TH) and, simultaneously, the image is entirely converted to a monochrome image.
3) Differentiation (EE) is used when target features outline components.
4) Binarization (TB) is performed for the entire monochrome image.
5) Reversion (RV) on binarized pixels is occasionally effective to enhance the components.
6) Fusion operations, expansion (EF) and contraction (CF), allow a reduction in noise, and occasionally, is performed repeatedly.

After these procedures are carried out, the image processed has been converted to a binarized image with target components defined. In the algorithm, we have adopted not the RGB color model, but the HSI model, because the latter is efficient for the segmentation of plants in fields. All operations are performed after each pixel value is converted from RGB to HSI. The smoothing algorithm is a median operator with a 3*3 mask of pixels, and it is applied only for the hue of the pixels. The thresholding has two different operators; one operates upon the hue and another upon the brightness of pixels. These operations substitute null for all bits of a pixel when the pixel value occurs within a range defined by minimum and maximum values. When the value is out of the range, they substitute unity for all bits of the pixel. For edge enhancement of components in images, a Sobel operator with a

Table 10.8 Filtering algorithms used as phenotypes

Manipulation	Algorithms	Symbols
Thresholding (hue)	Point processing	TH
Thresholding (brightness)	Point processing	TB
Smoothing	Median operator	SM
Edge enhancement	Sobel operator	EE
Contraction	4-neighbor fusion	CF
Expansion	4-neighbor fusion	EF
Reversion	Point processing	RV

10.6 Image Processing

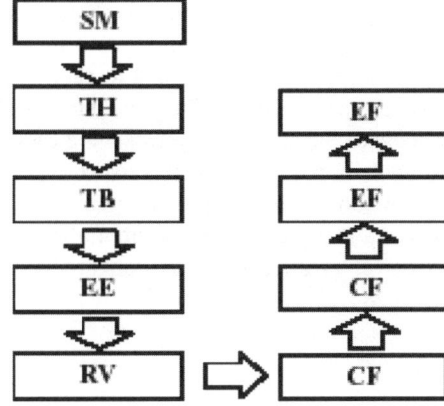

Fig. 10.20 Flow chart of a typical procedure for enhancing targets components

Fig. 10.21 Locus of genes defined on a chromosome

0 - 5	6 - 9	10 - 15	16 - 19	20 - 22	23 - 25
TH (minimum)	TH (range)	TB (minimum)	TB (range)	Toggle of TH	Toggle of SM

26 - 28	29 - 31	32 - 34	35 - 37	38 - 40	41 - 43
Toggle of CF	Toggle of CF	Toggle of EF	Toggle of EF	Toggle of EE	Toggle of RV

3*3 mask of pixels is used. The operator applied to the brightness value substitutes null for the saturation of pixels to convert the images to monochrome ones. Fusion operators search the four neighboring pixels. The contraction replaces a given pixel with a black one if the neighbors contain more than one black pixel. The expansion, on the other hand, replaces the given pixel with a white one if the neighbors contain more than one white pixel. Before the genetic operations are performed, an objective image, compared with processed images for fitness evaluation, must be provided as a binary drawing image. Target components in the image are represented with white pixels and the remainders are with black ones.

10.6.2.2 Genetic Algorithms

Chromosomes of the current GA consist of 44 binary strings, assigned to 12 genotypes as shown in Fig. 10.21. Phenotypes corresponding to the genotypes consist of on off states of the operations mentioned above and parameters for the operations concerning thresholding levels. The minimum thresholding levels on the hue and the brightness coordinates, ranging from 0.0 to 1.0, are encoded with 6 bits. Range of their minimum thresholding levels to the maximum ones is encoded with 4 bits in the same manner. Genotypes of the on-off states are encoded with 3 bits; a decimal value from 0 to 3 is defined as an "off" state of the operation and a value of more than 4 as a state of "on". Such redundant encoding allows sharp changes, caused by one bit reversion, to be avoided.

Figure 10.22 shows the flow diagram to search for appropriate procedures of the segmentation based on GAs. Conventional genetic operations called "simple GA" are used; crossover at the same points of two neighbor chromosomes, random mutation, and ranking depending on fitness evaluation and selection. At the beginning of the GA operation, chromosomes of a certain population size are generated, initialized with random strings. The crossover occurs at the points of certain string length determined at random, and then, each chromosome is mutated with a certain probability per a string. The each chromosome is interpreted as a sequence of filtering operations and their parameters. Subsequently, a clone of the original image is processed using the each sequence. After the fitness between the objective image and the processed ones is evaluated, the chromosomes are ranked and selected dependent on the degree of the fitness. The procedure from the crossover to the ranking is performed iteratively until appropriate procedures are obtained.

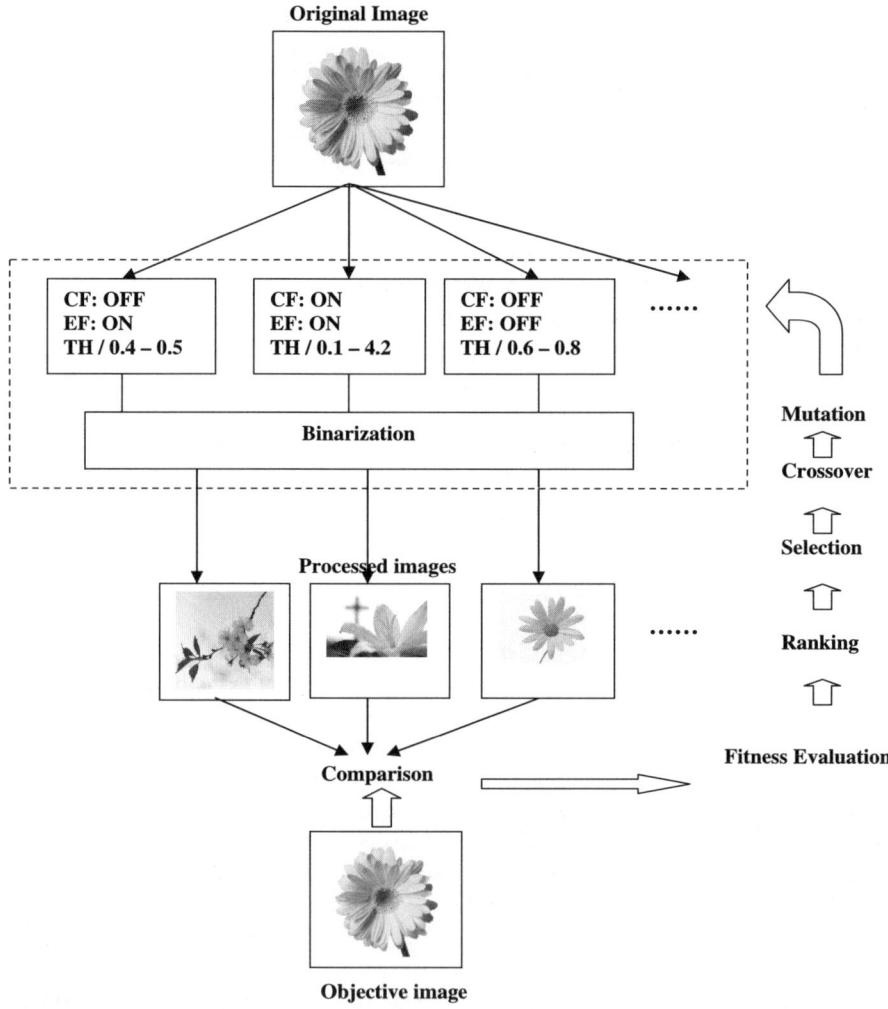

Fig. 10.22 Schematic of the knowledge acquisition system, combining GAs and operations for image

10.6 Image Processing

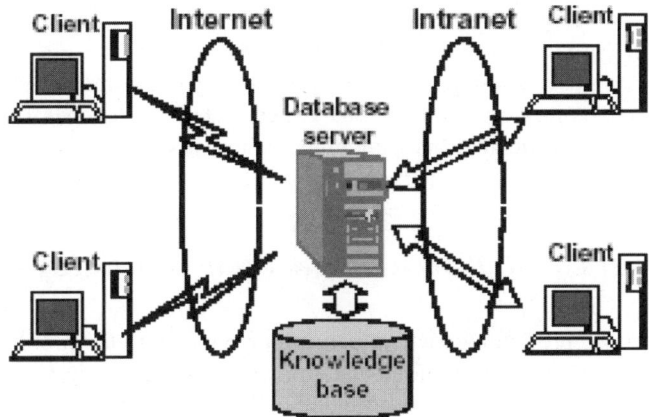

Fig. 10.23 Schematics of the knowledge base for image processing based on three-tiered database

Evaluation and selection play important roles in GAs because they determine the GA's performance of searching for solutions. The function for fitness evaluation is defined by the rate of correspondence between a processed image and an objective one, given as a successfully processed one. In detail, equivalence of each pixel, located in the same coordinate of both images, is verified as shown in the following formula:

$$f = P_{fit}/(P_{fit} + P_{unfit}) \qquad (10.18)$$

where P_{fit} is the number of pixels equivalent between images and P_{unfit} is the number in disagreement. After chromosomes are ranked according to their fitness, chromosomes of the population size, with high fitness in the rank, are selected as parents of the next generation.

10.6.2.3 Database Configuration

The knowledge base consists of three-tiered database architecture as shown in Fig. 10.23. The database can be developed using any database server and a middleware is required for controlling connection from client PCs to the database via the Internet. Since the connection is managed using a user ID and a password embedded in the software, the database is secured from illegal accesses. A table of the database consists of URL of images, description on features of targets segmented, and procedures obtained for acceptable segmentation.

In conventional image processing systems based on knowledge base, all the data for image processing is gotten ready beforehand. On the other hand, our knowledge base has been made without data at the beginning, and is increasing its knowledge as the software is used for processing various images. Figure 10.24 illustrates combination in processing by GA search, represented as the block arrows, with that by the knowledge base, represented as the normal ones.

Thus, this proposed approach can be applied to any color images and this architecture has an advantage over conventional expert system approaches, implementing all the knowledge for image processing in the system ahead.

10.6.3 Object Localization in Images Using Genetic Algorithm

In this section, we present a genetic algorithm application to the problem of object registration (i.e., object detection, localization and recognition) in a class of medical images containing various types of blood cells. The genetic algorithm approach taken here is seen to be most appropriate for this type of image, due to the characteristics of the objects.

One of the most frequently arising problems in the processing of (still) images is that of object registration. It arises in images containing objects, possibly overlapping, against a more-or-less uniform background. Objects may belong to one or more types or classes. Class identifying differences typically refer to the object morphology or shape, dimensions, color, opaqueness, surface texture and location / direction characteristics. The aims of digital processing of an object image are numerous: Object detection, localization, recognition and classification constitute major goals. Furthermore, more detailed object characterization in terms of size, color, direction, scaling, shift or rotation might be of interest for specific applications. Finally, search of an image for the existence or not of a specific object prototype (under a given degree of flexibility as to the similarity level required in the match) is often of importance. Common in all the problems mentioned above is the processing of the images digitally, through an appropriate software package, either general-purpose or custom developed for the application at hand.

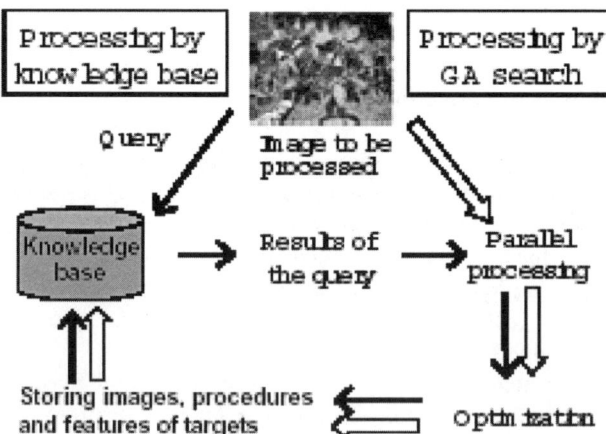

Fig. 10.24 Flow diagram of image processing by two different algorithms

10.6 Image Processing

Digital image processing is a mature field that offers to the researcher a variety of approaches. Given a field application, however, choice of the most suitable method or approach has not yet been fully automated. In this section, we present an application of the genetic algorithms approach to the problem of localization of objects in medical images of blood cells, taken via a microscope. The problem arises invariably in all blood or serum analysis medical contexts, and as such it has early received an intense research interest. Although there certainly exist automated solutions, the issue of quality along with the critical nature of the results, often necessitate manual / visual treatment by the human expert on a microscope.

The genetic algorithms approach is proposed here, because, as it will become clear through the results obtained, it was seen to be well suited to the morphology of the objects in the images treated.

A genetic algorithm is a non-linear optimization method that seeks the optimal solution of a problem via a non-exhaustive search among randomly generated solutions. Randomness is controlled through a set of parameters, thus turning genetic algorithms into exceptionally flexible and robust alternatives to conventional optimization methods. Genetic algorithms suffer a few disadvantages: they are not suitable for real time applications and take long to converge to the optimal solution. Convergence time cannot be predicted either. Nevertheless, they have become a strong optimization tool, while current research focuses on their combination with fuzzy logic and neural network techniques.

Genetic algorithms imitate natural evolutionary procedures for the production of successive generations of a population. In its simplest form, a genetic algorithm consists of three (3) mechanisms:

(i) *parent selection*
(ii) *genetic operation* for the production of descendants (offspring), and
(iii) *replacement of parents* by their descendants.

Parent selection process follows one of the selection processes of roulette, classification, constant situation, proportional forms or elitist choice. The *genetic operations* of (i) crossover and (ii) mutation combine parents to produce offspring of improved characteristics (getting higher grade by the evaluation function). *Parent replacement* strategies include (i) generational replacement and (ii) steady state reproduction.

10.6.4 Problem Description

Blood cell microscope images, such as the sample shown in Fig. 10.25, show cells of two different classes (possibly overlapping) against a uniform background. Class A is represented by bigger and usually more deformed cells whereas class B is represented by cells looking generally more normal and more uniform in shape and size. Cell color or grayscale can also be exploited; yet it is unreliable by itself, due to the various cell coloring techniques usually applied on the sample before it is placed in the microscope. In the present context, we will not go into the medical interpretation of the image, i.e. the diagnosis of certain pathologies connected to the presence or count or percentage of class A or class B cells, as this does not affect the technical problem addressed—although it renders the obtained results critical. Referring to Fig. 10.25, this section aims to address the following problems:

1) Detection of class A cells,
2) Percentage of the class A cells surface in the image, and
3) Registration of class A cells (coordinates and size).

Although this could be considered as an image segmentation problem, it is claimed that the genetic algorithms approach taken here is far more efficient in terms of processing time, while it yields high correct recognition scores.

10.6.5 Image Preprocessing

The histogram of the grayscale scale image is employed in order to obtain a grayscale threshold value Th, below which fall class A cells only. The sample histogram is shown in Fig. 10.26 (a), exhibits three major areas of grayscales, corresponding—from darker to lighter scale—to: (i) class A cell pixels, (ii) class B cell pixels and (iii) background pixels. Threshold value Th is set to the local minimum of the histogram curve, lying between the first two peaks mentioned above. The image is threshold by Th, thus producing a binary (black and the first two problems (detection of class A cells and calculation of their % area in the image) are straightforward if we use the binary image.

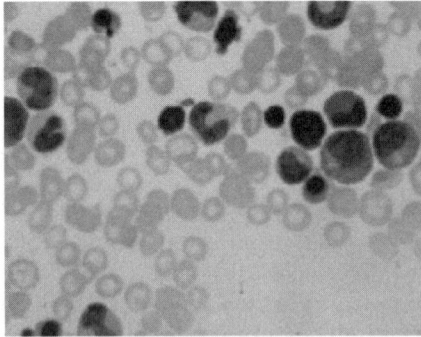

Fig. 10.25 Sample blood cell microscope image showing two classes of cells in a uniform background

10.6 Image Processing

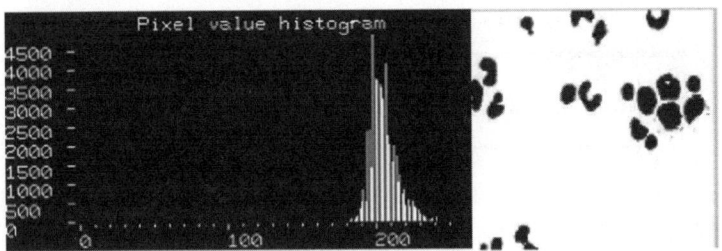

Fig. 10.26 (a) Histogram of the grayscale image in Fig. (10.25), (b) Binary version of Fig. (10.25) with threshold $Th = 110$

10.6.6 The Proposed Genetic Algorithm Approach

The genetic algorithm is repeatedly applied to the image as many times as the number of class A objects (bigger than a threshold area of TB pixels) it contains. Of course, an appropriate stopping rule is necessary, because the number of class A objects is originally unknown. Within each of the above repetitions, the genetic algorithm generates a succession of T generations, each consisting of N chromosomes. Each chromosome contains three (3) genes, namely, the 2-D plane coordinates of the center of an object (circle) and the radius of it. The first generation is generated randomly, whereas every next one is based on the following choices:

(i) Chromosomes are binary encoded, with 9, 10 and 4 bits for the 1st, 2nd and 3rd gene, respectively.
(ii) Parent pairs are selected by the roulette rule.
(iii) The genetic operations include 3-point crossover for the 1st and 2nd gene and 1-point crossover for the 3rd gene, with crossover probability Pc and arithmetic (bit) mutation, uniform across genes, with mutation probability Pm.
(iv) Generalized replacement is employed, combined with an elite strategy using a number of Pe elite chromosomes directly copied to the next generation.
(v) No schema theory is employed.

Once a new generation is produced, its N binary chromosomes are decoded and evaluated by the fitness function. This function assigns a numerical "grade" to each chromosome, which is used for the parent selection and genetic operations of the next generation. When the T-th generation is reached, iteration stops and the chromosome of the T-th generation with the highest grade is considered as a solution (localized circular object).

Repetition stops when the area of the image designated by such a solution is found to contain less than 40% of class A pixels—meaning that essentially there remain no more significant class A objects.

Critical for the success of the genetic algorithm is the choice of the evaluation (fitness) function. Indeed this is the only means of communication between the genetic evolutionary process and its environment (i.e., the problem it seeks to solve). When chromosomes of the current generation are graded by the fitness function, the

Table 10.9 Genetic algorithm parameters variation

Nr.	Parameter	Range
1	T	[10, 50, 90]
2	N	[10, 50, 70]
3	Pc	[10%, 50%, 80%]
4	Pm	[4%, 8%, 50%]
5	Pe	[1, 5, 20]

genetic algorithm gains feedback from the environment so as to adjust its evolution towards an improved next generation. For the problem at hand, we have employed the straight-forward option of a fitness function which counts the class A pixels contained in the area of the original image designated by the (center, radius) pair of a given chromosome. In that sense, chromosomes (circular objects) highly overlapping with class A objects in the image get a higher grade.

To implement this approach the parameters like Th, TB, T, N, Pm, Pc, Pe, etc should be initialized. These should be adjusted using prior information about the specific family of images, for optimal performance. These parameters can be varied as shown in Table 10.9.

A sample blood cell image with superimposed results is shown in Fig. 10.27. Circular objects localized by the genetic algorithm are marked with a white circle. This is a particularly successful experiment, as 20 out of 20 (100%) class A objects are localized. Major parameter choices are $N = 50$ chromosomes, $T = 50$ generations, $Pc = 80\%$, $Pm = 8\%$ and number of elite chromosomes $Pe = 5$.

However, not all parameter choices yield analogous results. Therefore the set of parameters can be varied according to Table 10.9, and the results can be examined for different images belonging to the same family. Thus, in this section we have dealt with a genetic algorithm approach to the problem of localization of objects belonging to a certain class, in blood cell microscope images.

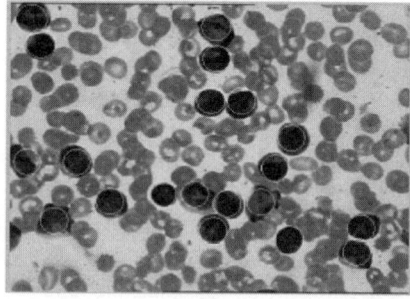

Fig. 10.27 Original image with superimposed genetic algorithm results marked with a white circle. 20 out of 20 (100%) of class A objects are localized correctly

10.7 Data Mining

10.7.1 A Genetic Algorithm for Feature Selection in Data-Mining

In this section, we look into discovering certain features and factors that are involved in large database. To exploit this data, data mining tools are required and a 2-phase approach using a specific genetic algorithm is employed.

This heuristic approach has been chosen as the number of features to consider is large. Consider a data which indicates for pairs of affected individuals of a same family their similarity at given points (locus) of their chromosomes. This is represented in a matrix where each locus is represented by a column and each pairs of individuals considered by a row. The objective is first to isolate the most relevant associations of features, and then to class individuals that have the considered similarities according to these associations.

For the first phase, the feature selection problem, we use a genetic algorithm (GA). To deal with this very specific problem, some advanced mechanisms have been introduced in the genetic algorithm such as sharing, random immigrant, dedicated genetic operators and a particular distance operator has been defined. Then, the second phase, a clustering based on the features selected during the previous phase, will use the clustering algorithm K-means, which is very popular in clustering.

10.7.1.1 GA for Feature Selection

The first phase of this algorithm deals with isolating the very few relevant features from the large set. This is not exactly the classical feature selection problem known in Data mining. Here, we have the idea that less than 5% of the features have to be selected. But this problem is close from the classical feature selection problem, and we will use a genetic algorithm as we saw they are well adapted for problems with a large number of features. Genetic algorithm considered here has different phases. It proceeds for a fixed number of generations. A chromosome, here, is a string of bits whose size corresponds to the number of features. A 0 or 1, at position i, indicates whether the feature i is selected (1) or not (0).

The Genetic Operators

These operators allow GAs to explore the search space. However, operators typically have destructive as well as constructive effects. They must be adapted to the problem.

We use a Subset Size-Oriented Common Feature Crossover Operator (SSOCF), which keeps useful informative blocks and produces offsprings which have the same distribution than the parents. Off- springs are kept, only if they fit better than the least good individual of the population. Features shared by the 2 parents are kept by offsprings and the non shared features are inherited by offsprings corresponding to the *ith* parent with the probability (*ni* - *nc/nu*) where *ni* is the number of selected features of the *ith* parent, *nc* is the number of commonly selected features across both mating partners and *nu* is the number of non-shared selected features (see Fig. 10.28).

The mutation is an operator which allows diversity. During the mutation stage, a chromosome has a probability *pmut* to mutate. If a chromosome is selected to mutate, we choose randomly a number *n* of bits to be flipped then *n* bits are chosen randomly and flipped.

A probabilistic binary tournament selection is taken. Tournament selection holds *n* tournaments to choose *n* individuals. Each tournament consists of sampling 2 elements of the population and choosing the best one with a probability $p \in [0.5, 1]$.

The Chromosomal Distance

Create a specific distance which is a kind of bit to bit distance where not a single bit i is considered but the whole window $(i-\sigma, i+\sigma)$ of the two individuals are compared. If one and only one individual has a selected feature in this window, the distance is increased by one.

The Fitness Function

The fitness function developed refers to the support notion, for an association, which, in data mining, denotes the number of times an association is met over the number of times at least one of the members of the association is met.

The function is composed of two parts. The first one favors for a small support a small number of selected features because biologists have in mind that associations will be composed of few features and if an association has a bad support, it is better to consider less features (to have opportunity to increase the support). The second part, the most important (multiplied by 2), favours for a large support a large number of features because if an association has a good support, it is generally composed of few features and then we must try to add other features in order to have a more complete association. What is expected is to favor good associations (in term of

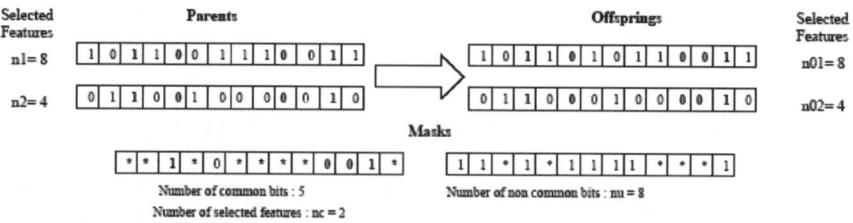

Fig. 10.28 The SSOCF crossover Operator

10.7 Data Mining

support) with as much as features as possible. This expression may be simplified, but we let it in this form in order to identify the two terms.

$$F = \left((1-S) \times \frac{\frac{T}{10} - 10 \times SF}{T}\right) + 2 \times \left(S \times \frac{\frac{T}{10} + 10 \times SF}{T}\right)$$

Where :
$$\begin{cases} \text{Support}: S = \frac{|A \cap B \cap C...|}{|A \cup B \cup C...|} \text{ where A, B, C ... are the selected features,} \\ T = \text{Total Number of features.} \\ SF = \text{Number of selected significant features (selected features that are not too close in term of the chromosomal distance).} \end{cases}$$
(10.19)

Sharing

To avoid premature convergence and to discover different good solutions (different relevant associations of features), we use a niching mechanism. Both crowding and sharing give good results and we choose to implement the fitness sharing. The objective is to boost the selection chance of individuals that lie in less crowded area of the search space. We use a niche count that measures of how crowded the neighborhood of a solution is. The distance D is the chromosomal distance adapted to our problem presented before. The fitness of individuals situating in high concentrated search space regions is degraded and a new fitness value is calculated and used, in place of the initial value of the fitness, for the selection.

The sharing fitness $f_{sh}(i)$ of an individual i, where n is the size of the population, $\alpha_{sh} = 1$ and $\sigma_{sh} = 3$), is:

$$f_{sh}(i) = \frac{F(i)}{\sum_{j=1}^{n} Sh(D(I_i, I_j))} \text{ where : } Sh(D(I_i, I_j))$$
$$= \begin{cases} 1 - \left(\frac{D(I_i, I_j)}{\sigma sh}\right)^{\alpha_{sh}} & \text{if } D(I_i, I_j) < \sigma_{sh} \\ 0 & \text{else} \end{cases}$$
(10.20)

Random Immigrant

Random Immigrant is a method that helps to maintain diversity in the population. It should also help to avoid premature convergence. Random immigrant is used as follows: if the best individual is the same during N generations, each individual of the population, whose fitness is under the mean, is replaced by a new randomly generated individual.

10.7.1.2 The Clustering Phase: Use of K-Means Algorithm

The k-means algorithm is an iterative procedure for clustering which requires an initial classification of the data. The k-means algorithm proceeds as follows: it computes the center of each cluster, then computes new partitions by assigning every object to the cluster whose center is the closest (in term of the Hamming distance) to that object. This cycle is repeated during a given number of iterations or until the assignment has not changed during one iteration. Since the number of features is now very small, we implement a classical k-means algorithm widely used in clustering, and to initialize the procedure we randomly select initial centers (Fig. 10.29).

Thus the approach proposed in the above section can be tested by employing it to any large databases based upon user's application.

10.7.2 Genetic Algorithm Based Fuzzy Data Mining to Intrusion Detection

The wide spread use of computer networks in today's society, especially the sudden surge in importance of e-commerce to the world economy, has made computer network security an international priority. Since it is not technically feasible to build a system with no vulnerabilities, intrusion detection has become an important area of research.

Intrusion detection approaches are commonly divided into two categories: misuse detection and anomaly detection. The misuse detection approach attempts to recognize attacks that follow intrusion patterns that have been recognized and reported by experts. Misuse detection systems are vulnerable to intruders who use new patterns of behavior or who mask their illegal behavior to deceive the detection system. Anomaly detection methods were developed to counter this problem. With the anomaly detection approach, one represents patterns of normal behavior, with the assumption that an intrusion can be identified based on some deviation from this normal behavior. When such a deviation is observed, an intrusion alarm is produced. Artificial intelligence (AI) techniques have been applied to both misuse detection and anomaly detection. Rule based expert systems have served as the basis for several systems including SRI's Intrusion Detection Expert System (IDES). These systems encode an expert's knowledge of known patterns of attack and system vulnerabilities as if-then rules. The acquisition of these rules is a tedious and error-prone process; this problem (known as the knowledge acquisition bottleneck

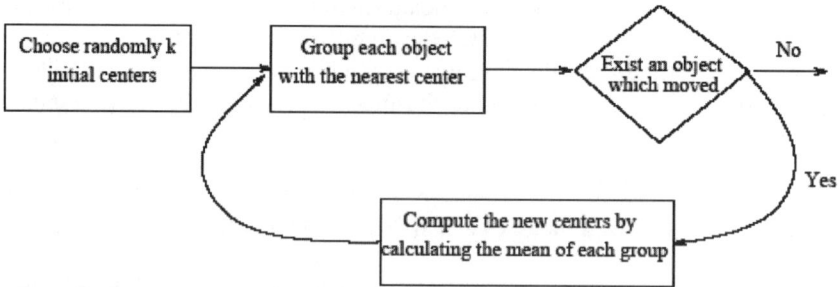

Fig. 10.29 The k-means algorithm: The states

10.7 Data Mining   371

in expert system literature) has generated a great deal of interest in the application of machine learning techniques to automate the process of learning the patterns. Examples include the Time-based Inductive Machine (TIM) for intrusion detection that learns sequential patterns and neural network-based intrusion detection systems. More recently, techniques from the data mining area (mining of association rules and frequency episodes) have been used to mine normal patterns from audit data.

Problems are encountered, however, if one derives rules that are directly dependent on audit data. An intrusion that deviates only slightly from a pattern derived from the audit data may not be detected or a small change in normal behavior may cause a false alarm. We have addressed this problem by integrating fuzzy logic with data mining methods for intrusion detection. Fuzzy logic is appropriate for the intrusion detection problem for two major reasons. First, many quantitative features are involved in intrusion detection. SRI's Nextgeneration Intrusion Detection Expert System (NIDES) categorizes security-related statistical measurements into four types: ordinal, categorical, binary categorical and linear categorical. Both ordinal and linear categorical measurements are quantitative features that can potentially be viewed as fuzzy variables. Two examples of ordinal measurements are the CPU usage time and the connection duration. An example of a linear categorical measurement is the number of different TCP/UDP services initiated by the same source host. The second motivation for using fuzzy logic to address the intrusion detection problem is that security itself includes fuzziness. Given a quantitative measurement, an interval can be used to denote a normal value. Then, any values falling outside the interval will be considered anomalous to the same degree regardless of their distance to the interval. The same applies to values inside the interval, i.e., all will be viewed as normal to the same degree. The use of fuzziness in representing these quantitative features helps to smooth the abrupt separation of normality and abnormality and provides a measure of the degree of normality or abnormality of a particular measure.

We describe a prototype intelligent intrusion detection system (IIDS) that is being developed to demonstrate the effectiveness of data mining techniques that utilize fuzzy logic. This system combines two distinct intrusion detection approaches: (1) anomaly based intrusion detection using fuzzy data mining techniques, and (2) misuse detection using traditional rule-based expert system techniques. The anomaly-based components look for deviations from stored patterns of normal behavior. The misuse detection components look for previously described patterns of behavior that are likely to indicate an intrusion. Both network traffic and system audit data are used as inputs. We are also using genetic algorithms to (1) tune the fuzzy membership functions to improve performance, and (2) select the set of features available from the audit data that provide the most information to the data mining component.

10.7.2.1 System Goals and Preliminary Architecture

Our long term goal is to design and build an intelligent intrusion detection system that is accurate (low false negative and false positive rates), flexible, not easily fooled by small variations in intrusion patterns, adaptive in new environments, modular with both misuse and anomaly detection components, distributed, and real-time. The architecture shown in Fig. 10.30 has been developed with these goals in mind.

The Machine Learning Component integrates fuzzy logic with association rules and frequency episodes to "learn" normal patterns of system behavior. This normal behavior is stored as sets of fuzzy association rules and fuzzy frequency episodes. The *Anomaly Intrusion Detection Module* extracts patterns for an observed audit trail and compares these new patterns with the "normal" patterns. If the similarity of the sets of patterns is below a specified threshold, the system alarms an intrusion. *Misuse Intrusion Detection Modules* use rules written in FuzzyCLIPS to match patterns of known attacks or patterns that are commonly associated with suspicious behavior to identify attacks. The use of fuzzy logic in both of these modules makes the rules of the system more flexible and less brittle. The machine learning component allows the system to adapt to new environments. The detection methods will be implemented as a set of intrusion detection modules. An intrusion detection module may address only one or even a dozen types of intrusions. Several intrusion detection modules may also cooperate to detect an intrusion in a loosely coupled way since these detection modules are relatively independent. Different modules may use different methods. For instance, one module can be implemented as a rule-based expert system and another module can be constructed as a neural network classi-

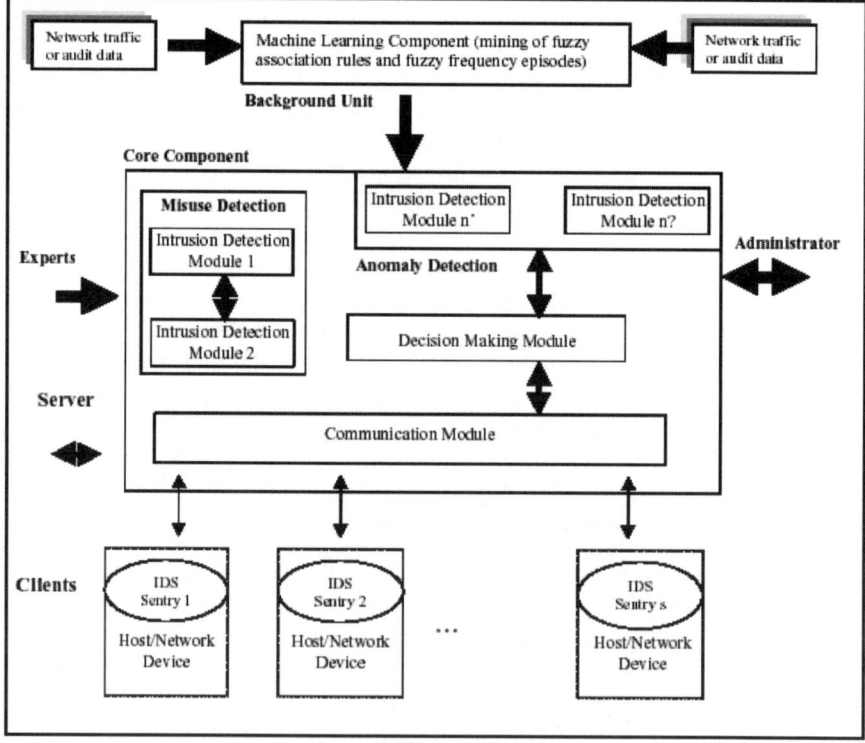

Fig. 10.30 Architecture of IIDS

10.7 Data Mining   373

fier. On the whole, this modular structure will ease future system expansion The *Decision-Making Module* will both decide whether or not to activate an intrusion detection module (misuse or anomaly) and integrate evaluation results provided by the intrusion detection modules. The Communication Module is the bridge between the intrusion detection sentries and the decision-making module. *Intrusion detection sentries* pre-process audit data and send results to the communication module. Feedback is returned to the sentries.

10.7.2.2 Anomaly Detection via Fuzzy Data Mining

We are combining techniques from fuzzy logic and data mining for our anomaly detection system. The advantage of using fuzzy logic is that it allows one to represent concepts that could be considered to be in more than one category (or from another point of view—it allows representation of overlapping categories). In standard set theory, each element is either completely a member of a category or not a member at all. In contrast, fuzzy set theory allows partial membership in sets or categories. The second technique, data mining, is used to automatically learn patterns from large quantities of data. The integration of fuzzy logic with data mining methods helps to create more abstract and flexible patterns for intrusion detection.

Fuzzy Logic

In the intrusion detection domain, we may want to reason about a quantity such as the number of different destination IP addresses in the last 2 seconds. Suppose one wants to write a rule such as

"**If** the number different destination addresses during the last 2 seconds was high **Then** an unusual situation exists".

Using traditional logic, one would need to decide which values for the number of destination addresses fall into the category high. As shown in Fig. 10.31 a, one would typically divide the range of possible values into discrete buckets, each representing a different set. The y-axis shows the degree of membership of each value in each set. The value 10, for example is a member of the set *low* to the degree 1 and a member of the other two sets, *medium* and *high*, to the degree 0. In fuzzy logic, a particular value can have a degree of membership between 0 and 1 and can be a member of more than one fuzzy set. In Fig. 10.31b, for example, the value 10 is a member of the set *low* to the degree 0.4 and a member of the set *medium* to the degree 0.75. In this example, the membership functions for the fuzzy sets are piecewise linear functions. Using fuzzy logic terminology, the number of destination ports is a fuzzy variable (also called a linguistic variable), while the possible values of the fuzzy variable are the fuzzy sets *low*, *medium*, and *high*. In general, fuzzy variables correspond to nouns and fuzzy sets correspond to adjectives.

In this section, FuzzyCLIPS are used to represent patterns using a rule-based system. FuzzyCLIPS, developed by the National Research Council of Canada, is a fuzzy extension of the popular CLIPS expert system shell developed by NASA.

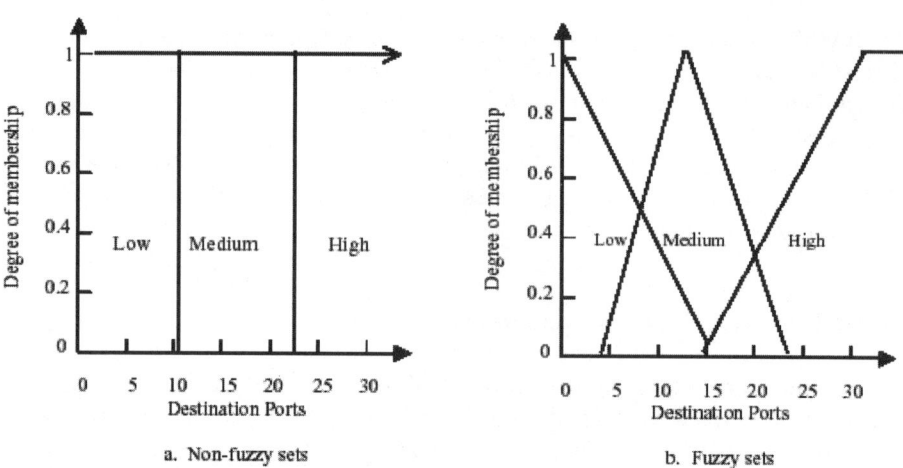

Fig. 10.31 Non-fuzzy and fuzzy representations of sets for quantitative variables. The x-axis is the value of a quantitative variable. The y-axis is the degree of membership in the sets *low*, *medium*, and *high*

FuzzyCLIPS provides several methods for defining fuzzy sets; we are using the three standard S, PI, and Z functions described by Zadeh. The graphical shapes and formal definitions of these functions are shown in Fig. 10.32. Each function is defined by exactly two parameters.

Using fuzzy logic, a rule like the one shown above could be written as **If** the DP = high **Then** an unusual situation exists where DP is a fuzzy variable and high is a fuzzy set. The degree of membership of the number of destination ports in the fuzzy set high determines whether or not the rule is activated.

Data Mining Methods

Data mining methods are used to automatically discover new patterns from a large amount of data. Two data mining methods, association rules and frequency episodes, have been used to mine audit data to find normal patterns for anomaly intrusion detection.

Association Rules

Association rules were first developed to find correlations in transactions using retail data. For example, if a customer who buys a soft drink (A) usually also buys potato chips (B), then potato chips are associated with soft drinks using the rule $A \rightarrow B$. Suppose that 25% of all customers buy both soft drinks and potato chips and that 50% of the customers who buy soft drinks also buy potato chips. Then the degree of support for the rule is $s = 0.25$ and the degree of confidence in the rule is $c = 0.50$. Agrawal and Srikant developed the fast Apriori algorithm for mining association rules. The Apriori algorithm requires two thresholds of *minconfidence* (representing

10.7 Data Mining

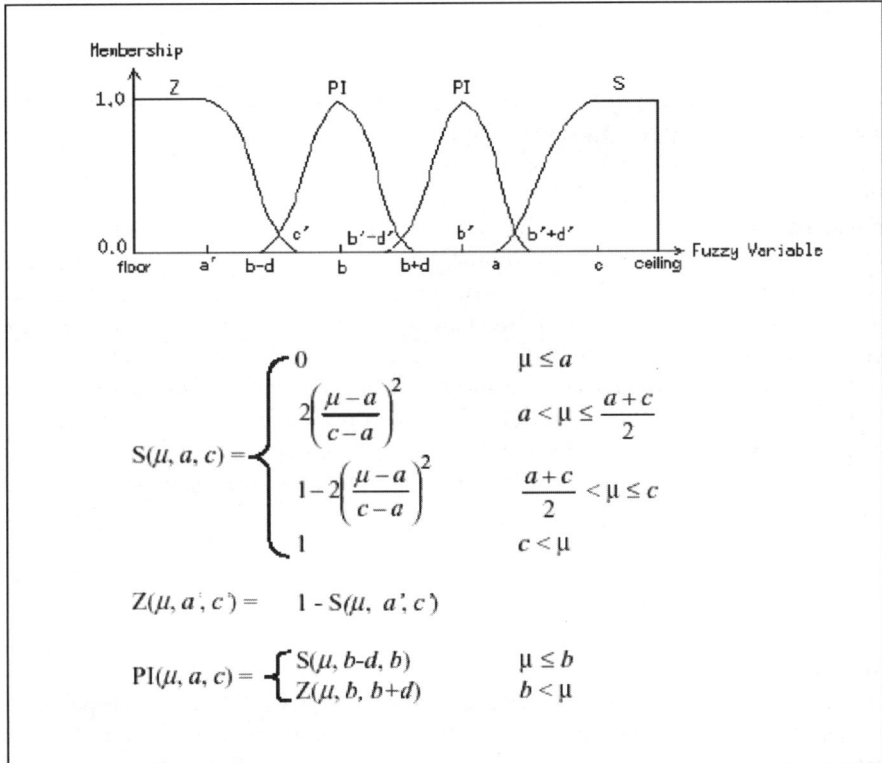

Fig. 10.32 Standard function representation of fuzzy sets

minimum confidence) and *minsupport* (representing minimum support). These two thresholds determine the degree of association that must hold before the rule will be mined.

Fuzzy Association Rules

In order to use the Apriori algorithm of Agrawal and Srikant for mining association rules, one must partition quantitative variables into discrete categories. This gives rise to the "sharp boundary problem" in which a very small change in value causes an abrupt change in category. Kuok, Fu, and Wong developed the concept of fuzzy association rules to address this problem. Their method allows a value to contribute to the support of more than one fuzzy set. We have modified the algorithm of Kuok, Fu, and Wong, by introducing a normalization factor to ensure that every transaction is counted only one time. An example of a fuzzy association rule mined by our system from one set of audit data is:

$$\{\text{SN} = LOW, \text{FN} = LOW\} \rightarrow \{\text{RN} = LOW\}, \qquad c = 0.924, s = 0.49 \tag{10.21}$$

where SN is the number of SYN flags, FN is the number of FIN flags and RN is the number of RST flags in a 2 second period.

When presented with a set of audit data, our system will mine a set of fuzzy association rules from the data. These rules will be considered a high level description of patterns of behavior found in the data. For anomaly detection, we mine a set of rules from a data set with no intrusions (termed a reference data set) and use this as a description of normal behavior. When considering a new set of audit data, a set of association rules is mined from the new data and the similarity of this new rule set and the reference set is computed. If the similarity is low, then the new data will cause an alarm. Figure. 10.33 shows results from one experiment comparing the similarities with the reference set of rules mined from data without intrusions and with intrusions. It is apparent that the set of rules mined from data with no intrusions (baseline) is more similar to the reference rule set than the sets of rules mined from data containing intrusions.

Misuse Detection Components

The misuse detection components are small rule-based expert systems that look for known patterns of intrusive behavior. The FuzzyCLIPS system allows us to implement both fuzzy and non-fuzzy rules. A simple example of a rule from the misuse detection component is given below: IF the number of consecutive logins by a user is greater than 3 THEN the behavior is suspicious Information from a number of misuse detection components will be combined by the decision component to determine if an alarm should be result.

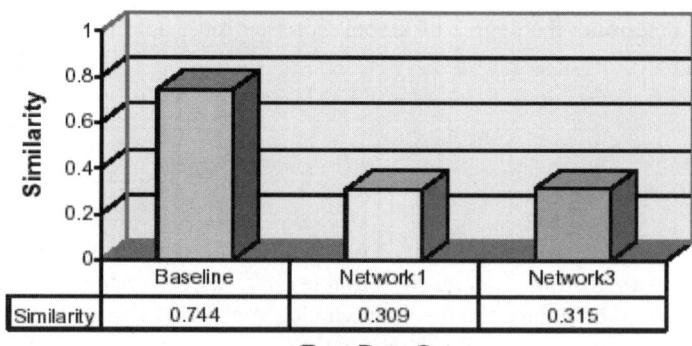

Fig. 10.33 Comparison of Similarities Between Training Data Set and Different Test Data Sets for Fuzzy Association Rules (minconfidence=0.6; minsupport=0.1 Training Data Set: reference (representing normal behavior) Test Data Sets: baseline (representing normal behavior), network1 (including simulated IP spoofing intrusions), and network3 (including simulated port scanning intrusions)

10.7 Data Mining

Genetic Algorithms

Genetic algorithms are search procedures often used for optimization problems. When using fuzzy logic, it is often difficult for an expert to provide "good" definitions for the membership functions for the fuzzy variables. We have found that genetic algorithms can be successfully used to tune the membership functions of the fuzzy sets used by our intrusion detection system. Each fuzzy membership function can be defined using two parameters as shown in Fig. 10.34. Each chromosome for the GA consists of a sequence of these parameters (two per membership function). An initial population of chromosomes is generated randomly where each chromosome represents a possible solution to the problem (an set of parameters). The goal is to increase the similarity of rules mined from data without intrusions and the reference rule set while decreasing the similarity of rules mined from intrusion data and the reference rule set. A fitness function is defined for the GA which rewards a high similarity of normal data and reference data while penalizing a high similarity of intrusion data and reference data. The genetic algorithm works by slowly "evolving" a population of chromosomes that represent better and better solutions to the problem.

Fitness Percentage

Figure 10.34 shows how the value of the fitness function changes as the GA progresses. The top line represents the fitness (or quality of solution) of the best individual in the population. We always retain the best individual from one generation to the next, so the fitness value of the best individual in the population never decreases. The middle line, showing the average fitness of the population, demonstrates that the overall fitness of the population continues to increase until it reaches

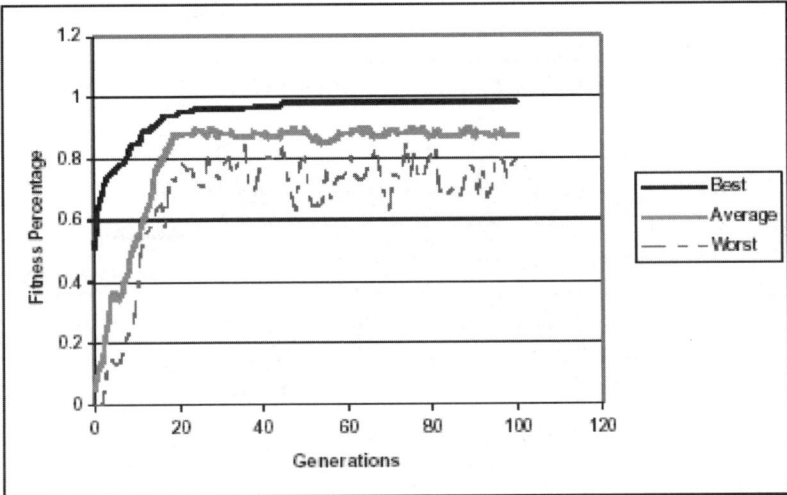

Fig. 10.34 The evolution process of the fitness of the population, including the fitness of the most fit individual, the fitness of the least fit individual and the average fitness of the whole population

a plateau. The lower line, the fitness of the least fit individual, demonstrates that we continue to introduce variation into the population using the genetic operators of mutation and crossover. Figure 10.35 demonstrates the evolution of the population of solutions in terms of the two components of the fitness function (similarity of mined ruled to the "normal" rules and similarity of the mined rules to the "abnormal" rules). This graph also demonstrates that the quality of the solution increases as the evolution process proceeds.

It is often difficult to know which items from an audit trail will provide the most useful information for detecting intrusions. The process of determining which items are most useful is called *feature selection* in the machine learning literature. We have conducted a set of experiments in which we are using genetic algorithms both to select the measurements from the audit trail that are the best indicators for different classes of intrusions and to "tune" the membership functions for the fuzzy variables. Figure 10.36 compares results when rules are mined (1) when there was no optimization and no feature selection, (2) when there was only optimization, and (3) when there was both optimization and feature selection. These results demonstrate that the GA can effectively select a set of features for intrusion detection while it tunes the membership functions. We have also found that the GA can identify different sets of features for different types of intrusions.

To conclude, we have integrated data mining techniques with fuzzy logic to provide new techniques for intrusion detection. Our system architecture allows us to support both anomaly detection and misuse detection components at both the individual workstation level and at the network level. Both fuzzy and non-fuzzy rules are supported within the system. We have also used genetic algorithms to tune the membership functions for the fuzzy variables used by our system to and select the most effective set of features for particular types of intrusions.

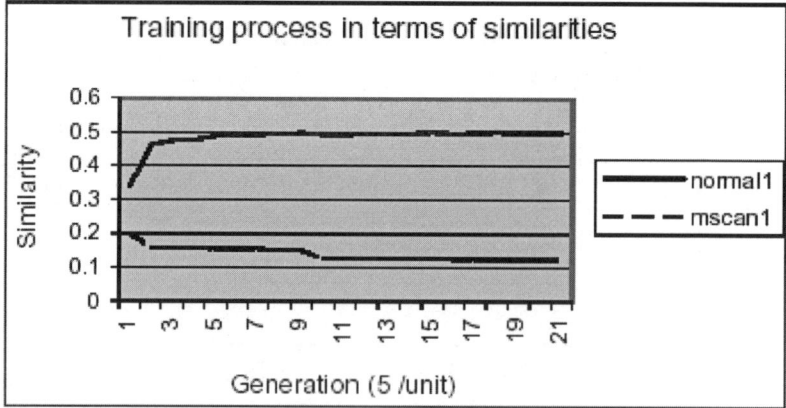

Fig. 10.35 The evolution process for tuning fuzzy membership functions in terms of similarity of data sets containing intrusions (mscan1) and not containing intrusions (normal1) with the reference rule set

10.7 Data Mining

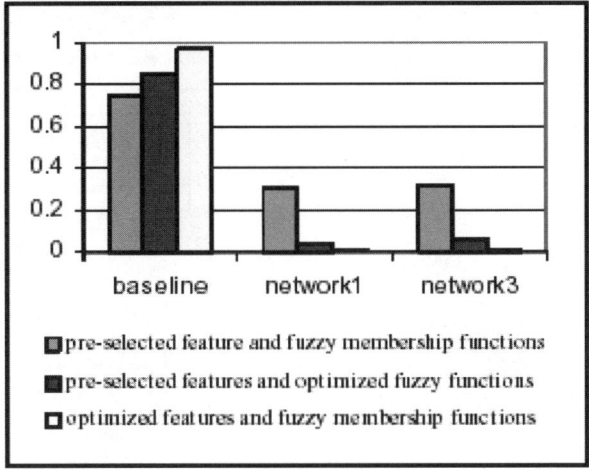

Fig. 10.36 Comparison of the similarity results using (1) features and fuzzy membership functions selected by the expert, (2) features selected by the expert and membership functions optimized by a GA, and (3) features selected by the GA and membership functions optimized by the GA

10.7.3 Selection and Partitioning of Attributes in Large-Scale Data Mining Problems Using Genetic Algorithm

This section presents the problems of reducing and decomposing large-scale concept learning problems in knowledge discovery in databases (KDD). The approach described here adapts the methodology of *wrappers* for performance enhancement and attribute subset selection to a genetic optimization problem. The fitness functions for this problem are defined in terms of classification accuracy given a particular supervised learning technique (or *inducer*). More precisely, the quality of a subset of attributes is measured in terms of empirical generalization quality (accuracy on cross validation data, or a continuation of the data in the case of time series prediction).

10.7.3.1 Attribute Selection, Partitioning, and Synthesis

The synthesis of a new group of attributes (also known as the *feature construction* problem) in inductive concept learning is an optimization problem. Its control parameters include the attributes used (i.e., which of the original inputs are relevant to distinguishing a particular target concept), how they are grouped (with respect to multiple targets), and how new attributes are defined in terms of *ground* (original) attributes. This synthesis and selection problem is a key initial step in *constructive induction*—the reformulation of a learning problem in terms of its inputs (attributes) and outputs (concept class descriptors).

Figure 10.37 illustrates the role of attribute selection (reduction of inputs) and partitioning (subdivision of inputs) in constructive induction (the "unsupervised" component of this generic KDD process). In this framework, the input consists of *heterogeneous* data (that originating from multiple sources). The performance ele-

Fig. 10.37 Attribute-based transformations in KDD

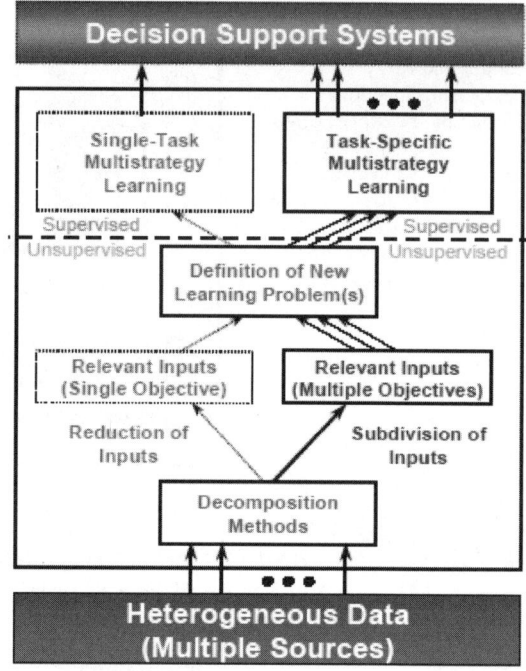

ment includes time series classification and other forms of pattern recognition that are important for decision support.

10.7.3.2 Attribute Partitioning in Constructive Induction

Attribute subset selection is the task of focusing a learning algorithm's attention on some subset of the given input attributes, while ignoring the rest. Here, subset selection is adapted to the systematic decomposition of concept learning problems in heterogeneous KDD. Instead of focusing a single algorithm on a single subset, the set of all input attributes is partitioned, and a specialized algorithm is focused on *each* subset. While subset selection is used to refinement of attribute sets in single-model learning, attribute partitioning is designed for multiple-model learning.

This approach adopts the role of feature construction in constructive induction: to formulate a new input specification from the original one. It uses subset partitioning to *decompose* a learning task into parts that are individually useful, rather than to *reduce* attributes to a single useful group. This permits new intermediate concepts to be formed by unsupervised learning (e.g., conceptual clustering or cluster formation using self-organizing algorithms). The newly defined problem or problems can then be mapped to one or more appropriate hypothesis languages (model specifications) as illustrated in Fig. 10.37. In the new system, the subproblem definitions obtained by partitioning of attributes also specify a mixture estimation problem. A

10.7 Data Mining

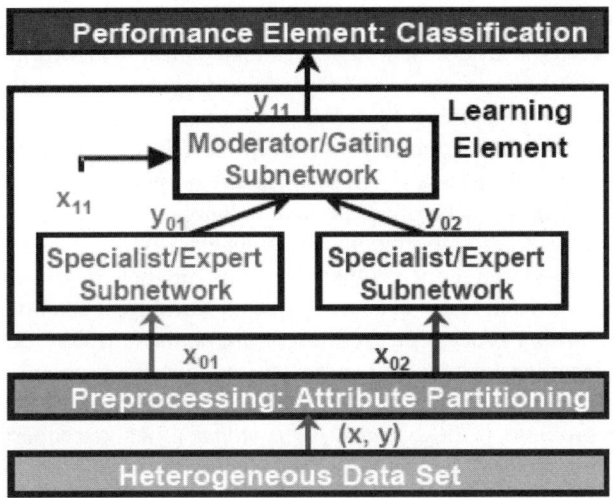

Fig. 10.38 The attribute partitioning approach

data fusion step, shown in Fig. 10.38, occurs after training of the models for all subproblems.

Together with attribute subset selection, attribute partitioning permits a concept learning problem to be refined for both increased classification accuracy and comprehensibility. The latter increases the utility of the model in systems that combine multiple models, such as hierarchical data fusion systems and large-scale multi-strategy data mining systems. Note that these systems may incorporate different type of concept learning algorithms, such as artificial neural networks. In our application, the multistrategy (hybrid) learning system is a GA wrapper that selects and configures probabilistic networks (especially temporal ANNs) and decision trees for KDD applications.

10.7.3.3 The Constructive Induction Problem and Supervised Concept Learning

In current practice, optimization problems in constructive induction are treated as a state space search. The primary difficulty encountered in applying search-based algorithms to synthesize attributes, select subsets of relevant attributes, or partition attributes into useful categories is the combinatorial complexity of uninformed search. The ability to constrain and control the search for useful attributes (or groups of them) is critical to making constructive induction viable. Toward this end, both domain knowledge and evaluation metrics have been applied in informed search algorithms (gradient and A*) for attribute subset selection and partitioning.

The definition of a concept learning problem consists of input *attributes* and *concept classes*. Each attribute is a function that maps an example, **x**, into a value. Conversely, a *classified* example can be defined as an object (a tuple of attribute values) whose type is the range of all combinations of these attributes together with a concept class, y. The task of an *inductive concept learning* algorithm is to produce a *concept description*, $y = g(\mathbf{x})$, that maps a newly observed example $\mathbf{x}$ to its class y. In inductive concept learning, therefore, the input (a training data set) consists of classified examples, and the output is a concept descriptor (a representation of the concept description such as a decision tree, classification rule base, linear separator, or classifier system). This classifier can then be applied to each new (unclassified) example to obtain a prediction (hypothesis) of its class.

Constructive induction is the problem of producing new descriptors of training examples (instances) and target classes in concept learning. It can be regarded as an *unsupervised learning* process that refines, or *filters*, the attributes (also referred to as *features* or *instance variables*) of some concept learning problem. The objective function of this process, called an attribute *filter* in the attribute subset selection and extraction problem, is the *expected performance* of a given supervised learning algorithm on the data set, restricted to the selected attributes. This expected performance measure can be based on any quantitative or qualitative analysis of the data set (including heuristic figures of merit), but the common trait of all attribute filters is that they operate independently of the induction algorithm (i.e., they ignore credit assignment based on actual supervised learning quality). The filter method can be used not only to *select* attributes, but to *compose* them using operators, such as the arithmetic operators $\{+, -, *, /\}$. The objective criterion is still based strictly on factors other than direct observation of supervised learning quality.

A more sophisticated variant, suitable for attribute selection, partitioning, or synthesis, casts the selection problem (for 0-1 subset membership, i.e., inclusion-exclusion; for subset membership; or for operator application order) as a multi-criterion optimization function. This function is defined subject to constraints of supervised learning performance: cross-validated classification accuracy and convergence time are most prevalent. This type of optimization is based on multiple runs of the supervised learning algorithm (concurrent across any population of candidate configurations, i.e., subsets, partitions, or synthetic attribute sets; serial among generations of candidates).

Because it takes the supervised learning algorithm into account and invokes it as a subroutine, this approach is referred to as the *wrapper* methodology. Wrappers can be used for both attribute reformulation (part of constructive induction) and other forms of parameter tuning in inductive learning. It is important to note that to date, attribute selection, partitioning, and synthesis wrappers have not been studied as genetic algorithms, although stochastic and heuristic search and optimization methods have been applied.

Composition of new attributes by such methods has been shown to increase accuracy of the classifiers produced by applying supervised learning algorithms to the reformulated data. The rationale is that concept learnability can be improved relative to given supervised learning algorithm through alternative representation. The step of transforming low-level attributes into useful attributes for supervised learning is known as *attribute synthesis* or, as is more common in the computational intelligence literature, *feature construction*. The complementary step to feature construction is *cluster definition*, the transformation of a given class definition into a more useful one.

10.7.3.4 Attribute Partitioning as Search

Both filters and wrappers for attribute selection and partitioning can be purely search-based or can incorporate constraint knowledge about operators, especially *which groups of attributes are coupled* (i.e., should be taken together for purposes of computing joint relevance measures discuss the use of such constraint knowledge in constructive induction. For example, in the automobile insurance KDD problem surveyed below, formulae are computed for *loss ratio* in automobile insurance customer evaluation. Only the number of exposures (units of customer membership) should be allowed as a denominator. Only certain attributes denoting loss paid (on accidents, for example) should be permitted as numerators, and these should always be summed. Similarly, *duration* attributes are a type of attribute that is always produced by taking the difference of two dates. This type of domain knowledge guided constructive induction drastically reduces the search space of candidate attributes from which the filter or wrapper algorithm must select.

The objective criterion for reformulation of a large-scale inductive learning problem in KDD is defined in terms of classification accuracy, and this leads naturally to the family of fitness functions and the scalability issues described below.

In scarch-bascd algorithms for attribute synthesis, constraint knowledge about operators has been shown to reduce the number of fitness evaluations for candidate attributes. This section shows how constraint knowledge about operators can be encoded in a fitness function. The purpose of this approach is to improve upon the non-genetic, search-based algorithm in terms of training sample efficiency. Several GA implementations of alternative (search-based and knowledge-based) attribute synthesis algorithms are surveyed, and their application to large-scale concept learning problems is addressed.

10.7.3.5 Methodology of Applying GAs to Constructive Induction Extending the Traditional Algorithm

This section briefly describes an encoding for attribute synthesis specifications for a simple GA with single-point crossover and a family of fitness functions that captures the objective criteria for wrapper systems.

Raymer *et al* use a masking GA, containing indicator bits for attributes to simultaneously extract and select attributes for a *k-nearest neighbor* (knn) supervised learning component. This masking GA is very similar to the state space encoding used by Kohavi *et al* for attribute subset selection, and is quite standard (e.g., forward selection and backward elimination algorithms in linear regression are described in similar fashion). Furthermore, the bit mask (inclusion-exclusion) encoding has an analogue in attribute partitioning that can be applied to encode pairwise sequential operations on attributes. Some related work on genetic search for feature selection permits replication of attributes by using a membership coding. The bit-mask coding is natural for attribute selection, but must be adapted for attribute partitioning. In the genetic wrapper for partitioning, two codings can be used. The first is a sparse n-by-n bit matrix encoding, where 1 in column j of row i denotes membership of the ith attribute in subset j. Empty subsets are permitted, but there can be no more than n. Also, in this design, membership is mutually exclusive (in a true partition, there is no overlap among subsets). The second coding uses numeric membership as in the state space representation, and is shown in Fig. 10.39; this is a more compact encoding but requires specialized crossover operators (corresponding to subset exchange) as well as mutation operators (corresponding to abstraction and refinement).

For an attribute selection, partitioning, or synthesis wrapper, the fitness function must always reflect the figure(s) of merit specified for the performance element of the KDD system. If this is a basic supervised concept learner that generates predictions, the fitness function should be based upon classification error (0-1, mean-squared error, or whatever loss function is actually used to evaluate the learner). This is not *necessarily* the same loss function as is used in the supervised learning algorithm (which may, for example, be based on gradient descent), but it frequently is. If the performance element is a classifier system, the fitness function for this wrapper should express the same criteria. Finally (and most important), the constraint knowledge for operator *preference* can be encoded as a penalty function and summed with the performance measure (or applied as a quick-rejection criterion). That is, if some operator is not permitted or not preferred, a penalty can be assessed that is either continuous or 0-1 loss.

10.7.3.6 Functional (Task-Level) Parallelism in Change of- Representation Search

As do simple GAs for most concept learning problems (supervised and unsupervised), genetic wrappers exhibit a high degree of functional (task-level) parallelism, as opposed to data parallelism (*aka* array or vector parallelism). This is doubly true for genetic attribute synthesis wrappers. With replication of the data across clus-

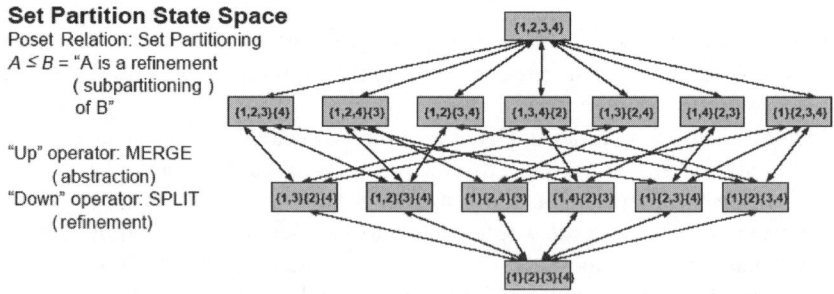

Fig. 10.39 A numeric encoding of individuals for attribute partitioning

10.7 Data Mining

ter nodes, the inter-task communication is limited to a specification string and the fitness value, with all of the computation for *one run* of the supervised learning algorithm being performed on a separate processor. The evaluation of each component of the specification (i.e., each synthetic attribute) can be also be functionally decomposed and parallelized. This approach, however, has a high internal data access overhead. Possible solutions include use of distributed shared memory and parallel I/O. Nevertheless, the break-even point for communications overhead is favorable, because the fitness function computations (for applications surveyed below) range from 5 minutes (for data sets on the order of 100 attributes and 25,000 exemplars) to 75 minutes (for data sets on the order of 400 attributes and 100,000 exemplars).

10.7.3.7 Applications of Genetic Constructive Induction in Large-Scale Data Mining

Record and Document Clustering (Information Retrieval)

The simple GA for attribute partitioning can be applied to knowledge discovery in very large databases. The purpose of constructive induction in these problems is to perform *change of representation* for supervised learning, thereby reducing the computational complexity of the learning problem given the transformed problem. For example self-organizing maps can be used to produce multiple, *intermediate* training targets (new, constructed attributes) that are used to define a new supervised learning problem. This technique has been used at NCSA (using manual and nongenetic methods such as Kohonen's self-organizing maps, or SOM) to cluster sales transaction records, insurance policy records, and claims data, as well as technical natural language reports (repair documents, warranty documents, and patent literature). In current research, the simple GA and more sophisticated genetic methods for attribute synthesis in record clustering (especially for repair documents and patent literature) are being evaluated in a Java-based infrastructure for large-scale KDD.

Supervised Learning for Insurance Policy Classification

Finally, another real-world application is multi-attribute risk assessment (prediction of expected financial loss) using insurance policy data. The input data is partitioned using a state space search over subdivisions of attributes (this approach is an extension of existing work on attribute subset selection The supervised learning task is represented as a discrete classification (concept learning) problem over continuous-valued input. It can be systematically decomposed by partitioning the input attributes (or fields) based on prior information such as *typing* of attributes (e.g., geographical, automobile specific demographics, driver-specific demographics, etc.). Preliminary experiments indicate that synthesis of *intra-type* attributes (such as paid loss, the sum of losses from different subcategories, and duration or membership, the difference between termination date and effective date of an insurance policy) and *inter-type* attributes (such as loss ratio) can be highly useful in supervised learning. This includes definition of new input attributes as well as intermediate target concepts.

10.8 Wireless Networks

10.8.1 Genetic Algorithms for Topology Planning in Wireless Networks

A wireless mesh network (WMN) is an attractive networking technology, providing with convenient access to the Internet as well as the spontaneous connection of mobile devices to each other. In WMNs, a number of studies have focused on the channel allocation problem because it is not easy to utilize the multi-channel and multi-radio characteristics of WMNs. Since the channel allocation in multi-channel multi-radio WMNs is an NP-Hard problem, most approaches design the network with a mathematical model and solve it with linear programming and some approximation algorithms.

On the other hand, where to deploy mesh routers is also a crucial matter in WMNs since it is directly related to the efficiency and deployment costs. This problem might also be solved by linear programming, however when the size of the target area becomes large, the linear programming method cannot handle this matter in a finite time. Therefore, in most cases, it is not an appropriate approach.

Genetic algorithm (GA) is introduced in deployment of mesh routers, in order to find a feasible solution to this matter in a finite and reasonable time. The target area is regarded with $n \times n$ grid and the mesh routers can be placed at the center of each rectangle, where n can be set as a larger number when more precision is needed. Each rectangle is assigned a sequence number where the top-left rectangle has the smallest number and the bottom-right the largest. And a binary-string is used in encoding scheme, in which i-th bit represents whether the rectangle with the sequence number i has a mesh router or not. Steady-state GA is used with tournament selection and toggling mutation with 0.0015 probability. In terms of the crossover operator, two-dimensional locus-based crossover is employed, where the schema (a series of dominant genes) of the parents is more likely to be passed down to the offspring. In terms of the fitness function, both the number of covered subscribers and the number of mesh routers are considered; the more the covered subscribers and the less the number of mesh routers, the better the fitness. Here, the strength of GA is that one can easily reflect restrictions such as obstacles or preference of the service providers by adjusting the fitness function; whereas in linear programming, it is usually quite hard to put such constraints into the mathematical model. Also, in this case a heuristic local optimization scheme is adopted, which is based on random toggling, to complement the slow convergence of genetic algorithms.

The performance of genetic algorithm can be implemented in planning mesh router deployment. The size of the target area is set 10,000 m $\times$ 10,000 m, and it is divided into 100 $\times$ 100 grids. And the transmission range of each mesh router is configured as 400 m. The location of the gateway to the Internet, and the expected subscriber vector can be arbitrarily configured, where the location of each expected subscriber is listed; insert 50 random entries there. The simulation result after 600 generations shows 250 mesh routers are sufficient for the given area; whereas a brief mathematical analysis shows 180 is the optimum.

Thus, genetic algorithm almost always found the solution close to the optimal topology within one hour, while linear programming required four to seven days to find the optimal topology for the much smaller (1/50 times) problem space. When the size of the target region grows, genetic algorithm will show much better performance over linear programming. This approach is also applicable to the channel allocation issue in WMNs.

10.8.2 Genetic Algorithm for Wireless ATM Network

Consider the example shown in Fig. 10.40, where cells A and B are connected to switch $s1$, and cells C and D are connected to switch $s2$. If the subscriber moves from cell B to cell A, switch $s1$ will perform a handoff for this call. This handoff is relatively simple and does not involve any location update in the databases that record the position of the subscriber. The handoff also does not involve any network entity other than switch $s1$. Now imagine that the subscriber moves from cell B to cell C. Then the handoff involves the execution of a fairly complicated protocol between switches $s1$ and $s2$. In addition, the location of the subscriber in the databases has to be updated. There is actually one more fact that makes this type of handoff difficult. If switch $s1$ is responsible for keeping the billing information about the call, then switch $s1$ cannot simply remove itself from the connection as a result of the handoff. In fact, the call continues to be routed through switch $s1$ (for billing purposes). The connection, in this case, is from cell C to switch $s2$, then to switch $s1$ and finally to the telephone network.

In this section, consider a group of cells and a group of switches in an ATM network (whose locations are fixed and known). The problem is to assign cells to switches in the ATM network in an optimum manner. We consider the topological design of a two-level hierarchical network. The upper-level network is a connected ATM network, and the lower-level network is a PCS network which is configured as an H-mesh (Fig. 10.40). The assumptions of the problem are stated as follows:

(1) The structures and positions of the ATM network and cell network are known.
(2) We assume that the cost of handoffs involving only one switch is negligible.
(3) Each cell in the cell network will be directly assigned and connected to only one switch in the ATM network.
(4) We assume that the number of calls that can be handled by each cell per unit time is equal to 1.
(5) The capacity of a switch, the number of cells that it can be assigned, is limited to a constant called Cap.

Fig. 10.40 Two-level hierarchical network. The handoff from *B* to *C* is more expensive than that from *B* to *A*

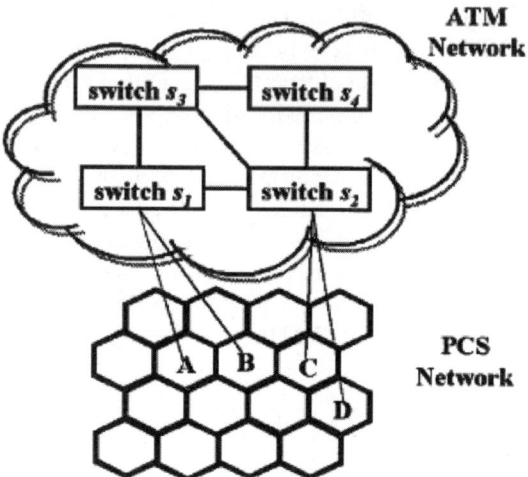

(6) The cost has two components. One is the cost of handoffs that involve two switches, and the other is the cost of **cabling** (or trucking).
(7) **Minimal switches assumption**: the number of switches assigned is assumed to be minimized.
(8) **Load balance assumption**: The load of assigned switches is assumed to be balanced. If this load balance assumption is satisfied, $m' = [n/Cap]$ switches need to be assigned, and the number of cells assigned to switches is $\lfloor n/m' \rceil$ or $\Box n/m' \rfloor$.

It is easy to see that finding an optimal solution to this problem is **NP-complete**, and that an exact search for optimal solutions is impractical due to exponential growth in execution time. Moreover, traditional heuristic methods and greedy approaches should trap in local optima. Genetic algorithms (GA) have been touted as a class of general-purpose search strategies that strike a reasonable balance between exploration and exploitation.

GA have been constructed as robust stochastic search algorithms for various optimization problems. GA searches by exploiting information sampled from different regions of the solution space. The combination of crossover and mutations helps GA escape from local optima. These properties of GA provide a good global search methodology for the two-level wireless ATM network design problem. In this section, we propose simple *GA* for optimal design for the two-level wireless ATM network problem.

10.8.2.1 Problem Description

The various notations used here are:

n total number of cells in the cell network
m total number of switches in the ATM network

10.8 Wireless Networks

$G(S, E)$	ATM network, where S is the set of switches and $E \subseteq S \times S$
$CG(C, L)$	cell network, where C is the set of cells and $L \subseteq C \times C$
(s_k, s_l)	edge between switches s_k and s_l in S
(c_i, c_j)	edge between cell c_i and c_j in C
(X_{s_k}, Y_{s_k})	coordinate of switch $s_k \in G, k = 1, 2, \ldots, m$
(X_{c_i}, Y_{c_i})	coordinate of cell $c_i \in S, i = 1, 2, \ldots, n$
d_{ki}	minimal cost between switches s_k and s_i in G
f_{ij}	cost per unit time of the handoffs that occur between cell c_i and c_j in CG, $i, j = 1, \ldots, n$
l_{ik}	cost of cabling per unit time and between cell $c_i \in CG$ and switch $s_k \in G, i = 1, \ldots, n; k = 1, \ldots, m$ and assume $l_{ik} = \sqrt{(X_{c_i} - X_{s_k})^2 + (Y_{c_i} - Y_{s_k})^2}$
w_{ij}	weight of edge $(c_i, c_j) \in CG$, where $w_{ij} = f_{ij} + f_{ji}, w_{ij} = w_{ji}$, and $w_{ii} = 0; i, j = 1, \ldots, n$
Cap	cell handling capacity of the switch
$m' = \lceil n/Cap \rceil$	number of switches that need to be assigned
α	ratio of the cost of cabling to that of handoff

10.8.2.2 Decision Variables

$$x_{ik} = \begin{cases} 1 & \text{if cell } c_i \text{ is assigned to switch } s_k \\ 0 & \text{otherwise} \end{cases}$$

$$z_{ijk} = x_{ik} x_{jk}, \text{ for } i, j = 1, \ldots, n \text{ and } k = 1, \ldots, m$$

i.e.,

$$z_{ijk} = \begin{cases} 1 & \text{if both cells } c_i \text{ and } c_j \text{ are connected to} \\ & \text{a common switch } s_k \\ 0 & \text{otherwise} \end{cases}$$

$$y_{ij} = \sum_{k=1}^{m} z_{ijk}, \text{ for } i, j = 1, \ldots, n$$

i.e.,

$$y_{ij} = \begin{cases} 1 & \text{if both cells } c_i \text{ and } c_j \text{ are connected to} \\ & \text{a common switch} \\ 0 & \text{otherwise} \end{cases}$$

Find variables x_{ik} which minimize,

$$\sum_{i=1}^{n}\sum_{k=1}^{m}l_{ik}x_{ik} + \alpha \sum_{i=1}^{n}\sum_{j=1}^{n}\sum_{k=1}^{m}\sum_{l=1}^{m}w_{ij}(1-y_{ij})x_{ik}y_{jl}d_{lk} \qquad (10.22)$$

subject to,

$$\sum_{i=1}^{n}x_{ik} \leq Cap, \ k=1,\ldots,m; \qquad (10.23)$$

$$\sum_{k=1}^{m}x_{ik} = 1, \ \text{for } i=1,\ldots,n; \qquad (10.24)$$

$$\left\lfloor \frac{n}{m'} \right\rfloor \leq \sum_{i=1}^{n}x_{ik} \leq \left\lceil \frac{n}{m'} \right\rceil, k=1,\ldots,m; \qquad (10.25)$$

$$x_{ik} \in \{0,1\}, \ \text{for } i=1,\ldots,n \text{ and } k=1,\ldots,m. \qquad (10.26)$$

If cells c_i and c_j are assigned to different switches, then a cost is incurred. If f_{ij} is the cost per unit time of handoffs that occurs between cells c_i and c_j, $(i, j = l, \ldots, n)$, then f_{ij} is proportional to the frequency of handoffs that occur between these cells which we assume is fixed and known.

The objective is to assign each cell to a switch so as to minimize (total cost) the sum of the cabling costs and handoff costs per unit time. The objective function (10.22) minimizes the total cost which is the sum of the cabling costs and handoffs costs per unit time. In (10.22), the first part is the total cabling costs between cells and switches; the second part is the cost of handoffs per unit time, and α is the ratio of the cost of cabling to that of and handoff costs. Constraint (10.23) ensures that the call handling capacity is limited to Cap. Constraint (10.24) ensures that each cell is assigned to exactly one switch. Constraint (10.25) ensures that the minimal switches assumption and load balance assumption can be satisfied. Constraint (10.26) is a binary and nonnegative constraint.

Consider the graph shown in Fig. 10.41 There are 10 cells in CG which should be assigned to 4 switches in S.

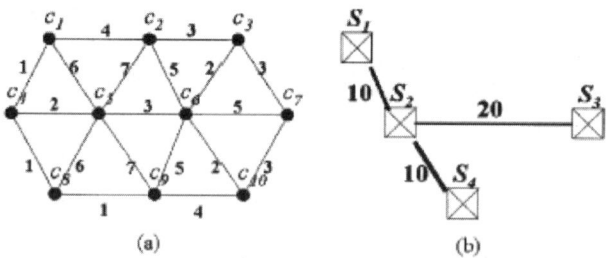

Fig. 10.41 An example of PCS and ATM networks: (a) Cell graph CG of PCS network, (b) ATM network

10.8 Wireless Networks

The weight of an edge between two cells is the cost per unit time of the handoffs that occur between them. Four switches are positioned at the center of the cell: c_1, c_5, c_7, and c_9. Assume the matrix CS of the distance between a cell and a switch is as follows:

$$CS = \{l_{ik}\}_{10 \times 4} = \begin{array}{c} c_1 \\ c_2 \\ c_3 \\ c_4 \\ c_5 \\ c_6 \\ c_7 \\ c_8 \\ c_9 \\ c_{10} \end{array} \begin{bmatrix} 0 & 1 & \sqrt{7} & 2 \\ 1 & 1 & \sqrt{3} & \sqrt{3} \\ 2 & \sqrt{3} & 1 & 2 \\ 1 & 1 & 3 & \sqrt{3} \\ 1 & 0 & 2 & 1 \\ \sqrt{3} & 1 & 1 & 1 \\ \sqrt{7} & 2 & 0 & \sqrt{3} \\ \sqrt{3} & 1 & \sqrt{7} & 1 \\ 2 & 1 & \sqrt{3} & 0 \\ \sqrt{7} & \sqrt{3} & 1 & 1 \end{bmatrix} \quad (10.27)$$

An initial assignment of example is shown in Fig. 10.42. Cells c_1, c_1, c_4, c_5 and c_8 are connected to switch s_2, and the others are connected to switch s_4.

In this section, we discuss the details of GA developed to solve the problem of optimum assignment of cells in PCSs to switches in the ATM network. The development of GA requires:

(1) a chromosomal coding scheme,
(2) genetic crossover operators,
(3) mutation operators,
(4) a fitness function definition,
(5) a replacement strategy,
(6) termination rules.

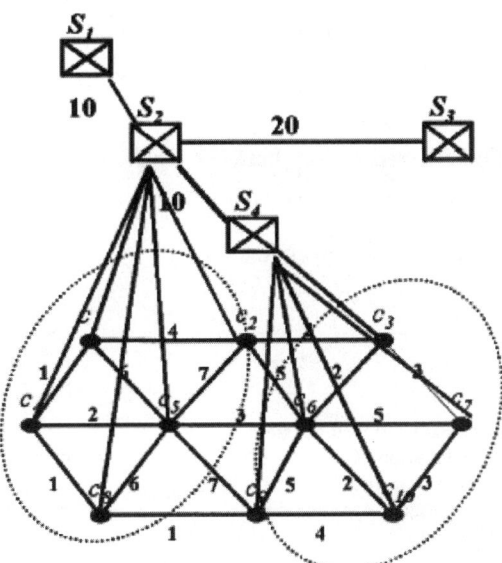

Fig. 10.42 An initial assignment for example. *Simple Genetic Algorithm (SGA) for Wireless ATM Network Design*

Chromosomal Coding

Since our problem involves representing of connections between cells and switches, we employ a coding scheme that uses positive integer numbers. Cells are labeled from one to n (the total number of cells); and switches are labeled from one to m (the total number of switches). The **cell-oriented representation** of the chromosome structure is shown in Fig. 10.43(a), where the ith cell belongs to the v_ith switch. For example, the chromosome of the example shown in Fig. 10.42 is shown in Fig. 10.37(b).

Genetic Crossover Operator

Two types of genetic operators were used to develop this algorithm:
(1) **simple single point crossover;**
(2) **the random cell swap operator (RCSO).**

The simple single point crossover is the traditional one. In RCSO, by randomly selecting two chromosomes (say $P1$ and $P2$) for crossover from previous generations and then using a random number generator, an integer value i is generated in the range $(1, n)$. This number is used as the crossover site. Let $v_i 1$ and $v_i 2$ be the value of the ith cell in $P1$ and $P2$, respectively. To create new offspring, RCSO employs two steps: first, all the characters between i and n of two parents are swapped and temporal chromosomes $C1$ and $C2$ are generated. Then, in all the characters in $C1$ and $C2$; the value $v_i 1$ ($v_i 2$) is change to $v_i 2$ ($v_i 1$). The following example provides a detailed description of the crossover operation (assume crossover site $i = 6$):

Parent $P1$
1 1 2 4 3 | 1 1 2 4 3;
Parent $P2$
3 1 4 2 3 | 3 1 2 3 2.

First, two substrings between 6 and 10 are swapped, and we have:

Temporal chromosome $C1$
1 1 2 4 3 | 3 1 2 3 2;

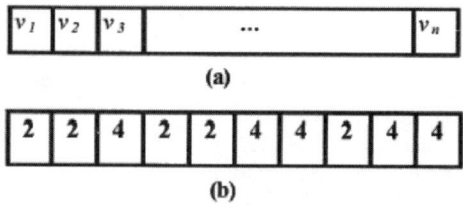

Fig. 10.43 (a) Cell-oriented representation of the chromosome structure, (b) Cell-oriented representation of Example

Temporal chromosome $C2$
3 1 4 2 3 | 1 1 2 4 3.

Then, every 1 is changed to 3, and every 3 is changed to 1 in both temporal chromosomes $C1$ and $C2$, and we have:

Offspring $O1$ **3 3** 2 4 **1** | **1 3** 2 **1** 2;
Offspring $O2$ **1** **3** 4 2 **1** | **3 3** 2 4 **1**.

Mutation

Two types of mutations were used to develop this algorithm:

(1) **The traditional mutation operation**: by randomly selecting a cell of a vector, the traditional mutation operation changes the value of the cell to a random number which is between 1 to m.
(2) **Multiple cell mutation**: by randomly selecting two random numbers k, l between 1 and m, multiple cell mutation change the value of each cell from k to l.

The following example provides a detailed description of multiple cells mutation (assume random numbers $k = 3$ and $l = 4$):

Before mutation 1 1 2 4 **3** 1 1 2 4 **3**;
After mutation 1 1 2 4 **4** 1 1 2 4 **4**.

Fitness Function Definition

Generally, GA use fitness functions to map objectives to costs to achieve the goal of an optimally designed two-level wireless ATM network. If cell ci is assigned to switch sk, then vi in the chromosome is set to be k. Let $d(v_i, v_j)$ be the minimal communication cost between switches sk and sl in G. An objective function value is associated with each chromosome, which is the same as the fitness measure mentioned above.

We use the following objective function:

$$\text{minimize} \sum_{i=1}^{n} l_{iv_i} + \alpha \sum_{i=1}^{n} \sum_{j=1}^{n} w_{ij} d_{(v_i, v_j)}. \tag{10.28}$$

While breeding chromosomes, GA does not require the chromosome to reflect a feasible solution. Thus, we need to attach a penalty to the fitness function in the event the solution is infeasible. Let n_k be the number of cells assigned to switch sk, and assume that n is a multiple of Cap. Sort switches in decreasing order according to the number of cells to be assigned. We rewrite the formulation above in an unconstrained form:

$$\text{minimize cost} = \sum_{i=1}^{n} l_{iv_i} + \alpha \sum_{i=1}^{n} \sum_{j=1}^{n} w_{ij} d_{(v_i, v_j)} + \Pi,$$

where

$$\Pi = \beta \left(\sum_{k=1}^{m'} |n_k - Cap| + \sum_{k=m'+1}^{m} |n_k| \right). \tag{10.29}$$

Π is the penalty measure associated with a chromosome, and β is the penalty weight.

Since the best-fit chromosomes should have a probability of being selected as parents that is proportional to their fitness, they need to be expressed in a maximization form. This is done by subtracting the objective from a large number $C_{\max}$. Hence, the fitness function becomes:

$$\text{maximize } C_{\max} - \left[\sum_{i=1}^{n} l_{iv_i} + \alpha \sum_{i=1}^{n} \sum_{j=l}^{n} w_{ij} d_{(v_i, v_j)} \right.$$
$$\left. + \beta \left(\sum_{k=1}^{m'} |n_k - Cap| + \sum_{k=m'+1}^{m} |n_k| \right) \right], \tag{10.30}$$

where $C_{\max}$ denotes the maximum value observed, so far, of the cost function in the population. Let cost be the value of the cost function for the chromosome; $C_{\max}$ can be calculated by the following iterative equation:

$$C_{\max} = \max \{C_{\max}, cost\}, \tag{10.31}$$

where $C_{\max}$ is initialized to zero.

Replacement Strategy

This subsection discusses a method used to create a new generation after crossover and mutation is carried out on the chromosomes of the previous generation. The most common strategies probabilistically replace the poorest performing chromosomes in the previous generation. The elitist strategy appends the best performing chromosome of a previous generation to the current population and thereby ensures that the chromosome with the best objective function value always survives to the next generation. The algorithm developed here combines both the concepts maintained above.

Each offspring generated after crossover is added to the new generation if it has a better objective function value than do both of its parents. If the objective function value of an offspring is better than that of only one of the parents, then we select a chromosome randomly from the better parent and the offspring. If the offspring is worse than both parents, then either of the parents is selected at random for the next generation. This ensures that the best chromosome is carried to the next generation while the worst is not carried to the succeeding generations.

Termination Rules

Execution of GA can be terminated using any one of the following rules:

R1: when the average and maximum fitness values exceed a predetermined threshold;
R2: when the average and maximum fitness values of strings in a generation become the same; or
R3: when the number of generations exceeds an upper bound specified by the user.

The best value for a given problem can be obtained from a GA when the algorithm is terminated using R2.

In this section, we have investigated the problem of obtaining the optimum design of the two-level wireless ATM network. Given cells and switches on an ATM network (whose locations are fixed and known), the problem is to assign cells to switches in an optimum manner. This problem has been modeled as a complex integer programming problem, and the optimal solution of this problem has been found to be **NPcomplete**. A stochastic search methods (SGA) based on a genetic approach have been proposed to solve this problem. Simulation can be performed considering a hexagonal system or any other system and the results can be observed indicating the robustness of Genetic Algorithm.

10.9 Very Large Scale Integration (VLSI)

10.9.1 Development of a Genetic Algorithm Technique for VLSI Testing

The objective of VLSI testing is to generate compact set of test vectors that has high coverage of manufacturing defects.

- Stuck at fault modeling is the widely used fault modeling method in VLSI Testing.
- Here nodes are assumed to be stuck at either "0" or "1", for the purpose fault modeling.
- Testing methodology for a digital circuit is shown in Fig. 10.44.

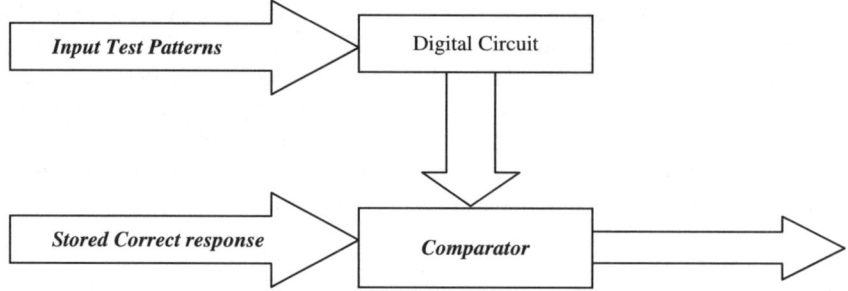

Fig. 10.44 Digital circuit

- Test vectors are encoded as *Binary bit stream*.
- Fitness function gives the *number of faults covered* by each test vector.

Consider the XOR circuit shown in the Fig. 10.45 below.
The description of given XOR Circuit is as follows:

- Number of primary inputs is "2" and primary output is "1".
- Each parent width is 2 bits.
- In this example, XOR circuit has 12 fault sites and 24 stuck at faults.
- For a fault free circuit, the output is "1" for a input vector [1,0].
- If the circuit has a stuck at "0" at "a", the output response is "0". So input vector [1,0] detects stuck at 0 [SA0] fault at "a".
- Like wise [1,0] can also detect SA0 fault at [a, c, d, g, h, z] and SA1 at [b, e, j].
- Thus [1,0] can detect 9 out of a total of 24 faults and it's fitness is 0.375 [i.e. 9/24].

The experimental circuit is as given below in Fig. 10.46. Table 10.10 shows the fault coverage of different test vectors.
The advantages of GA in VLSI Testing is as follows:

- Concept is easy to understand and separate from the application.
- Easy to exploit previous or alternate solutions.
- They are adaptive and learn from experience.
- They are efficient for complex programs.
- They are easy to parallelize as they have intrinsic parallelism.
- As they are inherently parallel, the computation can be easily distributed.

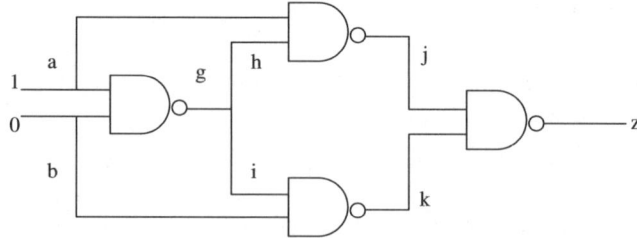

Fig. 10.45 XOR circuit

10.9 Very Large Scale Integration (VLSI)

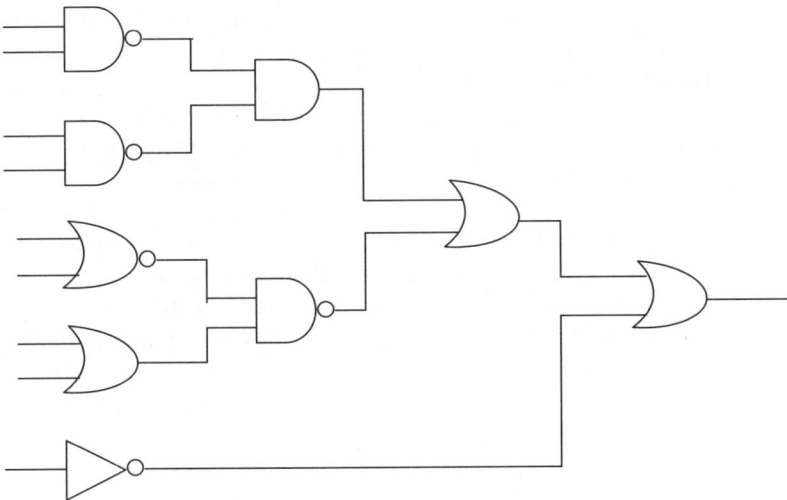

Fig. 10.46 Experimental circuit

10.9.2 VLSI Macro Cell Layout Using Hybrid GA

Genetic algorithms have proven to be a well-suited technique for solving selected combinatorial optimization problems. The blindness of the algorithm during the search in the space of encoding must be abandoned, because this space is discrete and the search has to reach feasible points after the application of the genetic operators. This can be achieved by the use of a problem specific genotype encoding, and hybrid, knowledge based techniques, which support the algorithm during the creation of the initial individuals and the following optimization process. In this section a novel hybrid genetic algorithm, which is used to solve macrocell placement problem is presented.

The design of VLSI *(very large scale integrated)* microchips is a process of many consecutive steps including specification, functional design, circuit design, physical design, and fabrication. Macro-cell layout generation is a task in the *physical design cycle*. The circuit is partitioned and the components are grouped in functional units,

Table 10.10 Fault coverage of different test vectors

Test Vector a b c d e f g h I	Stuck at 0 faults detected	Stuck at 1 faults detected	Fitness Value	Cumulative sum of fault coverage %
1 1 0 1 0 0 1 0 1	a, b, g, I, 3, 4	e, f, 1, 5, 6, 7, 8, 9	0.40	40
1 0 1 1 1 0 0 1 1	e, 7, 8, 9	3	0.14	54
1 0 1 1 0 1 0 1 1	f, c, d, h [I]	2 [3, 5, 6, 7, 8, 9]	0.34	68
1 0 1 0 0 0 1 1 1	1, 2, 6 [8]	b, d	0.17	82
1 1 0 1 0 0 0 0 1	[7, 8, 9]	g, h, 4	0.17	90
1 1 0 1 0 0 1 1 0	5 [9]	I	0.08	95
0 1 0 1 0 0 1 1 1	[1, 2, 6, 8, 9]	a, c	0.20	100

the *macro-cell*s. These cells can be described as rectangular blocks with *terminals* (pins) along their borders. These terminals have to be connected by *signal nets*, along which power or signals (e.g., clock ticks) are transmitted between the various units of the chip. A net can connect two or more terminals, and some nets must be routed to *pads* at the outer border of the layout, since they are involved in the I/O of the chip. The layout defines the positions of the cells (Fig. 10.47).

The major objectives are chip area minimization and interconnection wire length minimization. Since the number of possible placements increases explosively with the number of blocks, even subsets of the problem have been shown to be NP-complete or NP-hard. In this section, a hybrid genetic algorithm with a genotype representation based on binary trees and the genetic operators that work directly on this tree structure is used.

10.9.3 Problem Description

Inputs of the placement problem are

- a set of blocks with fixed geometries and fixed pin positions
- a set of nets specifying the interconnections between pins of blocks
- a set of pads (external pins) with fixed positions
- a set of user constraints, e.g., block positions/orientations, critical nets, if any

Given the inputs, the objective of the problem is to find the positions and orientations of each block, so that the chip area and interconnection wire length between blocks are minimized while satisfying all the given constraints. We take wire length into account simultaneously in the optimization process. Since it is impossible to calculate the exact wire length at this stage where detailed routing has not yet been carried out, we estimate the length of each net as one-half of the perimeter of the bounding box of the net.

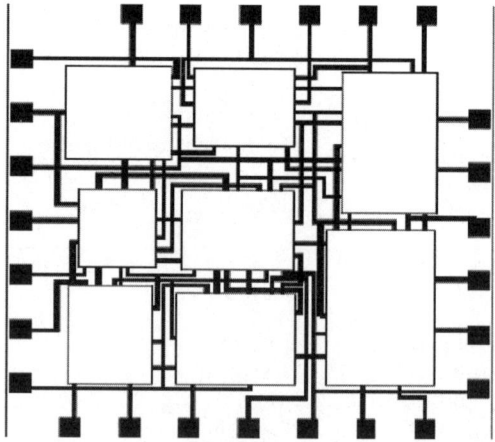

Fig. 10.47 The schematic representation of a VLSI macro-cell layout, which shows the position of eight cells, the routes for the signal nets, and the I/O pads

10.9 Very Large Scale Integration (VLSI)

The objective function, which measures the quality of the resulting placement, can be expressed as follows,

$$E = 1/(C1\, ChipArea + C2\, WireLength) \qquad (10.32)$$

where $C1$, $C2$ are the corresponding weights.

10.9.4 Genetic Layout Optimization

10.9.4.1 The Hybrid Genetic Algorithm

A Hybrid Genetic Algorithm is designed to use heuristics for improvement of offspring produced by crossover. Initial population is randomly generated. The offspring is obtained by crossover between two parents selected randomly. The layout improvement heuristics RemoveSharp and LocalOpt are used to ring the offspring to a local maximum. If fitness of the layout of the offspring thus obtained is greater than the fitness of the layout of any one of the parents then the parent with lower fitness is removed from the population and the offspring is added to the population. If the fitness of the layout of the offspring is lesser than that of both of its parent then it is discarded. For mutation a random number is generated within one and if it is less than the specified probability of the mutation operator a layout is randomly selected and removed from the population. Its layout is randomized and then added to the population. The algorithm works as below:

Step 1 : Initialize population randomly
Step 2 : Apply **RemoveSharp** algorithm to all layouts in the initial population
  Apply **LocalOpt** algorithm to all layouts in the initial population
Step 3 : Select two parents randomly
  Apply **Crossover** between parents and generate an offspring
  Apply **RemoveSharp** algorithm to offspring
  Apply **LocalOpt** algorithm to offspring
  If Fitness(offspring) > Fitness (any one of the parents) then replace the weaker parent by the offspring
Step 4 : **Mutate** any one randomly selected layout from population
Step 5 : Repeat steps 3 and 4 until end of specified number of iterations.

10.9.4.2 Genotype Representation

The phenotypic representation for the placement problems is basically the pattern that describes the position of the blocks. Binary slicing trees are well suited to represent placement patterns and have already been used in genetic algorithms. During recombination, partial arrangements of blocks are transmitted from parents to offspring. The corresponding operation is the inheritance of subtrees from the parents. Encoding the tree in a string complicates this operation, since the string needs to be decoded into the slicing tree to execute the recombination, then recoded into an offspring chromosome afterwards. There is no reason for using a string encoding except for the analogy to the natural evolution process, where the genetic information is encoded in a DNA string.

When directly using the slicing tree as the genotype representation, further decoding or encoding the tree when applying genetic operators is avoided. The genotype is encoded as a binary slicing tree, which defines the relative placement of the cells (Fig. 10.48). It is composed in a bottom-up fashion. In each inner node two *blocks* (in the lowest level these are single cells) are joined to a *meta-block* (partial placement). In each meta-block the orientations of the combined blocks are fixed (Fig. 10.49). Therefore every tree describes several possible shapes for the corresponding layout, which enormously improves the performance of the GA. Blocks or sub-patterns in a tree defining a layout, is always stacked vertically upon each other. The pattern characterized by the right successor of an inner tree node is always positioned on top of the pattern characterized by its left successor when combining both parts into a pattern or meta-block.

10.9.4.3 Genetic Operators

During the optimization process the placement of the blocks has to be changed. The genetic operators directly work on the tree-structure by combining subtrees of parents (crossover) and modifying the tree of an individual (mutation). The crossover operator takes two individuals (parent) out of which one offspring is composed by combining two subtrees, one from each parents. Unfortunately, these parts usually do not add up to a complete layout. After the combination of the two subtrees the redundant blocks are deleted and the missing blocks have to be added at random positions to the tree to ensure that the offspring finally represents a correct layout. Mutation operator modifies either by exchanging simple blocks or a block (leaf) with a meta-block (subtree) or by exchanging two meta-blocks. These cases

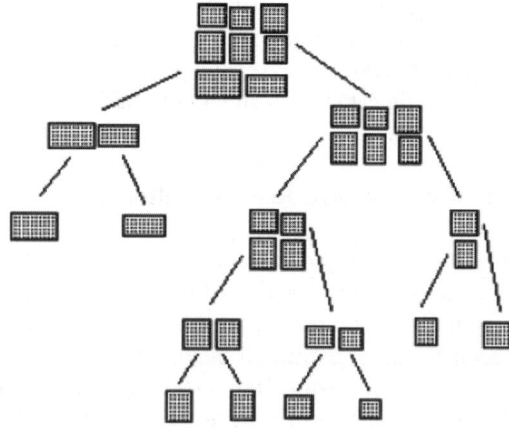

Fig. 10.48 The genotype

10.9 Very Large Scale Integration (VLSI)

Fig. 10.49 The composition of a meta-block

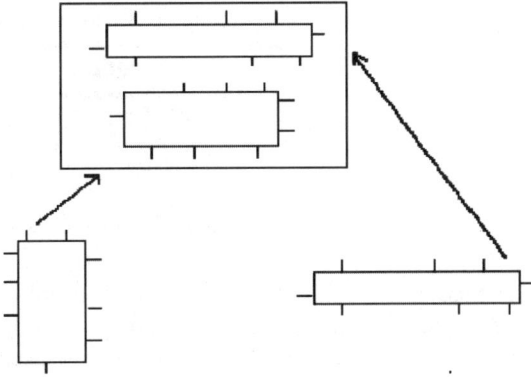

represent the exchange of two cells, a cell with a partial layout, and the exchange of two partial layouts on the layout surface.

10.9.4.4 Simulation

The hybrid genetic algorithm for the layout generation problem was tested on real life circuits chosen from a benchmark suite that was released for design workshops in the early 90s and is often referenced in the literature as the *MCNC benchmarks*. They were originally maintained by *North Carolina's Microelectronics, Computing and Networking Center*, but are now located at the *CAD Benchmarking Laboratory (CBL)* at North Carolina State University. These benchmarks are standard problems in macro-cell layout, and the characteristics of the circuits are shown in Table 10.11.

Using the following parameters, the heuristics developed are analyzed; the best performance of HGA is found when the values of the parameters are set as below:

> RemoveSharp (m) : 5
> LocalOpt(q) : 6
> Probability of Population size : 50
> Mutation operator : 0.02
> Number of Iterations : 10000

Table 10.11 The benchmark circuits for the macro-cell layout generation problem

Benchmark	Cells
Apte	9
Xerox	10
Hp	11

Table 10.12 Results of HGA

Benchmark	HGA
Apte	47.30
Xerox	20.30
Hp	9.39

The classical layout problem discussed in problem description; in which wirelength is also a factor in fitness value is analyzed. But due to technological progress, technologies like Over-The-Cell routing are now used so there is no need to determine wirelength and to add routing space (through which wires are routed) to the layout. Therefore in the fitness function introduced in problem description, C1 is assigned one, and C2 is assigned zero.

The result of using hybrid GA is shown in Table 10.12.

In this section an approach has been presented to incorporate domain knowledge into a genetic algorithm, which is supposed to compute near-optimal solutions to VLSI Placement problem. The feasibility of the approach has been demonstrated by presenting performance results for benchmark instances. It can be found that the implementation of the two newly introduced heuristics result in near optimal solutions in all cases. These heuristics are simple, straightforward and easy to implement when compared to other algorithms. This approach promises to be a useful tool in VLSI Design Automation.

10.10 Summary

GAs can even be faster in finding global maxima than conventional methods, in particular when derivatives provide misleading information. We should not forget, however, that, in most cases where conventional methods can be applied, GAs are much slower because they do not take auxiliary information like derivatives into account. In these optimization problems, there is no need to apply a GA which gives less accurate solutions after much longer computation time. The enormous potential of GAs lies elsewhere—in optimization of non-differentiable or even discontinuous functions, discrete optimization, and program induction. Thus due to these reasons genetic algorithm is found to be used in a variety of applications as discussed in this chapter. Apart from these applications dealt in this chapter, GAs can be applied to production planning, air traffic problems, automobile, signal processing, communication networks, environmental engineering and so on.

Chapter 11
Introduction to Particle Swarm Optimization and Ant Colony Optimization

11.1 Introduction

In this chapter, a brief introduction is given to Particle Swarm Optimization (PSO) and Ant Colony Optimization (ACO). Optimization is the process to find a best optimal solution for the problem under consideration. Particle Swarm Optimization and Ant Colony Optimization achieve finding an optimal solution for the search problems using the social behavior of the living organisms. Particle swarm optimization is a form of swarm intelligence and Ant colony optimization is a population-based metaheuristic that can be used to find approximate solutions to difficult optimization problems. The chapter gives an overview of basic concepts and fucntional operation of Particle Swarm Optimization and Ant Colony Optimization.

11.2 Particle Swarm Optimization

Particle swarm optimization (PSO) is a population based stochastic optimization technique developed by Dr. Eberhart and Dr. Kennedy in 1995, inspired by social behavior of bird flocking or fish schooling. PSO shares many similarities with evolutionary computation techniques such as Genetic Algorithms (GA). The system is initialized with a population of random solutions and searches for optima by updating generations. However, unlike GA, PSO has no evolution operators such as crossover and mutation. In PSO, the potential solutions, called particles, fly through the problem space by following the current optimum particles.

In past several years, PSO has been successfully applied in many research and application areas. It is demonstrated that PSO gets better results in a faster, cheaper way compared with other methods. Another reason that PSO is attractive is that there are few parameters to adjust. One version, with slight variations, works well in a wide variety of applications. Particle swarm optimization has been used for approaches that can be used across a wide range of applications, as well as for specific applications focused on a specific requirement.

11.2.1 Background of Particle Swarm Optimization

Particle swarm optimization (PSO) is a form of swarm intelligence and is inspired by bird flocks, fish schooling and swarm of insects. The flock of birds, fish schooling and swarm of insects is as shown in Fig. 11.1.

Consider Fig. 11.1 and imagine a swarm of insects or a school of fish. If one sees a desirable path to go (e.g., for food, protection, etc.) the rest of the swarm will be able to follow quickly even if they are on the opposite side of the swarm. On the other hand, in order to facilitate felicitous exploration of the search space, typically one wants each particle to have a certain level of "craziness" or randomness in their movement, so that the movement of the swarm has a certain explorative capability: the swarm should be influenced by the rest of the swarm but also should independently explore to a certain extent.

This is performed by particles in multidimensional space that have a position and a velocity. These particles are flying through hyperspace (i.e., $\Re^n$) and have two essential reasoning capabilities: their memory of their own best position and knowledge of the swarm's best, "best" simply meaning the position with the smallest objective value. Members of a swarm communicate good positions to each other and

Fig. 11.1 Social behavior

11.2 Particle Swarm Optimization

adjust their own position and velocity based on these good positions. There are two main ways this is done:

- a global best that is known to all and immediately updated when a new best position is found by any particle in the swarm
- "neighborhood" bests where each particle only immediately communicates with a subset of the swarm about best positions

Each particle keeps track of its coordinates in the problem space which are associated with the best solution (fitness) it has achieved so far. (The fitness value is also stored.) This value is called *pbest*. Another "best" value that is tracked by the particle swarm optimizer is the best value, obtained so far by any particle in the neighbors of the particle. This location is called *lbest*. when a particle takes all the population as its topological neighbors, the best value is a global best and is called *gbest*.

The particle swarm optimization concept consists of, at each time step, changing the velocity of (accelerating) each particle toward its *pbest* and *lbest* locations (local version of PSO). Acceleration is weighted by a random term, with separate random numbers being generated for acceleration toward *pbest* and *lbest* locations.

11.2.2 Operation of Particle Swarm Optimization

Consider Swarm of particles is flying through the parameter space and searching for optimum. Each particle is characterized by,

$$\text{Position vector} \ldots \ldots x_i(t)$$
$$\text{Velocity vector} \ldots \ldots v_i(t)$$

as shown in Fig. 11.2.

During the process, each particle will have its individual knowledge *pbest*, i.e., its own best-so-far in the position and social knowledge *gbest* i.e., *pbest* of its best neighbor as shown in Fig. 11.3.

Performing the velocity update, using the formula given below,

$$v_i(t+1) = \alpha \, v_i + c_1 \times rand \times (pbest(t) \text{-} x_i(t)) + c_2 \times rand \times (gbest(t) \text{-} x_i(t))$$
$$(11.1)$$

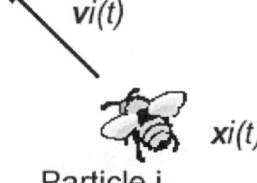

Fig. 11.2 A particle with position vector and velocity vector

Fig. 11.3 Particle with *pbest* and *gbest*

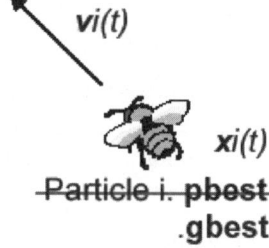

where α is the inertia weight that controls the exploration and exploitation of the search space. c_1 and c_2, the cognition and social components respectively are the acceleration constants which changes the velocity of a particle towards the *pbest* and *gbest*. *rand* is a random number between 0 and 1. Usually c_1 and c_2 values are set to 2. The velocity update is based on the parameters as shown in Fig. 11.4.

Now, performing the position update,

$$X_i(t+1) = X_i(t) + V_i(t+1) \tag{11.2}$$

The position update process is as shown in Fig. 11.5

The above process discussed is repeated for each and evry particle considered in the computation and the best optimal solution is obtained.

PSO utilizes several searching points like genetic algorithm (GA) and the searching points gradually get close to the optimal point using their pbests and the gbest. The first term of RHS of (11.1) is corresponding to diversification in the search procedure. The second and third terms of that are corresponding to intensification in the search procedure. Namely, the method has a well balanced mechanism to utilize diversification and intensification in the search procedure efficiently. The original PSO can be applied to the only continuous problem. However, the method can be expanded to the discrete problem using discrete number position and its velocity easily.

The above feature can be explained as follows. The RHS of (11.1) consists of three terms. The first term is the previous velocity of the agent. The second and third

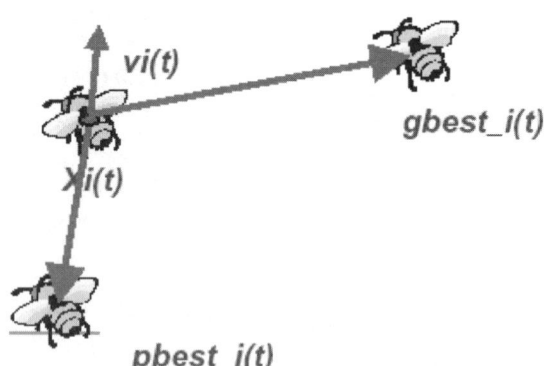

Fig. 11.4 Parameters for velocity update

Fig. 11.5 Position update using $x_i(t)$ and $v_i(t+1)$

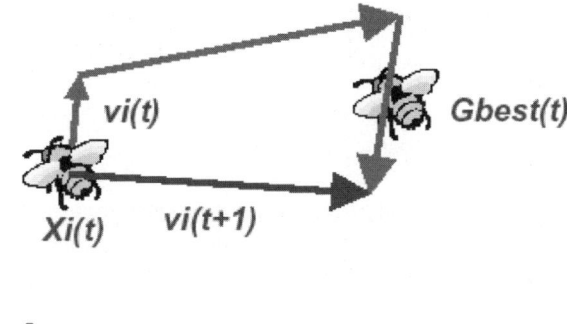

terms are utilized to change the velocity of the agent. Without the second and third terms, the agent will keep on "flying" in the same direction until it hits the boundary. Namely, it tries to explore new areas and, therefore, the first term is corresponding to diversification in the search procedure. On the other hand, without the first term, the velocity of the "flying" agent is only determined by using its current position and its best positions in history. Namely, the agents will try to converge to the their pbests and/or gbest and, therefore, the terms are corresponding to intensification in the search procedure.

11.2.3 Basic Flow of Particle Swarm Optimization

The basic operation of PSO is given by,

    Step 1: Initialize the *swarm* from the solution space
    Step 2: Evaluate *fitness* of individual particles
    Step 3: Modify *gbest, pbest* and *velocity*
    Step 4: Move each *particle* to a new *position*
    Step 5: Goto step 2, and repeat until convergence or stopping condition is satisfied

The pseudo code of the procedure is as follows

```
For each particle
    Initialize particle
END
Do
    For each particle
        Calculate fitness value
```

                If the fitness value is better than the best fitness value
                    (pbest) in history set current value as the new pbest
            End
            Choose the particle with the best fitness value of all the
                particles as the gbest
            For each particle
                Calculate particle velocity according equation (11.1)
                Update particle position according equation (11.2)
            End

        While maximum iterations or minimum error criteria is not attained
Particles' velocities on each dimension are clamped to a maximum velocity V_{max}. If the sum of accelerations would cause the velocity on that dimension to exceed V_{max}, which is a parameter specified by the user. Then the velocity on that dimension is limited to V_{max}.

The basic flowchart of Particle Swarm optimization is as shown in Fig. 11.6. In that Repository refers to the memory location of each particle.

11.2.4 Comparison Between PSO and GA

The strength of GAs is in the parallel nature of their search. A GA implements a powerful form of hill climbing that preserves multiple solutions, eradicates unpromising solutions, and provides reasonable solutions. Through genetic operators, even weak solutions may continue to be part of the makeup of future candidate solutions. The genetic operators used are central to the success of the search. All GAs require some form of recombination, as this allows the creation of new solutions that have, by virtue of their parent's success, a higher probability of exhibiting a good performance. In practice, crossover is the principal genetic operator, whereas

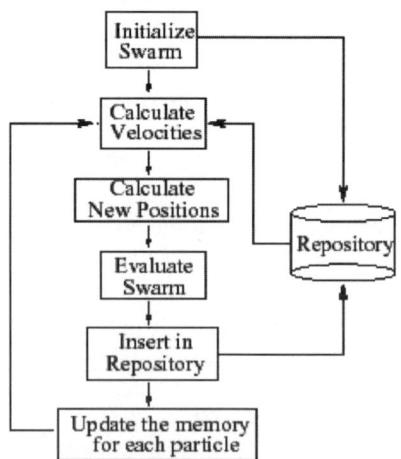

Fig. 11.6 Basic flowchart of PSO

11.2 Particle Swarm Optimization

mutation is used much less frequently. Crossover attempts to preserve the beneficial aspects of candidate solutions and to eliminate undesirable components, while the random nature of mutation is probably more likely to degrade a strong candidate solution than to improve it. Another source of the algorithm's power is the implicit parallelism inherent in the evolutionary metaphor. By restricting the reproduction of weak candidates, GAs eliminate not only that solution but also all of its descendants. This tends to make the algorithm likely to converge towards high quality solutions within a few generations.

Most of evolutionary techniques have the following procedure:

1. Random generation of an initial population
2. Reckoning of a fitness value for each subject. It will directly depend on the distance to the optimum.
3. Reproduction of the population based on fitness values.
4. If requirements are met, then stop. Otherwise go back to 2.

From the procedure, one can learn that PSO shares many common points with GA. Both algorithms start with a group of a randomly generated population; both have fitness values to evaluate the population. Both update the population and search for the optimum with random techniques. Both systems do not guarantee success. However, PSO does not have genetic operators like crossover and mutation. Particles update themselves with the internal velocity. They also have memory, which is important to the algorithm. Compared with genetic algorithms (GAs), the information sharing mechanism in PSO is significantly different. In GAs, chromosomes share information with each other. So the whole population moves like a one group towards an optimal area. In PSO, only gbest (or lbest) gives out the information to others. It is a one -way information sharing mechanism. The evolution only looks for the best solution. Compared with GA, all the particles tend to converge to the best solution quickly even in the local version in most cases.

Particle Swarm Optimization shares many similarities with evolutionary computation (EC) techniques in general and GAs in particular. All three techniques begin with a group of a randomly generated population, all utilize a fitness value to evaluate the population. They all update the population and search for the optimum with random techniques. A large inertia weight facilitates global exploration (search in new areas), while a small one tends to assist local exploration. The main difference between the PSO approach compared to EC and GA, is that PSO does not have genetic operators such as crossover and mutation. Particles update themselves with the internal velocity; they also have a memory that is important to the algorithm. Compared with EC algorithms (such as evolutionary programming, evolutionary strategy and genetic programming), the information sharing mechanism in PSO is significantly different. In EC approaches, chromosomes share information with each other, thus the whole population moves like one group towards an optimal area. In PSO, only the "best" particle gives out the information to others. It is a one-way information sharing mechanism; the evolution only looks for the best solution. Compared with ECs, all the particles tend to converge to the best solution quickly even in the local version in most cases. Compared to GAs, the advantages of PSO are that PSO is easy to implement and there are few parameters to adjust.

11.2.5 Applications of PSO

PSO has been successfully applied in many areas: function optimization, artificial neural network training, fuzzy system control, and other areas where GA can be applied. The various application areas of Particle Swarm Optimization include:

- Power Systems operations and control
- NP-Hard combinatorial problems
- Job Scheduling problems
- Vehicle Routing Problems
- Mobile Networking
- Modeling optimized parameters
- Batch process scheduling
- Multi-objective optimization problems
- Image processing and Pattern recognition problems

and so on. Currently, several researchers are being carried out in the area of particle swarm optimization and hence the application area also increases tremendously.

11.3 Ant Colony Optimization

Ant Colony Optimization (ACO) is a population-based, general search technique for the solution of difficult combinatorial problems, which is inspired by the pheromone trail laying behavior of real ant colonies. In ACO, a set of software agents called artificial ants search for good solutions to a given optimization problem. To apply ACO, the optimization problem is transformed into the problem of finding the best path on a weighted graph. The artificial ants (hereafter ants) incrementally build solutions by moving on the graph. The solution construction process is stochastic and is biased by a pheromone model, that is, a set of parameters associated with graph components (either nodes or edges) whose values are modified at runtime by the ants.

The first member of ACO class of algorithms, called Ant System, was initially proposed by Colorni, Dorigo and Maniezzo. The main underlying idea, loosely inspired by the behavior of real ants, is that of a parallel search over several constructive computational threads based on local problem data and on a dynamic memory structure containing information on the quality of previously obtained result. The collective behavior emerging from the interaction of the different search threads has proved effective in solving combinatorial optimization (CO) problems.

11.3.1 Biological Inspiration

In the 40s and 50s of the 20th century, the French entomologist Pierre-Paul Grass observed that some species of termites react to what he called "significant stimuli".

11.3 Ant Colony Optimization

He observed that the effects of these reactions can act as new significant stimuli for both the insect that produced them and for the other insects in the colony. Grass used the term stigmergy to describe this particular type of communication in which the "workers are stimulated by the performance they have achieved".

The two main characteristics of stigmergy that differentiate it from other forms of communication are the following.

- Stigmergy is an indirect, non-symbolic form of communication mediated by the environment: insects exchange information by modifying their environment; and
- Stigmergic information is local: it can only be accessed by those insects that visit the locus in which it was released (or its immediate neighborhood).

Stigmergy is an indirect and asynchronous form of communication in which the insects manipulate the environment to transport information to the other insects, which then respond to the change. The insects therefore do not have to be at the same place at the same time as the others to communicate with them. In many ant species colonies, stigmergy refers to the deposition of pheromone by ants while they are moving. Other ants can then smell the deposited pheromone and have a natural tendency to follow the laid trail. This constitutes an asynchronous and indirect communication scheme, where one ant communicates with other ants wherever they are, and it is how positive feedback is created. A little pheromone on a path might lead other ants to follow the same path, depositing even more pheromone, which can lead to a positive feedback effect, if the selected path is good (leading to food) thus recruiting even more ants to follow the path.

The main elements of a biological stigmergic system are shown in Fig. 11.7:

- The insect as the acting individual.
- The pheromone as an information carrier, used to create a dissipation field.
- The environment as a display and distribution mechanism for information.

Examples of stigmergy can be observed in colonies of ants. In many ant species, ants walking to and from a food source deposit on the ground a substance called pheromone. Other ants perceive the presence of pheromone and tend to follow paths where pheromone concentration is higher. Through this mechanism, ants are able to transport food to their nest in a remarkably effective way. The basic behavior of the ants is discussed as follows:

Real ants are capable of finding shortest path from a food source to the nest without using visual cues. Also, they are capable of adapting to changes in the environment, for example finding a new shortest path once the old one is no longer

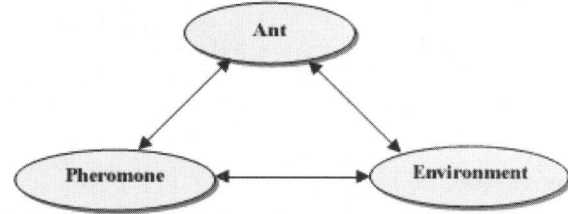

Fig. 11.7 Elements of a stigmeric system

Fig. 11.8 Movement of ant in a straight line

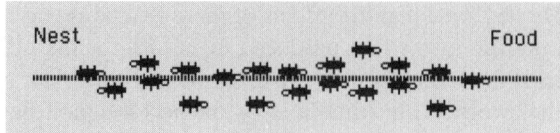

feasible due to a new obstacle. Consider the following Fig. 11.8 in which ants are moving on a straight line, which connects a food source to the nest:

It is well known that the main means used by ants to form and maintain the line is a pheromone trail. Ants deposit a certain amount of pheromone while walking, and each ant probabilistically prefers to follow a direction rich in pheromone rather than a poorer one. This elementary behavior of real ants can be used to explain how they can find the shortest path, which reconnects a broken line after the sudden appearance of an unexpected obstacle, has interrupted the initial path (Fig. 11.9).

In fact, once the obstacle has appeared, those ants, which are just in front of the obstacle, cannot continue to follow the pheromone trail and therefore they have to choose between turning right or left. In this situation we can expect half the ants to choose to turn right and the other half to turn left. The very same situation can be found on the other side of the obstacle (Fig. 11.10).

It is interesting to note that those ants which choose, by chance, the shorter path around the obstacle will more rapidly reconstitute the interrupted pheromone trail compared to those which choose the longer path. Hence, the shorter path will receive a higher amount of pheromone in the time unit and this will in turn cause a higher number of ants to choose the shorter path. Due to this positive feedback (autocatalytic) process, very soon all the ants will choose the shorter path (Fig. 11.11).

The most interesting aspect of this autocatalytic process is that finding the shortest path around the obstacle seems to be an emergent property of the interaction between the obstacle shape and ants distributed behavior: Although all ants move at approximately the same speed and deposit a pheromone trail at approximately the same rate, it is a fact that it takes longer to contour obstacles on their longer side than on their shorter side which makes the pheromone trail accumulate quicker on the shorter side. It is the ants' preference for higher pheromone trail levels, which makes this accumulation still quicker on the shorter path.

Deneubourg et al. thoroughly investigated the pheromone laying and following behavior of ants. In an experiment known as the "double bridge experiment", the nest of a colony of Argentine ants was connected to a food source by two bridges

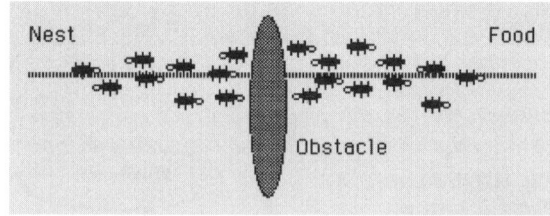

Fig. 11.9 Obstacle on ant paths

11.3 Ant Colony Optimization

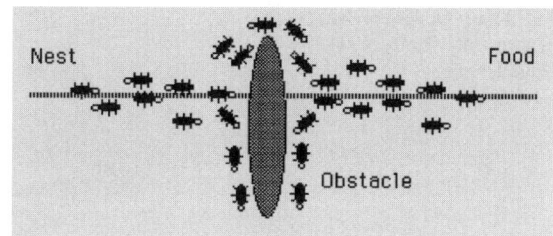

Fig. 11.10 Behavior of ants to obstacle

of equal lengths, as shown in Fig. 11.12. In such a setting, ants start to explore the surroundings of the nest and eventually reach the food source. Along their path between food source and nest, Argentine ants deposit pheromone. Initially, each ant randomly chooses one of the two bridges. However, due to random fluctuations, after some time one of the two bridges presents a higher concentration of pheromone than the other and, therefore, attracts more ants. This brings a further amount of pheromone on that bridge making it more attractive with the result that after some time the whole colony converges toward the use of the same bridge.

This colony-level behavior, based on autocatalysis, that is, on the exploitation of positive feedback, can be used by ants to find the shortest path between a food source and their nest. Goss et al. considered a variant of the double bridge experiment in which one bridge is significantly longer than the other, as shown in Fig. 11.13. In this case, the stochastic fluctuations in the initial choice of a bridge are much reduced and a second mechanism plays an important role: the ants choosing by chance the short bridge are the first to reach the nest. The short bridge receives, therefore, pheromone earlier than the long one and this fact increases the probability that further ants select it rather than the long one. Goss et al. developed a model of the observed behavior: assuming that at a given moment in time m_1 ants have used the first bridge and m_2 the second one, the probability p_1 for an ant to choose the first bridge is:

$$p_1 = \frac{(m_1 + k)^h}{(m_1 + k)^h + (m_2 + k)^h}$$

where parameters k and h are to be fitted to the experimental data—obviously $p_2 = 1 - p_1$.

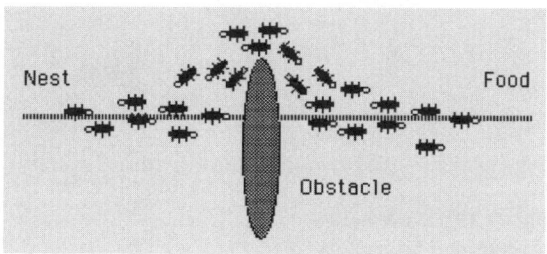

Fig. 11.11 Ants choosing shorter path

Fig. 11.12 Double bridge experiment—Bridges of equal length

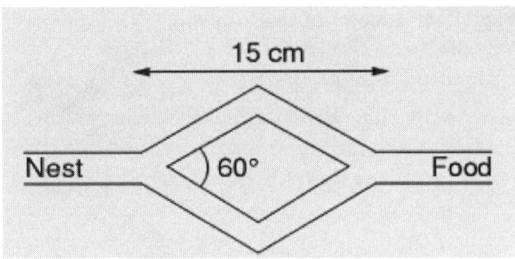

11.3.2 Similarities and Differences Between Real Ants and Artificial Ants

Most of the ideas of ACO stem from real ants. In particular, the use of a colony cooperating individuals, an artificial pheromone trail for local stigmergetic communication a sequence of local moves to find shortest paths and a stochastic decision policy using local information. Researchers have used the most ideas from real ants behavior in order to build Ant System (AS). There exist some differences and similarities between real and artificial ants which could be stated as follows:

11.3.2.1 Similarities

- Colony of cooperating individuals–Both real ant colonies and ant algorithms are composed of a population, or colony of independent individual agents. They globally cooperate in order to find a good solution to the task under consideration. Although the complexity of each artificial ant is such that it can build a feasible solution (as a real ant can find somehow a path between the nest and the food), high quality solutions are the result of the cooperation among the individuals of the whole colony.
- Pheromone trail and stigmergy–Like real ants, artificial ants change some aspects of their environment while walking. Real ants deposit a chemical substance called pheromone on the visited state. Artificial ants will change some numerical information of the problem state, locally stored, when that state is visited. Based on analogy, these information and changes could be called an artificial pheromone trail. Ant system algorithms assume that a local pheromone trail is the single

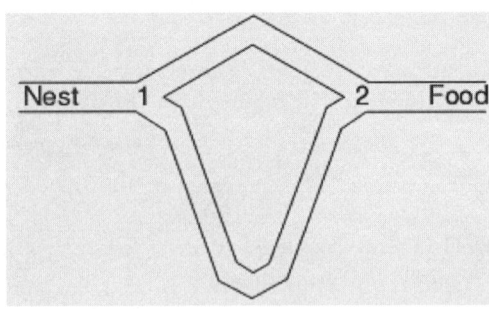

Fig. 11.13 Double bridge experiment—Bridges of varying length

way of communication among artificial ants. AS algorithms include artificial pheromone evaporation in form of reduction in the artificial pheromone trail over time as in nature. Pheromone evaporation in nature and in AS algorithms are important because it will allow ant colony to slowly forget the history and direct the searching process in new directions. Artificial pheromone evaporation could be helpful to move the searching process toward new regions and to avoid stacking in local extremes.
- Local moves and the shortest path searching–Despite real ants are walking through adjacent states and artificial ants are jumping from one to another adjacent state of the considered problem, both walking and jumping have the same purpose, which is finding the shortest path between the origin and the destination.
- Transition policy–Both real ants and artificial ones will build solutions by applying decision making procedures to move through adjacent states. Decision making procedures could be based on some probabilistic rules or probabilities could be calculated based on approximate reasoning rules. In both cases, the transition policy will use local information that should be local in the space and time sense. The transition policy is a function of local state information represented by problem specifications (this could be equivalent to the terrain's structure that surrounds the real ants) and the local modification of the environment (existing pheromone trails) introduced by ants that have visited the same location.
- Deposited amount of pheromone–Amount of pheromone that an artificial ant will deposit is mostly a function of the quality of the discovered solution. In nature, some ants behave in a similar way, the deposited amount of pheromone is highly dependent on the quality of the discovered food source.

11.3.2.2 Differences

The differences are as follows:

- Artificial ants live in a discrete world. All their moves are jumps from one discrete state to another adjacent one.
- Artificial ants have memory, they could remember states that have been visited already (tabu lists in the model).
- Pheromone deposit methodology is significantly different between real and artificial ants. Timing in pheromone laying is problem dependent and often does not have similarities with the real ants pheromone deposit methodology.
- To improve overall performance, AS algorithms could be enriched with some additional capabilities that cannot be found in real ant colonies. Most AS contains some local optimization techniques to improve solutions developed by ants.

11.3.3 Characteristics of Ant Colony Optimization

The characteristics of ant colony optimization are as follows:

- **Natural algorithm** since it is based on the behavior of real ants in establishing paths from their colony to source of food and back.

- **Parallel and distributed** since it concerns a population of agents moving simultaneously, independently and without a supervisor.
- **Cooperative** since each agent chooses a path on the basis of the information, pheromone trails laid by the other agents, which have previously selected the same path. This cooperative behavior is also autocatalytic, i.e., it provides a positive feedback, since the probability of choosing a path increases with the number of agents that previously chose that path.
- **Versatile** that it can be applied to similar versions of the same problem; for example, there is a straightforward extension from the traveling salesman problem (TSP) to the asymmetric traveling salesman problem (ATSP).
- **Robust** that it can be applied with minimal changes to other combinatorial optimization problems such as quadratic assignment problem (QAP) and the job-shop scheduling problem (JSP).

11.3.4 Ant Colony Optimization Algorithms

The model proposed by Deneubourg and co-workers for explaining the foraging behavior of ants was the main source of inspiration for the development of ant colony optimization. In ACO, a number of artificial ants build solutions to the considered optimization problem at hand and exchange information on the quality of these solutions via a communication scheme that is reminiscent of the one adopted by real ants.

Different ant colony optimization algorithms have been proposed. The original ant colony optimization algorithm is known as Ant System and was proposed in the early 90s. Since then, a number of other ACO algorithms were introduced. Table 11.1 gives a list of successful variants of Ant Colony Optimization Algorithms. Figure 11.14 gives a narrow overview on Ant Colony Optimization Algorithms.

ACO is a class of algorithms, whose first member, called Ant System, was initially proposed by Colorni, Dorigo and Maniezzo. The main underlying idea, loosely inspired by the behavior of real ants, is that of a parallel search over several

Table 11.1 Development of various Ant Colony Optimization Algorithms

ACO algorithm	Authors	Year
Ant System	Dorigo, Maniezzo & Colomi	1991
Elitist AS	Dorigo	1992
Ant-Q	Gambardella & Dorigo	1995
Ant Colony System	Dorigo & Gambardella	1996
$\mathcal{MMAS}$	Stützle & Hoos	1996
Rank-based AS	Bullnheimer, Hartl & Strauss	1997
ANTS	Maniezzo	1998
Best-Worst AS	Cordón, et al.	2000
Hyper-cube ACO	Blum, Roli, Dorigo	2001

11.3 Ant Colony Optimization

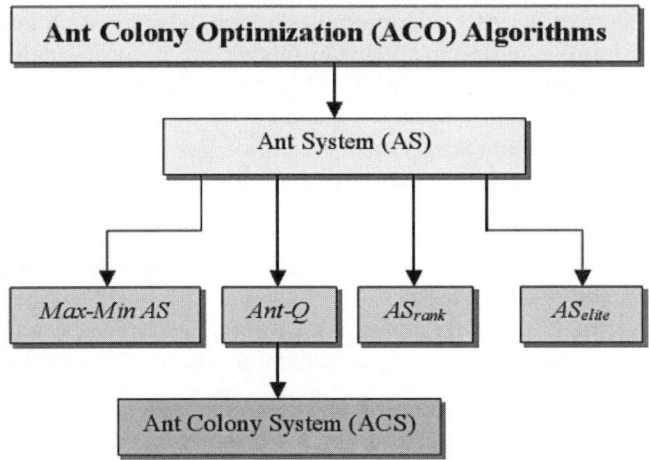

Fig. 11.14 Overview of ACO algorithms

constructive computational threads based on local problem data and on a dynamic memory structure containing information on the quality of previously obtained result. The collective behavior emerging from the interaction of the different search threads has proved effective in solving combinatorial optimization (CO) problems.

The notation is used as follows: A combinatorial optimization problem is a problem defined over a set $C = c_1, \ldots, c_n$ of basic components. A subset S of components represents a solution of the problem; $F \subseteq 2^C$ is the subset of feasible solutions, thus a solution S is feasible if and only if $S \in F$. A cost function z is defined over the solution domain, $z : 2^C \to R$, the objective being to find a minimum cost feasible solution S^*, i.e., to find $S^* : S^* \in F$ and $z(S^*) \leq z(S), \forall S \in F$.

Given this, the functioning of an ACO algorithm can be summarized as follows. A set of computational concurrent and asynchronous agents (a colony of ants) moves through states of the problem corresponding to partial solutions of the problem to solve. They move by applying a stochastic local decision policy based on two parameters, called trails and attractiveness. By moving, each ant incrementally constructs a solution to the problem. When an ant completes a solution, or during the construction phase, the ant evaluates the solution and modifies the trail value on the components used in its solution. This pheromone information will direct the search of the future ants.

Furthermore, an ACO algorithm includes two more mechanisms: trail evaporation and, optionally, daemon actions. Trail evaporation decreases all trail values over time, in order to avoid unlimited accumulation of trails over some component. Daemon actions can be used to implement centralized actions which cannot be performed by single ants, such as the invocation of a local optimization procedure, or the update of global information to be used to decide whether to bias the search process from a non-local perspective.

More specifically, an ant is a simple computational agent, which iteratively constructs a solution for the instance to solve. Partial problem solutions are seen as states. At the core of the ACO algorithm lies a loop, where at each iteration, each ant moves (performs a step) from a state ι to another one ψ, corresponding to a more complete partial solution. That is, at each step σ, each ant k computes a set $A_k^\sigma(t)$ of feasible expansions to its current state, and moves to one of these in probability. The probability distribution is specified as follows. For ant k, the probability $p_{\tau\psi}^k$ of moving from state ι to state ψ depends on the combination of two values:

- the attractiveness $n_{\iota\psi}$ of the move, as computed by some heuristic indicating the a priori desirability of that move;
- the trail level $\tau_{\iota\psi}$ of the move, indicating how proficient it has been in the past to make that particular move: it represents therefore an a posteriori indication of the desirability of that move.

Trails are updated usually when all ants have completed their solution, increasing or decreasing the level of trails corresponding to moves that were part of "good" or "bad" solutions, respectively. The general framework just presented has been specified in different ways by the authors working on the ACO approach. A brief introduction is given for Ant System (AS) and Ant Colony System (ACS) to traveling salesman problem.

11.3.4.1 Ant System

Ant System is the first ACO algorithm proposed in the literature. Ant System applied to traveling Sales Man problem is discussed here. Its main characteristic is that, at each iteration, the pheromone values are updated by all the m ants that have built a solution in the iteration itself. The pheromone τ_{ij} associated with the edge joining cities i and j, is updated as follows:

$$\tau_{ij} \leftarrow (1-\rho) \cdot \tau_{ij} + \sum_{k=1}^{m} \Delta \tau_{ij}^k, \qquad (11.3)$$

where ρ is the evaporation rate, m is the number of ants, and $\Delta \tau_{ij}^k$ is the quantity of pheromone laid on edge (i, j) by ant k:

$$\Delta \tau_{ij}^k = \begin{cases} Q/L_k & \text{if ant } k \text{ used edge } (i, j) \text{ in its tour,} \\ 0 & \text{otherwise,} \end{cases} \qquad (11.4)$$

Where Q is a constant, and L_k is the length of the tour constructed by ant k.

In the construction of a solution, ants select the following city to be visited through a stochastic mechanism. When ant k is in city i and has so far constructed the partial solution s^P probability of going to city j is given by:

11.3 Ant Colony Optimization

$$p_{ij}^k = \begin{cases} \dfrac{\tau_{ij}^\alpha \cdot \eta_{ij}^\beta}{\sum_{c_{ij} \in N(s^P)} \tau_{ij}^\alpha \cdot \eta_{ij}^\beta} & \text{if } c_{ij} \in N(s^P), \\ 0 & \text{otherwise,} \end{cases} \quad (11.5)$$

where $N(s^P)$ is the set of feasible components; that is, edges (i, 1) where l is the city not yet visited by the ant k. The parameters α and β control the relative importance of the pheromone versus the heuristic information η_{ij}, which is given by,

$$\eta_{ij} = \frac{1}{d_{ij}}, \quad (11.6)$$

where d_{ij} is the distance between cities i and j.

11.3.4.2 Ant Colony System

The most interesting contribution of ACS is the introduction of a local pheromone update in addition to the pheromone update performed at the end of the construction process (called offline pheromone update). The local pheromone update is performed by all the ants after each construction step. Each ant applies it only to the last edge traversed:

$$\tau_{ij} = (1 - \varphi) \cdot \tau_{ij} + \varphi \cdot \tau_0, \quad (11.7)$$

where $\varphi \in (0, 1)$ is the pheromone decay coefficient, and τ_0 is the initial value of the pheromone.

The main goal of the local update is to diversify the search performed by subsequent ants during an iteration: by decreasing the pheromone concentration on the traversed edges, ants encourage subsequent ants to choose other edges and, hence, to produce different solutions. This makes it less likely that several ants produce identical solutions during one iteration. The offline pheromone update is applied at the end of each iteration by only one ant, which can be either the iteration-best or the best-so-far. However, the update formula is:

$$\tau_{ij} \leftarrow \begin{cases} (1 - \rho) \cdot \tau_{ij} + \rho \cdot \Delta \tau_{ij} & \text{if } (i, j) \text{ belongs to best tour,} \\ \tau_{ij} & \text{otherwise.} \end{cases} \quad (11.8)$$

where $\Delta \tau_{ij} = 1/L_{best}$, where L_{best} can be either L_{ib} or L_{bs}. L_{best} is the length of the tour of the best ant. This may be (subject to the algorithm designer decision) either the best tour found in the current iteration—*iteration-best*, L_{ib}—or the best solution found since the start of the algorithm—*best-so-far*, L_{bs}—or a combination of both.

Another important difference between ACS and AS is in the decision rule used by the ants during the construction process. In ACS, the so-called *pseudorandom proportional rule* is used: the probability for an ant to move from city i to city j depends on a random variable q uniformly distributed over [0, 1], and a parameter q_0; if $q \leq q_0$, then $j = \arg\max_{c_{ij} \in N(s^p)} \{\tau_{il} \eta_{il}^{\beta}\}$ otherwise (11.5) is used.

11.3.4.3 Basic Flow of ACO

The basic operational flow in Ant Colony Optimization is as follows:

Step 1: Represent the solution space by a construction graph
Step 2: Initialize ACO parameters
Step 3: Generate random solutions from each ant's random walk
Step 4: Update pheromone intensities
Step 5: Goto Step 3, and repeat until convergence or a stopping condition is satisfied.

The generic ant algorithm is given as shown in Fig. 11.15,
    The generalized flowchart for Ant Colony Optimization algorithm is as shown in Fig. 11.16.
    The step by step procedure to solve combinatorial optimization problems using ACO in a nutshell is:

- Represent the problem in the form of sets of components and transitions or by means of a weighted graph that is travelled by the ants to build solutions.
- Appropriately define the meaning of the pheromone trails, i.e., the type of decision they bias. This is a crucial step in the implementation of an ACO algorithm. A good definition of the pheromone trails is not a trivial task and it typically requires insight into the problem being solved.
- Appropriately define the heuristic preference to each decision that an ant has to take while constructing a solution, i.e., define the heuristic information associated to each component or transition. Notice that heuristic information is crucial for good performance if local search algorithms are not available or can not be applied.
- If possible, implement an efficient local search algorithm for the problem under consideration, because the results of many ACO applications to NP-hard

---

Step 1: Initilization
  – Initialize the pheromone trail
Step 2: Iteration
  – For each Ant Repeat
    – Solution construction using the current pheromone trail
    – Evaluate the solution constructed
    – Update the pheromone trail
  – Until stopping criteria

Fig. 11.15 A generic ant algorithm

11.3 Ant Colony Optimization

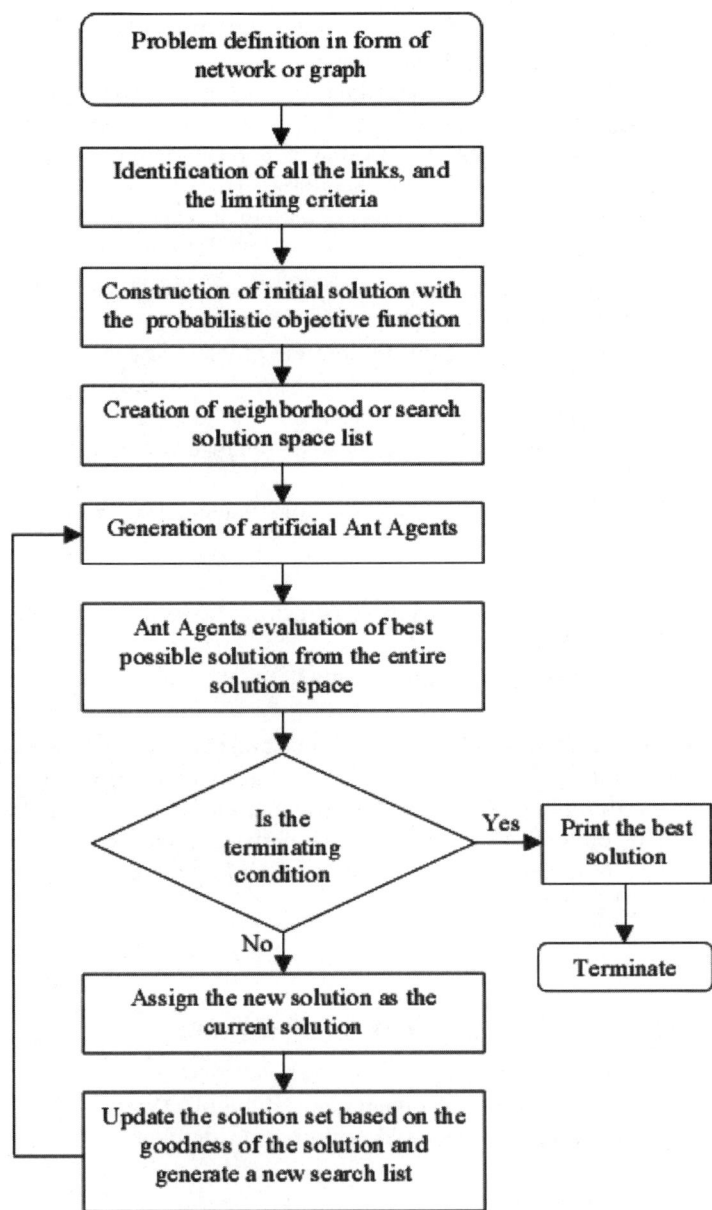

Fig. 11.16 Basic flowchart of Ant Colony Optimization Algorithm

combinatorial optimization problems show that the best performance is achieved when coupling ACO with local optimizers.
- Choose a specific ACO algorithm and apply it to the problem being solved, taking the previous aspects into consideration.
- Tune the parameters of the ACO algorithm. A good starting point for parameter tuning is to use parameter settings that were found to be good when applying the ACO algorithm to similar problems or to a variety of other problems.

It should be clear that the above steps can only give a very rough guide to the implementation of ACO algorithms. In addition, the implementation is often an iterative process, where with some further insight into the problem and the behavior of the algorithm; some initially taken choices need to be revised. Finally, we want to insist on the fact that probably the most important of these steps are the first four, because a poor choice at this stage typically can not be made up with pure parameter fine-tuning.

An ACO algorithm iteratively performs a loop containing two basic procedures, namely:

- A procedure specifying how the ants construct/modify solutions of the problem to be solved;
- A procedure to update the pheromone trails.

The construction/modification of a solution is performed in a probabilistic way. The probability of adding a new item to the current partial solution is given by a function that depends on a problem-dependent heuristic and on the amount of pheromone deposited by ants on the trail in the past. The updates in the pheromone trail are implemented as a function that depends on the rate of pheromone evaporation and on the quality of the produced solution.

11.3.5 Applications of Ant Colony Optimization

There are numerous successful implementations of the ACO meta-heuristic applied to a number of different combinatorial optimization problems. The applications include:

- Traveling Salesman Problem, where a salesman must find the shortest route by which he can visit a given number of cities, each city exactly once.
- Quadratic Assignment Problem, the problem of assigning n facilities to n locations so that the costs of the assignment are minimized.
- Job-Shop Scheduling Problem, where a given set of machines and set of job operations must be assigned to time intervals in such a way that no two jobs are processed at the same time on the same machine and the maximum time of completion of all operations is minimized.

11.3 Ant Colony Optimization

- Vehicle Routing Problem, the objective is to find minimum cost vehicle routes such that:

  (a) Every customer is visited exactly once by exactly one vehicle;
  (b) For every vehicle the total demand does not exceed the vehicle capacity;
  (c) The total tour length of each vehicle does not exceed a given limit;
  (d) Every vehicle starts and ends its tour at the same position.

- Shortest Common Super sequence Problem, where—given a set of strings over an alphabet—a string of minimal length that is a super sequence of each string of the given set has to be found (a super sequence S of string A can be obtained from A by inserting zero or more characters in A).
- Graph-Coloring Problem, which is the problem of finding a coloring of a graph so that the number of colors used is minimal.
- Sequential Ordering Problem, which consists of finding a minimum weight Hamiltonian path 2 on a directed graph with weights on the arcs and on the nodes, subject to precedent constraints among the nodes.
- Connection-Oriented Network Routing, where all packets of the same session follow the same path selected by a preliminary setup phase.
- Connectionless Network Routing where data packets of the same session can follow different paths (Internet-type networks).

Table 11.2 gives an overall view for the application areas of ant colony optimization and researchers who performed it.

Table 11.2 Applications of ACO algorithms

PROBLEM TYPE	PROBLEM NAME	AUTHORS	YEAR
ROUTING	TRAVELING SALESMAN	DORIGO ET AL.	1991, 1996
		DORIGO & GAMBARDELLA	1997
		STÜTZLE & HOOS	1997, 2000
	VEHICLE ROUTING	GAMBARDELLA ET AL.	1999
		REIMANN ET AL	2004
	SEQUENTIAL ORDERING	GAMBARDELLA & DORIGO	2000
ASSIGNMENT	QUADRATIC ASSIGNMENT	STÜTZLE & HOOS	2000
		MANIEZZO	1999
	COURSE TIMETABLING	SOCHA ET AL.	2002, 2003
	GRAPH COLORING	COSTA & HERTZ	1997
SCHEDULING	PROJECT SCHEDULING	MERKLE ET AL.	2002
	TOTAL WEIGHTED TARDINESS	DEN BESTEN ET AL.	2000
		MERKLE & MIDDENDORF	2000
	OPEN SHOP	BLUM	2005
SUBSET	SET COVERING	LESSING ET AL.	2004
	l-CARDINALITY TREES	BLUM & BLESA	2005
	MULTIPLE KNAPSACK	LEGUIZAMÓN & MICHALEWICZ	1999
	MAXIMUM CLIQUE	FENET & SOLNON	2003
OTHER	CONSTRAINT SATISFACTION	SOLNON	2000, 2002
	CLASSIFICATION RULES	PARPINELLI ET AL.	2002
		MARTENS ET AL.	2006
	BAYESIAN NETWORKS	CAMPOS, ET AL.	2002
	PROTEIN FOLDING	SHMYGELSKA & HOOS	2005
	PROTEIN-LIGAND DOCKING	KORB ET AL.	2006

11.4 Summary

In this chapter, the basic concepts of Particle Swarm Optimization and Ant Colony Optimization are discussed. PSO and ACO algorithms operate on the social behavior of the living organisms. PSO system combines local search methods with global search methods, attempting to balance exploration and exploitation. Thus, PSO is an extremely simple algorithm that seems to be effective for optimizing a wide range of functions. Also, PSO is attractive is that there are only few parameters to adjust. Particle swarm optimization has been used for approaches that can be used across a wide range of applications, as well as for specific applications focused on a specific requirement. Ant Colony Optimization has been and continues to be a fruitful paradigm for designing effective combinatorial optimization solution algorithms. After more than ten years of studies, both its application effectiveness and its theoretical groundings have been demonstrated, making ACO one of the most successful paradigms in the metaheuristic area.

Review Questions

1. Give the history of the development of swarm intelligence.
2. Define: particles and ants.
3. How are social behaviors of living organisms helpful in developing optimization techniques?
4. Explain in detail on the operation of Particle Swarm Optimization.
5. Mention the advantages and applications of Particle Swarm optimization.
6. Discuss the behavior of real ants and compare it with artificial ants.
7. List the various types of ant colony optimization algorithms.
8. With a neat flowchart, explain the algorithm of Ant Colony Optimization
9. State the various applications of ACO.
10. Compare and Contrast—Genetic Algorithm, Particle Swarm Optimization and Ant Colony Optimization.

Exercise Problems

1. Write a MATLAB program to implement particle swarm optimization and ant colony optimization for traveling salesman problem.
2. Implement a vehicle routing problem using the concept of particle swarm optimization.
3. Write a computer program to perform ant colony optimization for a Graph Coloring Problem.
4. Ant colony optimization algorithm is best suited for protein-folding problems—Justify
5. Develop a C program for performing image segmentation using particle swarm optimization.

Bibliography

1. Dawkins, R. (1989). The Selfish Gene - New Ed. Oxford University Press, Great Britain.
2. Fraser, A. P. (1994). Genetic Programming in C++. Technical report 040, University of Salford.
3. Goldberg, D. E. & Smith, R. E. (1987) Nonstationary Function Optimization using Genetic Algorithms with Diploidy and Dominance. In J.J Grefenstette, editor, Proceedings of the Second International Conference on Genetic Algorithms, 59–68. Lawrence Erlbaum Associates.
4. Hadad B. S. & Eick C. F. (1997) Supporting Polyploidy in Genetic Algorithms Using Dominance Vectors. In P.J. Angeline et al. (eds.), Proceedings of the Sixth International Conference on Evolutionary Programming, 223–234. Berlin: Springer-Verlag.
5. Koza, J.R. (1992). Genetic Programming: On the Programming of Computers by Means of Natural Selection. Cambridge, MA:MIT Press.
6. Merrell, D. J. (1994) The Adaptive Seascape: The Mechanism for Evolution. University of Minnesota Press, Minneapolis.
7. Vekaria K. & Clack C. (1997) Genetic Programming with Gene Dominance. In J. Koza (editor). Late Breaking Papers at the Genetic Programming 1997 Conference, 300. Stanford CA:Stanford University Bookstore.
8. Emma Collingwood, David Corne and Peter Ross, "Useful Diversity via Multiploidy," IEEE International Conference on Evolutionary Computing, Nagoya, Japan, 1996.
9. Richard Dawkins, The Extended Phenotype, Oxford University Press, 1982.
10. Tomofumi Hikage, Hitoshi Hemmii, and Katsunori Shimohara, "Diploid Chromosome and Progressive Evolution Model for Real-Time hardware Evolution," Fourth European Conference on Artificial Life, 1997.
11. Daniel Hillis, "Co-evolving parasites improve simulated evolution as an optimization procedure," Artificial Life II, Addison Wesley, 1992.
12. Young-il Kim, et. al., "Winner take all strategy for a Diploid Genetic Algorithm.", The first Asian Conference on simulated Evolution and Learning, 1996.
13. Orazio Miglino, Stefano Nolfi, and Domenico Parisi. "Discontinuity in evolution: how different levels of organization imply pre-adaptation."Technical Report, Institute of Psychology, National Research Council of Italy. 1993.
14. Melanie Mitchell, An Introduction to Genetic Algorithms, MIT Press, 1996.
15. P. Osmera, V. Kvasnicka, J. Pospichal, "Genetic Algorithms with Diploid Chromosomes," Mendel '97, PC-DIR Brno, 1997, ISBN 80-214-0884-7, pp. 111–116.
16. Smith, R.E., & Goldberg, D.E. (1992). Diploidy and dominance in artificial genetic search. Complex Systems, 6(3). 251–285.
17. Deborah Stacey, "Diploidy and Dominance," online course notes for "Topics in Artificial Intelligence." University of Guelph, Ontario, Canada: http://hebb.cis.uoguelph.ca/~deb/27662/Lectures/diploidy.html
18. J.D. Bagley. The Beh.aviour of Adaptive Systems Which Employ Genetic and Correlation Algorithms. PhD thesis, University of Michigan, 1967.
19. Dipankar Dasgupta and Douglas R. McGregor. Using structured genetic algorithms for solving deceptive problems. Technical re- port, University of Strathclyde Department of Computer Science, 1993.

20. Darrell Whitley. The GENITOR algorithm and selection pressure. In J. D. Schaffer, editor, Proceedings of the Third International Conference on Genetic Algorithms, pages 161–121. San Mateo: Morgan Kaufmann,1989.
21. Yukiko Yoshida and Nobue Adachi. A diploid genetic algorithm for preserving population diversity - pseudo-meiosis ga. In Manner Davidor, Schwefel, editor, Parallel Problem Solving from Nature: PPSN 111, pages 36–45. Springer-Verlag, 1994.
22. Beasley, D., Bull, D. R., and Martin, R. R. (1993). A sequential technique for multimodal function optimization, Evolutionary Computation, volume 1, number 1, MIT Press, MA.
23. Cavicchio, D. J. (1970). Adaptive search using simulated evolution. Ph.D. thesis, University of Michigan, Ann Arbor, MI.
24. Cedeño, W. (1995). The multi-niche crowding genetic algorithm: analysis and applications. UMI Dissertation Services, 9617947.
25. Cedeño, W. and Vemuri, V. (1996). Database design with genetic algorithms. D. Dasgupta and Z. Michalewicz (eds), Evolutionary Algorithms in Engineering Applications, Springer Verlag, 3/97.
26. Cedeño, W., Vemuri, V., and Slezak, T. (1995). Multi-Niche crowding in genetic algorithms and its application to the assembly of DNA restriction-fragments. Evolutionary Computation, 2:4, 321–345.
27. Cedeño, W. and Vemuri, V. (1992). Dynamic multimodal function optimization using genetic algorithms. In Proceedings of the XVIII Latin-American Informatics Conference, Las Palmas de Gran Canaria, Spain: University of Las Palmas, 292–301.
28. Cobb, H. J. and Grefenstette, J. J. (1993). Genetic algorithms for tracking changing environments. In S. Forrest (ed.) Proceedings of the Fifth International Conference on Genetic Algorithms. Morgan Kaufmann Publishers San Mateo, California, 523–530.
29. Dasgupta, D. & McGregor, D. R. (1992). Non-stationary function optimization using the structured genetic algorithm. In R. Manner and B. Manderick (eds.), Parallel Problem Solving from Nature, 2. Amsterdam: North Holland, 145–154.
30. De Jong, K. A. (1975). An analysis of the behaviour of a class of genetic adaptive systems. Doctoral dissertation, University of Michigan. Dissertation Abstracts International 36(0), 5140B. (University Microfilms No. 76–9381).
31. Deb, K. and Goldberg, D. E. (1989). An investigation of niche and species formation in genetic function optimization, In J. D. Schaffer (Ed.), Proceedings of the Third International Conference on Genetic Algorithms. San Mateo, CA: Morgan Kaufmann, 42–50.
32. Goldberg, D. E. (1989). Genetic Algorithms in Search, Optimization & Machine Learning. Reading MA: Addison-Wesley.
33. Goldberg, D. E., & Richardson, J. (1987). Genetic algorithms with sharing for multimodal function optimization. In J. J. Grefenstette (Ed.), Proceedings of the Second International Conference on Genetic Algorithms. Hillsdale, NJ: Lawrence Erlbaum Associates, 41–49.
34. Goldberg D. E. & Smith R. E. (1987). Non-stationary function optimization using genetic algorithms with dominance and diploidy. In J. J. Grefenstette (Ed.), Proceedings of the Second International Conference on Genetic Algorithms. Hillsdale, NJ
35. Grefenstette, J.J. (1992). Genetic algorithms for changing environments. In R. Manner and B. Manderick (eds.), Parallel Problem Solving form Nature, 2. Amsterdam: North Holland, 137–144.
36. Harik, G. R. (1995). Finding multimodal solutions using restricted tournament selection. In L. J. Eshelman (ed.), Proceedings of the Sixth International Conference on Genetic Algorithms. San Mateo, CA:Morgan Kaufmann Publishers, 24–31.
37. Holland, J. H. (1975). Adaptation in natural and artificial systems, Ann Arbor MI: The University of Michigan Press.
38. Mahfoud, S. W. (1992). Crowding and preselection revisited. In R. Männer & B. Manderick (Eds.), Proceedings of Parallel Problem Solving from Nature 2. New York, NY: Elsevier Science B. V., 27–36.
39. Maresky, J., Davidor, Y., Gitler, D., Aharoni, G., and Barak, A. (1995). Selectivelydestructive restart. In L. J. Eshelman (ed.), Proceedings of the Sixth International Conference on Genetic Algorithms. San Mateo, CA:Morgan Kaufmann Publishers.

40. Ng, K. P. & Wong, K. C. (1995). A new diploid scheme and dominance change mechanism for non-stationary function optimization. In L. J. Eshelman (ed.), Proceedings of the Sixth International Conference on Genetic Algorithms. San Mateo, CA:Morgan Kaufmann Publishers.
41. Nix, A. & Vose, M. D. (1992). Modeling genetic algorithms with Markov chains, Annals of Mathematics and Artificial Intelligence 5, 79–88.
42. Spears, W. M. (1994). Simple subpopulation schemes, in Proceedings of the 94 Evolutionary Programming Conference, San Diego, CA.
43. Syswerda, G. (1989). Uniform crossover in genetic algorithms. In J. D. Schaffer (Ed.), Proceedings of the Third International Conference on Genetic Algorithms. San Mateo, CA: Morgan Kaufmann, 2–9.
44. Whitley, D. (1988). GENITOR: a different genetic algorithm. In Proceedings of the Rocky Mountain Conference on Artificial Intelligence. Denver Colorado, 118–130.
45. KwanWoo Kim, Mitsuo Gen, MyoungHun Kim, "Adaptive Genetic Algorithms for Multi-Resource Constrained Project Scheduling Problem with Multiple Modes", International Journal of Innovative Computing, Information and Control ICIC pp.1349–4198,Volume 2, Number 1, February 2006
46. Gen, M. and R. Cheng, Genetic Algorithm and Engineering Optimization, John Wily and Sons, New York, 2000.
47. Bouleimen, K. and H. Lecocq, A new e.cient simulated annealing algorithm for the resource constrained project scheduling problem and its multiple mode version, European Journal of Operational Research, vol.144, pp.268–281, 2003.
48. Heilmann, R., A branch-and-bound procedure for the multi-mode resource-constrained project scheduling problem with minimum and maximum time lags, European Journal of Operational Research, vol.144, pp.348–365, 2003.
49. Mak, K. L., Y. S. Wong and X. X. Wang, An adaptive genetic algorithm for manufacturing cell formation, International Journal of Manufacturing Technology, vol.16, pp.491–497, 2000.
50. Michalewicz, Z., Genetic Algorithm + Data Structure = Evolution Programs, Third Edition, Springer-Verlag, New York, 1996.
51. Ozdamar, L., A genetic algorithm approach to a general category project scheduling problem, IEEE Transactions on Systems, Man, and Cybernetics -Part C: Applications and Reviews, vol.29, no.1, pp.44–59, 1999.
52. Reyck, B. D. and W. Herroelen, The multi-mode resource-constrained project scheduling problem with generalized precedence relations, European Journal of Operational Research, vol.119, pp.538–556, 1999.
53. im, K., Y. Yun, J. Yoon, M. Gen and G. Yamazaki, Hybrid genetic algorithm with adaptive abilities for resource-constrained multiple project scheduling, Computer in Industry, vol.56, pp.143–160, 2004.
54. Wei-Guo Zhang, Wei CHEN, and Ying-Luo Wang, "The Adaptive Genetic Algorithms for Portfolio Selection Problem", IJCSNS International Journal of Computer Science and Network Security, Vol.6 No.1, January 2006.
55. J. H. Holland, Adaptation in Natural and Artificial Systems. University of Michigan Press. Ann Arbor, 1975.
56. W. E. Hart, The role of development in genetic algorithms. In D. Whitley and M. Vose (Eds.), Foundations of Genetic Algorithms 3. Morgan Kaufmann. 1994.
57. David B.Fogel. "An Introduction to Simulated Evolutionary Optimization". IEEE Trans Neural Networks, 1994, Jan.5(1).
58. Yusen Xia, Baoding Liu, Shouyang Wang, K.K Lai.A new model for portfolio selection with order of expected returns. Computers and Operations Research 2000, 27:409–22.
59. Yusen Xia, Shouyang Wang, Xiaotie Deng. A compromise solution to mutual funds portfolio selection with transaction costs. European Journal of Operations Research, 2001,134: 564–581.
60. Strinivas M , Patnaik L M . Adaptive Probabilities of Crossover and Mutation In Genetic Algorithms. IEEE Trans.Syst. Man and Cybernetics , 1994,24(4):656–667.

61. H. Markowitz, Analysis in portfolio choice and capital markets, Oxford, Basil Blackwell, 1987.
62. Marco Dorigo and Maria Gambardella - "Ant Colony System: A Cooperative Learning Approach To Traveling Salesman Problem" – 1997
63. Darrell Whitley, Timothy Startweather and D'Ann Fuquay - "Scheduling Problems And Traveling Salesman: The Genetic Edge Recombination Operator" – 1989
64. M K Pakhira, A Hybrid Genetic Algorithm using Probabilistic Selection,IE(I) Journal - CP, Vol 84, May 2003,pp. 23–30
65. Lienig J. (1997) A Parallel Genetic Algorithm for Performance Driven VLSI Routing. IEEE Transactions on Evolutionary Computation Vol. I. No.1 :29–39
66. Mazumder P., Rudnick E. (1999) Genetic Algorithm for VLSI Design, Layout and Automation. Addison-Wesley Longman Singapore Pte. Ltd., Singapore.
67. Schnecke V., Vornberger O (1996) A Genetic Algorithm for VLSI Physical Design Automation :In Proceedings of Second Int. Conf. on Adaptive Computing in Engineering Design and Control, ACEDC '96 26–28 Mar 1996, University of Plymouth, U.K., pp 53–58
68. Schnecke V., Vornberger O (1996) An Adaptive Parallel Genetic Algorithm for VLSILayout Optimization :In Proceedings of 4th Int. Conf. on Parallel Problem Solving from Nature (PPSN IV) 22–27 Sep 1996, Springer LNCS 1141, pp 859–868
69. Schnecke V., Vornberger O (1997) Hybrid Genetic Algorithms for Constrained Placement Problems. IEEE Transactions on Evolutionary Computation. Vol. I. No.4. :266- 277
70. Laurence D. Merkle, George H. Gates, Jr., Gary B. Lamont, and Ruth Pachter. Application of the parallel fast messy genetic algorithm to the protein structure prediction problem. Proceedings of the Intel Supercomputer Users' Group Users Conference, pages 189–195, 1994.
71. Kenneth M. Merz and Scott M. Le Grand, editors. The Protein Folding Problem and Tertiary Structure Prediction. Springer, New York, 1994.
72. Steven R. Michaud. Solving the Protein Structure Prediction Problem with Parallel Messy Genetic Algorithms. Master's thesis, Air Force Institute of Technology, Wright Patterson AFB, March 2001.
73. Steven R. Michaud, Jesse B. Zydallis, Gary Lamont, and Ruth Pachter, Scaling a genetic algorithm to medium sized peptides by detecting secondary structures with an analysis of building blocks. In Matthew Laudon and Bart Romanowicz, editors, Proceedings of the First International Conference on Computational Nanoscience, pages 29–32, Hilton Head, SC, March 2001.
74. Steven R. Michaud, Jesse B. Zydallis, David M. Strong, and Gary Lamont. Load balancing search algorithms on a heterogeneous cluster of pcs. In Proceedings of the Tenth SIAM Conference on Parallel Processing for Scientific Computing (PP01), Portsmouth, VA, March 2001.
75. Amer DRAA, Hichem TALBI, Mohamed BATOUCHE, A Quantum-Inspired Genetic Algorithm for Solving the Nqueens Problem", 7th ISPS'Algiers May 2005,pp.145–152.
76. Erbas, Cengiz., Sarkeshik, Sayed and Tanik, Murat M. (1992) "Different Perspectives of the N-Queens Problem," In Proceedings of ACM 1992 Computer Science Conference, Kansas City, MO, March 3–5.
77. Watkins, John J. (2004). Across the Board: The Mathematics of Chess Problems. Princeton: Princeton University Press. ISBN 0-691-11503-6.
78. K. Han and J. Kim, "Quantum-inspired evolutionary algorithm for a class of combinatorial optimization". IEEE transactions on evolutionary computation, vol. 6, no. 6, December 2002.
79. P. W. Shor, "Quantum Computing," Documenta Mathematica, vol. Extra Volume ICM, pp. 467- 486, 1998.
80. H.Talbi, A.Draa And M.Batouche, "A Quantum-Inspired Genetic Algorithm for Multi-Source Affine Image Registration", In the proceedings of the International Conference on Image Analysis and Recognition (ICIAR'04), Porto, September 2004, Springer-Verlag Press, LNCS 3211 pp. 147–154.
81. K.-H. Han and J.-H. Kim, "Genetic Quantum Algorithm and its Application to Combinatorial Optimization Problem," in Proceedings of the 2000 Congress on Evolutionary Computation, IEEE Press, pp.1354–1360, July 2000.

Bibliography

82. Möller, B, Graf, W, and Stransky, W (2004) Fuzzy-Optimization of Structures, In: Proceedings of ICCES04, edited by S.N. Atluri and S.J.N. Tadeu. Tech Science Press, Madeira, pages 1765–1770.
83. Möller, B, Beer, M, Graf, W, and Stransky, W (2000) Dynamic Structural Analysis Considering Fuzziness, In: 4th Euromech Solid Mechanics Conference, edited by M. Potier-Ferry and L. S. Toth. Euromech, Metz, pages 616.
84. Thomas Bernard, Markoto Sajidman, and Helge-Björn Kuntze," A New Fuzzy-Based Multiobjective Optimization Concept for Process Control Systems", Fuzzy Days 2001, LNCS 2206, pp. 653–670, 2001.
85. Ackermann, J.: Robuste Regelung. Springer, Heidelberg 1993
86. Fonseca, C.M.; Fleming, P.J.: Multiobjective optimization and multiple constraint handling with evolutionary algorithms - part I: a unified formulation. IEEE Trans. Syst. Man & Cybernetics A, 28 (1), pp. 26–37, 1998
87. Ng, W.Y: Interactive Multi-Objective Programming as a Framework for Computer-Aided Control System Design, volume 132 of Lect. Notes Control & Inf. Sci. Springer-Verlag, Berlin, 1989
88. Zakian, V.; Al-Naib, U.: Design of dynamical and control systems by the method of inequalities. Proc. IEE, 120(11), pp. 1421–1427, 1973
89. Bellman, R.E.; Zadeh, L.A.: Decision Making In A Fuzzy Environment, Management Science, 17 (1970), S. 141–163
90. Rommelfanger, H.: Fuzzy Decision Support Systeme, Springer, Heidelberg 1994
91. Sajidman, M.; Kuntze, H.-B.: Integration of Fuzzy Control and Model Based Concepts for Disturbed Industrial Plants with Large Dead-Times. Proc. 6th IEEE Int. Conf. on Fuzzy Systems (FUZZ IEEE'97), Barcelona (Spain), July 1–5, 1997.
92. C.Carlsson and R.Fuller, Interdependence in fuzzy multiple objective programming, Fuzzy Sets and Systems, 65(1994) 19–29.
93. C.Carlsson and R.Fuller, Fuzzy if-then rules for modeling interdependencies in FMOP problems, in: Proceedings of EUFIT'94 Conference, September 20–23, 1994 Aachen, Germany, Verlag der Augustinus Buchhandlung, Aachen, 1994 1504–1508.
94. C.Carlsson and R.Full'er, Fuzzy reasoning for solving fuzzy multiple objective linear programs, in: R.Trappl ed., Cybernetics and Systems '94, Proceedings of the Twelfth European Meeting on Cybernetics and Systems Research, World Scientific Publisher, London, 1994, vol.1, 295–301.
95. C.Carlsson and R.Full'er, Multiple Criteria Decision Making: The Case for Interdependence, Computers & Operations Research, 22(1995) 251–260.
96. C.Carlsson and R.Full'er, Fuzzy multiple criteria decision making: Recent developments, Fuzzy Sets and Systems, 78(1996) 139–153.
97. C.Carlsson and R.Full'er, Optimization with linguistic values, TUCS Technical Reports, Turku Centre for Computer Science, No. 157/1998.
98. R.Felix, Relationships between goals in multiple attribute decision-making, Fuzzy Sets and Systems, 67(1994) 47–52.
99. M.Inuiguchi, H.Ichihashi and H. Tanaka, Fuzzy Programming: A Survey of Recent Developments, in: Slowinski and Teghem eds., Stochastic versus Fuzzy Approaches to Multiobjective Mathematical Programming under Uncertainty, Kluwer Academic Publishers, Dordrecht 1990, pp 45–68
100. A. Kusiak and J.Wang, Dependency analysis in constraint negotiation, IEEE Transactions on Systems, Man, and Cybernetics, 25(1995) 1301- 1313.
101. Y.-J.Lai and C.-L.Hwang, Fuzzy Multiple Objective Decision Making: Methods and Applications, Lecture Notes in Economics and Mathematical Systems, Vol. 404 (Springer-Verlag, New York, 1994).
102. M.K. Luhandjula, Fuzzy optimization: an appraisal, Fuzzy Sets and Systems,30(1989) 257–282.
103. Y. Tsukamoto, An approach to fuzzy reasoning method, in: M.M. Gupta, R.K. Ragade and R.R. Yager eds., Advances in Fuzzy Set Theory and Applications (North-Holland, New-York, 1979).

104. R.R. Yager, Constrained OWA aggregation, Fuzzy Sets and Systems, 81(1996) 89–101.
105. H.-J.Zimmermann, Methods and applications of fuzzy mathematical programming, in: R.R.Yager and L.A.Zadeh eds., An Introduction to Fuzzy Logic Applications in Intelligent Systems, Kluwer Academic Publisher, Boston, 1992 97–120.
106. D. Dubois and H. Prade, Ranking fuzzy numbers in the setting of possibility theory. Inform. Sci. 30, (1983) 183 - 224.
107. D. Dubois and H. Prade, Possibility Theory: An Approach to Computerized Processing of Uncertainty. Plenum Press, New York - London, 1988.
108. M. Inuiguchi, H. Ichihashi and Y. Kume, Some properties of extended fuzzy preference relations using modalities. Inform. Sci. 61, (1992) 187 - 209.
109. M. Inuiguchi and M. Sakawa, Possible and necessary optimality tests in possibilistic linear programming problems, Fuzzy Sets and Systems 67 (1994), 29–46.
110. M. Inuiguchi, J. Ramik, T. Tanino and M. Vlach, Satisficing solutions and duality in interval and fuzzy linear programming. Fuzzy Sets and Systems 135 (2003), 151–177.
111. M. Inuiguchi, Enumeration of all possibly optimal vertices with possible optimality degreesin linear programming problems with a possibilistic objective function, Fuzzy Optimization and Decision Making, 3, (2004), 311–326.
112. J. Ramik and M. Vlach, Generalized Concavity in Fuzzy Optimization and Decision Analysis. Kluwer Acad. Publ., Dordrecht - Boston - London, 2002.
113. J. Ramik, Duality in Fuzzy Linear Programming: Some New Concepts and Results. Fuzzy Optimization and Decision Making, Vol.4, (2005), 25–39.
114. Francisco Herrera, Luis Magdalena, Introduction: Genetic Fuzzy Systems, International Journal Of Intelligent Systems, Vol. 13, 887–890 1998.
115. David Shaw, John Miles and Alex Gray, Genetic Programming within Civil Engineering, Organisation of the Adaptive Computing in Design and Manufacture 2004 Conference. April 20–22, 2004, Engineers House, Clifton, Bristol, UK., pp.
116. Koza J.R.,Genetic Programming: On the programming of computers by means of natural selection, Cambridge MA: MIT Press, ISBN 0-262-11170-5, 1992.
117. Banzhaf W et al, Genetic Programming- An introduction (On the automatic evolution of computer programs and its applications), Morgan Kaufmann Publishers, ISBN 1-55860-510-X, 1998.
118. Montana D.J, "Strongly typed genetic programming", Evolutionary computation, 3(2), 1995, pp199–230.
119. Radcliffe N.J and Surry P.D, "Formal memetic algorithms", Lecutre Notes in Computer Science 865, 1994.
120. Ashour A.F et al, "Empirical modelling of shear strength of RC deep beams by genetic programming", Computers and Structures, Pergamon, 81 (2003), pp331–338.
121. Hong YS and Bhamidimarri R, "Evolutionary self-organising modelling of a municipal wastewater treatment plant", Water Research, 37(2003), pp1199–1212.
122. Roberts S.C. and Howard D, "Detection of incidents on motorways in low flow high speed conditions by genetic programming", Cagnoni S et al (eds): EvoWorkshops 2002, LNCS 2279, Springer-Verlag, 2002, pp245–254.
123. Dorado J et al, "Prediction and modelling of the flow of a typical urban basin through genetic programming", Cagnoni S et al (eds): EvoWorkshops 2002, LNCS 2279, Springer-Verlag, 2002, pp190–201.
124. Howard D and Roberts SC, "The prediction of journey times on motorways using genetic programming", Cagnoni S et al (eds): EvoWorkshops 2002, LNCS 2279, Springer-Verlag, 2002, pp210–221.
125. Ishino Y and Jin Y, "Estimate design intent: a multiple genetic programming and multivariate analysis based approach", Advanced Engineering Infomatics, 16(2002), pp107–125.
126. Babovic V et al, "A data mining approach to modelling of water supply assets", Urban Water, 4(2002), pp401–414.
127. Kojima F. et al, "Identification of crack profiles using genetic programming and fuzzy inference", Journal of Materials Processing Technology, Elsevier, 108 (2001), pp263–267.

128. Whigham P.A. and Crapper P.F, "Modelling rainfall-runoff using genetic programming", Mathematical and Computer Modelling, 33(2001), pp707–721.
129. Lee D.G et al, "Genetic programming model for long-term forecasting of electric power demand", Electric power systems research, Elsevier, 40, 1997, pp17–22.
130. Montana D.J. and Czerwinski S, "Evolving control laws for a network of traffic signals", Proceedings of the First Annual Conference: Genetic Programming, July 28–3, 1996. Stanford University, pp333–338.
131. Köppen M and Nickolay B, "Design of image exploring agent using genetic programming", Proceedings of IIZUKA'96 Japan, 1996, pp549–552.
132. Yang Y. and Soh C.K, "Automated optimum design of structures using genetic programming", Computers and Structures, Pergamon, 80 (2002), pp1537–1546.
133. Yang J and Soh C.K, "Structural optimization by genetic algorithms with tournament selection", Journal of Computing in Civil Engineering, July 1997, pp195–200.
134. Diada J.M et al, "Visualizing tree structures in genetic programming", Lecture Notes in Computer Science 2724, 2003, pp1652–1664.
135. Wernert E.A and Hanson A.J, "Tethering and reattachment in collaborative virtual environments", Proceedings of IEEE Virtual Reality 2000, IEEE Computer Society Press, 2000, pp292.
136. Bulfin, R. & Liu, C. (1985). Optimal allocation of redundant components for large systems. IEEE Trans on Reliability, 34, 241–247.
137. Campbell, J. & Painton, L. (1996). Optimization of reliability allocation strategies through use of genetic algorithms. Proceedings of 6th Symposium on Multidisciplinary Design and Optimization, (pp. 1233–1242).
138. Chern, M. (1992). On the computational complexity of reliability redundancy allocation in a series system. Operations Research Letters, 11, 309–315.
139. Coit, D. & Smith, A. (1998). Redundancy allocation to maximize a lower percentile of the system time-to-failure distribution. IEEE Trans on Reliability, 47(1), 79–87.
140. Coit, D. & A. Smith (1996): Solving the redundancy allocation problem using a combined neural network/GA approach; Computers & Operations Research, 23.
141. Fyffe, D. E., Hines, W. W. & Lee, N. K. (1968). System reliability allocation and a computational algorithm. IEEE Trans on Reliability, 17, 64–69.
142. Gen, M., Ida, K. & Lee, J. U. (1990). A computational algorithm for solving 0-1 goal programming with GUB structures and its application for optimization problems in system reliability. Electronics and Communications in Japan, Part 3, 73, 88–96.
143. Ida, K., Gen M. & Yokota, T. (1994). System reliability optimization with several failure modes by genetic algorithm. In: Proceedings of 16th International Conference on Computers and Industrial Engineering, (pp 349–352).
144. Kulturel-Konak, S., A. Smith, & B. Norman (2004): Multi-Objective Tabu Search Using a Multinomial Probability Mass Function, European Journal of Operational Research.
145. MacQueen J. (1967): Some methods for classification and analysis of multivariate observations. In L. M. LeCam and J. Neyman, editors, Proceedings of the Fifth Berkeley Symposium on Mathematical Statistics and Probability, volume 1, 281–297.
146. Nakagawa, Y. & Miyazaki, S. (1981). Surrogate constraints algorithm for reliability optimization problems with two constraints. IEEE Trans on Reliability, R-30, 175- 180.
147. Painton, L. & Campbell, J. (1995). Genetic algorithms in optimization of system reliability. IEEE Trans on Reliability, 44(2), 172–178.
148. Rousseeuw Peter J. (1987): Silhouettes: A graphical aid to the interpretation and validation of cluster analysis. Journal of computational and applied mathematics, 20, 53–65.
149. Rousseeuw P., Trauwaert E. and Kaufman L. (1989): Some silhouette-based graphics for clustering interpretation. Belgian Journal of Operations Research, Statistics and Computer Science, 29, No. 3.
150. Srinivas, N. and K. Deb: Multi-objective Optimization Using Nondominated Sorting in Genetic Algorithms. Journal of Evolutionary Computation, 2(3).
151. S. Chopra, E.R. Gorres, M.R. Rao, Solving the Steiner tree problem on a graph using branch and cut, Oper. Res. Soc. Am. J. Comput. 4 (3), pp. 320–335, 1992

152. R.M. Karp, Reducibility among combinatorial problems, Complexity of Computer Computations, Plenum Press, New York, 1976
153. M. Gerla, L.Kleinrock, On the Topological Design of Distributed Computer Networks, IEEE Trans.Commun., 25 (1), Jan 1977
154. Grover, W.D., Venables, B.D., Sandham, J., and Milne, Performance of the Self-Healing Network Protocol with Random Individual Link Failure Time, in Proceedings of ICC 1991, vol. pp. 660–666, 1991.
155. Sakauchi, H., Nishimura, Y., and Hasegawa, S., A Self-Healing Network with an Economical Spare-Channel Assignment, in Proceedings of IEEE Globecom '90, pp. 438- 443, Dec., 1990.
156. Wu, T., and Lau, C., A Class of Self-Healing Ring Architectures for SONET Network Applications, in Proceedings of IEEE Globecom '90, pp. 444–452, Dec., 1990.
157. Wu, T., Kolar, D. J., and Cardwell, R.H., Survivable Network Architectures for Broadband Fiber Optic Networks: Models and Performance Comparisons, in IEEE Journal of Lightwave Technology, vol. 6, no. 11, pp. 1698–1709, Nov. 1988.
158. Nathan, Sri, Multi-Ring Topology Based Network Planning, Nortel Internal Document.
159. Frank, H., Frisch, I.T. and Chou, W., Topological Considerations in the design of theARPA computer network. in Conf. Rec., 1970 Spring Joint Computer Conference, AFIPS Conference Proceedings, vol. 36. Montvale, NJ: AFIPS Press, 1970.
160. Steiglitz, K., Weiner, P., and Kleitman, D. J., The design of minimum cost survivable networks. IEEE Transactions on Circuit Theory, pp. 455–460, Nov. 1969.
161. Frank, H., Frisch, I.T., Chou, W. and Van Slyke, R., Optimal design of centralized computer networks, Networks vol. 1, pp. 43–57, 1971.
162. Gerla, M. and Kleinrock, L. On the Topological Design of Distributed Computer Networks. IEEE Transactions on Communications, vol. 25, no. 1, January, 1977.
163. Gendreau, M., Labbe, M. and Laporte, G., Efficient heuristics for the design of ring networks, in Telecommunication Systems, vol. 4, pp 177–188, 1995.
164. Davis, L., Cox, A., and Qiu, Y., A Genetic Algorithm for Survivable Network Design in Proceedings of the Fifth International Conference on Genetic Algorithms, Morgan Kauffman, 1993, pp 408–415.
165. Davis, L., and Coombs, S., Genetic Algorithms and Communication Link Speed Design: Theoretical Considerations in Proceedings of the Second International Conference on Genetic Algorithms, Lawrence Erlbaum, 1987, pp 252–256.
166. Coombs, S., and Davis, L., Genetic Algorithms and Communication Link Speed Design: Constraints and Operators in Proceedings of the Second International Conference on Genetic Algorithms, Lawrence Erlbaum, 1987, pp 257–260.
167. Cormen, T.H., Leiserson, C.E. and Rivest, R.L., Introduction to Algorithms. McGraw Hill, 1990.
168. Starkweather, T., McDaniel, S., Mathias, K., Whitley, D. and Whitley, C., A Comparison of Genetic Sequencing Operators, in Proceedings of the Fourth International Conference on Genetic Algorithms, Morgan Kauffman, 1991, pp. 69–76.
169. Orvosh, D., and Davis, L., Shall We Repair? Genetic Algorithms, Combinatorial Optimization and Feasibility Constraints in Proceedings of the Fifth International Conference on Genetic Algorithms, Morgan Kauffman, 1993, pp. 650.
170. Mann, J., Kapsalis, A. and Smith, G.D., The GAmeter Toolkit in Applications of Modern Heuristic Methods, Rayward-Smith, V. J. (ed.), Alfred Walker, pp. 195–209, 1995.
171. Wasem, O. J., An Algorithm for Designing Rings for Survivable Fiber Networks, IEEE Transactions on Reliability, Vol. 40, No. 4, pp. 428–439, Oct. 1991.
172. Wasem, O. J., Optimal Topologies for Survivable Fiber Optic Networks Using SONET Self-Healing Rings, Proceedings of IEEE Globecom '91, pp. 2032–2038, Nov. 1991.
173. Slevinsky, J. B., Grover, W. D. and MacGregor, M. H., An Algorithm for Survivable Network Design Employing Multiple Self-healing Rings, Proceedings of IEEE Globecom '93, pp. 1586–1573, Nov. 1993.
174. Wasem, O. J., Wu, T. H. and Cardwell, R. H., Survivable SONET Networks – Design Methodology, IEEE JSAC, Vol. 12, pp. 205–212, Jan. 1994.

175. Gardner, L. M., Haydari, M., Shah, J., Sudborough, I. H. and Xia, C., Techniques for Finding Ring Covers in Survivable Networks, Proceedings of IEEE Globecom '94, pp. 1862- 1866, Nov. 1994.
176. Boot, J., Wever, H. W. and Zwinkels, A. M. E., Planning an SDH Network, Proceedings of the Sixth International Network Planning Symposium, pp. 143–148, Sep. 1994.
177. Poppe, F. and Demeester, P., An Integrated Approach to the Capacitated Network Design Problem, Proceedings of the Fourth International Conference on Telecommunication Systems, pp. 391–397, Mar. 1996.
178. Baldwin, J. M., A new factor in evolution, American Naturalist, vol. 30, pp. 441–451, 1896.
179. Whitley, D., Gordon, S. and Mathias, K., Lamarckian evolution, the Baldwin effect and function optimization. In Parallel Problem Solving from Nature - PPSN III. In Y. Davidor, H.P. Schwefel, and R. Manner, editors, pp. 6–15. berlin: Springer-Verlag, 1994.
180. Belew, R.K., McInerney, J. and Schraudolph, N.N., Evolving networks: Using the Genetic Algorithm with connectionist learning. In Proceedings of the Second Artificial Life Conference, pp. 511–547, Addison-Wesley, 1991.
181. P. K. Nanda, P. Kanungo, D. P. Muni," Parallel Genetic Algorithm Based Crowding Scheme Using Neighbouring Net Topology,"Proc. Of 6th International conference on Information technology, dec 2003, Bhubaneswae, pp 583–585.
182. T. Back, D. B. Fogel and T. Michalewicz, (Ed.):Evolutionary Computation _; Basic Algorithms and operators, Institute of Physics publishing, Bristol and Philadelphia; 2000.
183. Samir W. Mahfoud, Simple Analytical Models of Genetic Algorithms for Multi modal Function Optimization, Proceedings of the Fifth International Conference on genetic Algorithms, 1993.
184. P. Kanungo, Parallel Genetic Algorithm Based Crowding Scheme for Cluster Analysis, M.E. thesis, Department of Electrical Engineering, R. E. C. Rourkela, Jan.,2001.
185. E. Cantu-Paz, A survey of Parallel Genetic Algorithms, Calculateurs Paralleles, Vol. 10, No. 2, 1998, pp.141–171.
186. Cantu-Paz,E.: Migration policies and takeover times in parallel Genetic Algorithms,Proceedings of the International Conference on Genetic and Evolutionary Computation, San Francisco, CA, 1999, pp.775–779.
187. P. K. Nanda, Bikash Ghose and T. N. Swain, Parallel genetic algorithm based unsupervised scheme for extraction of power frequency signals in the steel industry, IEE Proceedings on vision,Image and Signal Processing, UK, Vol.149, No. 4,2002, pp.204–210.
188. E. Cantu-Paz: Designing Efficient and Accurate Parallel Genetic Algorithms, parallel Genetic Algorithms, Ph.D. dissertation, Illinois Genetic Algorithm Laboratory, UIUC, USA; 1999.
189. Aggarwal, K. K., Rai, S., (1981). Reliability evaluation in computer communication networks, IEEE Transactions on Reliability, R-30 (1).
190. Aggarwal, K. K., Chopra, Y. C., Bajwa, J. S., (1982). Reliability evaluation by network decomposition, IEEE Transactions on Reliability, R-31 (4), 355–358.
191. Biegel, J. E., Davern, J. J., (1990). Genetic algorithms and job shop scheduling, Computers and Industrial Engineering, 19 (1–4), 81–91.
192. Boorstyn, R. R., Plank, H., (1977). Large scale network topological optimization, IEEE Transactions on Communications, Com-25 (1), 29–37.
193. Cavers, J. K., (1975). Cutset manipulations for communication network reliability estimation, IEEE Transactions on Communications, Com-23 (6).
194. Coit, D. W., Smith, A. E., (1994). Use of genetic algorithm to optimize a combinatorial reliability design problem, Proceeding of the Third IIE Research Conference, 467–472.
195. Colbourn, C. J., (1987). The Combinatorics of Network Reliability, Oxford University Press.
196. Cohoon, J., Hedge, S. U., Martin, W. N., Richards, D. S., (1991). Distributed genetic algorithms for the floorplan design problem, IEEE Transactions on Computer Aided Design, 10 (4), 483–492.
197. Goldberg, D. E., (1989). Genetic Algorithms in Search, Optimization and Machine Learning, Addison-Wesley.
198. Hopcroft, J., Ullman, J., (1973). Set merging algorithms, SIAM Journal of Computers, 2, 296–303.

199. Jan, R. H., (1993). Design of reliable networks, Computers and Operations Research, 20, 25–34.
200. Jan, R. H., Hwang, F. J., Cheng, S. T., (1993). Topological optimization of a communication network subject to a reliability constraint, IEEE Transactions on Reliability, 42 (1), 63–70.
201. Muhlenbein, H., Schleuter, M. G., Kramer, O., (1988). Evolution algorithms in combinatorial optimization, Parallel Computing, 7, 65–85.
202. Nakawaza, H., (1981). Decomposition method for computing the reliability of complex networks, IEEE Transactions on Reliability, R-30 (3).
203. Rai, S., (1982). A cutset approach to reliability evaluation in communication networks, IEEE Transactions on Reliability, R-31 (5).
204. Roberts, L. G., Wessler, B. D., (1970). Computer network development to achieve resource sharing, AFIPS Conference Proceedings, 36. Montvale, NJ: AFIPS Press, 543–599.
205. Smith, A. E., Tate, D. M., (1993). Genetic optimization using a penalty function, Proceedings of the Fifth International Conference on Genetic Algorithms, 499–505.
206. Yeh, M. S., Lin, J. S., Yeh, W. C., (1994). A new Monte Carlo method for estimating network reliability, Proc 16th International Conference on Computers & Industrial Engineering, 723–726.
207. James C. Werner, Mehmet E. Aydin, Terence C. Fogarty," Evolving genetic algorithm for Job Shop Scheduling problems," Proceedings of ACDM 2000 PEDC, Unviersity of Plymouth, UK
208. Aarts, E.H.L., Van Laarhoven, P.J.M., Lenstra, J.K. and Ulder, N.L.J., (1994). A computational study of local search algorithms for job shop scheduling, ORSA Journal on Computing 6, pp. 118–125.
209. Aiex, R.M., Binato S. and Resende, M.G.C. (2001). Parallel GRASP with Path-Relinking for Job Shop Scheduling, AT&T Labs Research Technical Report, USA. To appear in Parallel Computing.
210. Adams, J., Balas, E. and Zawack., D. (1988). The shifting bottleneck procedure for job shop scheduling, Management Science, Vol. 34, pp. 391–401.
211. Applegate, D. and Cook, W., (1991). A computational study of the job-shop scheduling problem. ORSA Journal on Computing, Vol. 3, pp. 149–156.
212. Baker, K.R., (1974). Introduction to Sequencing and Scheduling, John Wiley, New York.
213. Bean, J.C., (1994). Genetics and Random Keys for Sequencing and Optimization, ORSA Journal on Computing, Vol. 6, pp. 154–160.
214. Beasley, D., Bull, D.R. and Martin, R.R. (1993). An Overview of Genetic Algorithms: Part 1, Fundamentals, University Computing, Vol. 15, No.2, pp. 58–69, Department of Computing Mathematics, University of Cardiff, UK.
215. Binato, S., Hery, W.J., Loewenstern, D.M. and Resende, M.G.C., (2002). A GRASP for Job Shop Scheduling. In: Essays and Surveys in Metaheuristics, Ribeiro, Celso C., Hansen, Pierre (Eds.), Kluwer Academic Publishers.
216. Brucker, P., Jurisch, B. and Sievers, B., (1994). A Branch and Bound Algorithm for Job-Shop Scheduling Problem, Discrete Applied Mathematics, Vol 49, pp. 105–127.
217. Carlier, J. and Pinson, E., (1989). An Algorithm for Solving the Job Shop Problem. Management Science, Feb, 35(29; pp.164–176.
218. Carlier, J. and Pinson, E., (1990). A practical use of Jackson's preemptive schedule for solving the jobshop problem. Annals of Operations Research, Vol. 26, pp. 269–287.
219. Cheng, R., Gen, M. and Tsujimura, Y. (1999). A tutorial survey of job-shop scheduling problems using genetic algorithms, part II: hybrid genetic search strategies, Computers & Industrial Engineering, Vol. 36, pp. 343–364.
220. Croce, F., Menga, G., Tadei, R., Cavalotto, M. and Petri, L., (1993). Cellular Control of Manufacturing Systems, European Journal of Operations Research, Vol. 69, pp. 498–509.
221. Croce, F., Tadei, R. and Volta, G., (1995). A Genetic Algorithm for the Job Shop Problem, Computers and Operations Research, Vol. 22(1), pp. 15–24.
222. Davis, L., (1985). Job shop scheduling with genetic algorithms. In Proceedings of the First International Conference on Genetic Algorithms and their Applications, pp. 136–140. Morgan Kaufmann.

Bibliography

223. Dorndorf, U. and Pesch, E., (1995). Evolution Based Learning in a Job Shop Environment, Computers and Operations Research, Vol. 22, pp. 25–40.
224. Fisher, H. and Thompson, G.L., (1963). Probabilistic Learning Combinations of Local Job-Shop Scheduling Rules, in: Industrial Scheduling, J.F. Muth and G.L. Thompson (eds.), Prentice-Hall, Englewood Cliffs, NJ, pp. 225–251.
225. French, S., (1982). Sequencing and Scheduling - An Introduction to the Mathematics of the Job-Shop, Ellis Horwood, John-Wiley & Sons, New York.
226. Garey, M.R. and Johnson, D.S., (1979). Computers and Intractability, W. H. Freeman and Co., San Francisco.
227. Giffler, B. and Thompson, G.L., (1960). Algorithms for Solving Production Scheduling Problems, Operations Research, Vol. 8(4), pp. 487–503.
228. Gray, C. and Hoesada, M. (1991). Matching Heuristic Scheduling Rules for Job Shops to the Business Sales Level, Production and Inventory Management Journal, Vol. 4, pp. 12–17.
229. Jackson, J.R., (1955). Scheduling a Production Line to Minimize Maximum Tardiness, Research Report 43, Management Science Research Projects, University of California, Los Angeles, USA.
230. Jain, A.S. and Meeran, S. (1999). A State-of-the-Art Review of Job-Shop Scheduling Techniques. European Journal of Operations Research, Vol. 113, pp. 390–434.
231. Jain, A. S., Rangaswamy, B. and Meeran, S. (1998). New and Stronger Job-Shop Neighborhoods: A Focus on the Method of Nowicki and Smutnicki (1996), Department of Applied Physics, Electronic and Mechanical Engineering, University of Dundee, Dundee, Scotland.
232. Johnson, S.M., (1954). Optimal Two and Three-Stage Production Schedules with Set-Up Times Included, Naval Research Logistics Quarterly, Vol. 1, pp. 61–68.
233. Laarhoven, P.J.M.V., Aarts, E.H.L. and Lenstra, J.K. (1992). Job shop scheduling by simulated annealing. Operations Research, Vol. 40, pp. 113–125.
234. Lawrence, S., (1984). Resource Constrained Project Scheduling: An Experimental Investigation of Heuristic Scheduling Techniques, GSIA, Carnegie Mellon University, Pittsburgh, PA.
235. Lenstra, J.K. and Rinnoy Kan, A.H.G., (1979). Computational complexity of discrete optimization problems. Annals of Discrete Mathematics, Vol. 4, pp. 121–140.
236. Lourenço, H.R. (1995). Local optimization and the job-shop scheduling problem. European Journal of Operational Research, Vol. 83, pp. 347–364.
237. Lourenço, H.R. and Zwijnenburg, M. (1996). Combining the large-step optimization with tabu-search: Application to the job-shop scheduling problem. In I.H. Osman and J.P. Kelly, editors, Metaheuristics: Theory and Apllications, pp. 219–236, Kluwer Academic Publishers.
238. Nowicki, E. and Smutnicki, C. (1996). A Fast Taboo Search Algorithm for the Job-Shop Problem, Management Science, Vol. 42, No. 6, pp. 797–813.
239. Perregaad, M. and Clausen, J., (1995). Parallel Branch-and-Bound Methods for the Job_shop Scheduling Problem, Working Paper, University of Copenhagen, Copenhagen, Denmark.
240. Resende, M.G.C., (1997). A GRASP for Job Shop Scheduling, INFORMS Spring Meeting, San Diego, California, USA.
241. Roy, B. and Sussmann, (1964). Les Problèmes d' ordonnancement avec contraintes dijonctives, Note DS 9 bis, SEMA, Montrouge.
242. Sabuncuoglu, I., Bayiz, M., (1997). A Beam Search Based Algorithm for the Job Shop Scheduling Problem, Research Report: IEOR-9705, Department of Industrial Engineering, Faculty of Engineering, Bilkent University, Ancara, Turkey.
243. Spears, W.M. and Dejong, K.A., (1991). On the Virtues of Parameterized Uniform Crossover, in Proceedings of the Fourth International Conference on Genetic Algorithms, pp. 230–236.
244. Storer, R.H., Wu, S.D. and Park, I., (1992). Genetic Algorithms in Problem Space for Sequencing Problems, Proceedings of a Joint US-German Conference on Operations Research in Production Planning and Control, pp. 584–597.
245. Storer, R.H., Wu, S.D., Vaccari, R., (1995). Problem and Heuristic Space Search Strategies for Job Shop Scheduling, ORSA Journal on Computing, 7(4), Fall, pp. 453–467.

246. Taillard, Eric D. (1994). Parallel Taboo Search Techniques for the Job Shop Scheduling Problem, ORSA Journal on Computing, Vol. 6, No. 2, pp. 108–117.
247. Vaessens, R.J.M., Aarts, E.H.L. and Lenstra, J.K., (1996). Job Shop Scheduling by local search. INFORMS Journal.
248. Wang, L. and Zheng, D. (2001). An effective hybrid optimisation strategy for job-shop scheduling problems, Computers & Operations Research, Vol. 28, pp. 585–596.
249. Williamson, D. P., Hall, L. A., Hoogeveen, J. A., Hurkens, C. A. J., Lenstra, J. K., Sevast'janov, S. V. and Shmoys, D. B. (1997) Short Shop Schedules, Operations Research, March - April, 45(2), pp. 288- 294.
250. Udhaya B. Nallamottu, Terrence L. Chambers, William E. Simon," Comparison of the Genetic Algorithm to Simulated Annealing Algorithm in Solving Transportation Location-allocation Problems With Euclidean Distances ",Proceedings of the 2002 ASEE Gulf-Southwest Annual Conference, The University of Louisiana at Lafayette, March 20 – 22, 2002.
251. Cooper, L. L., (1964), Heuristic Methods For Location-Allocation Problems, Siam Rev., 6, 37–53.
252. Cooper, L. L., (1972), The Transportation-Location Problems, Oper..Res., 20, 94–108. Gonzalez-Monroy, L. I., Cordoba, A., (2000), Optimization of Energy Supply Systems: Simulated Annealing Versus Genetic Algorithm, International Journal of Modern Physics C, 11 (4), 675 – 690.
253. Liu C.M., Kao, R. L., Wang, A.H., (1994), Solving Location-Allocation Problems with Rectilinear Distances by Simulated Annealing, Journal of The Operational Research Society, 45, 1304–1315.
254. Chowdury, H. I., Chambers, T. L., Zaloom, V., 2001, "The Use of Simulated Annealing to Solve Large Transportation-Location Problems With Euclidean Distances," Proceedings of the International Conference on Computers and Industrial Engineering (29th ICC&IE), Montreal, Canada, October 31 - November 3, 2001.
255. K. Guptaa, A. K. Bhuniab, "An Application of real-coded Genetic Algorithm (RCGA) for integer linear programming in Production-Transportation Problems with flexible transportation cost", AMO-Advanced Modeling and Optimization, Volume 8, Number 1, 2006,pp.73–98.
256. Arsham, H., (1992) Post optimality analysis of the transportaton problem. Journal of the Operational Research Society, vol.43, pp. 121 – 139.
257. Arshmam H, Khen AB.(1989) A simplex type algorithm for general transportation problems : an alternative to stepping-stone. Journal of the Operational Research Society; vol.40, pp.581–590.
258. Charness A, Copper WW.(1954) The steping stone method for explaining linear Programming calculation in transportation problem. Management Science; vol. 1 pp. 49 - 69.
259. Dantzig, GB.(1963) Linear programming and extentions. Princeton , NJ; Prinston University Press.
260. Davis L.(1991) Handbook of Genetic Algorithms. Van Nostrand Reinhold, Newyork.
261. Deb K. (1995) Optimization for Engineering Design-Algorithms and Examples. Prentice Hall of India,New Delhi .
262. Forest S. (1993). Proceedings of 5-th international conference on Genetic Algorithms. Margen Kaufmann, California. Goldberg DE (1989) Genetic Algorithms: Search, Optimization and Machine Learning. Addison Wesley.
263. Hitchock FL.(1941) The distribution of a product from several sources to numerous locations. Journal of Mathematical Physics,vol 20, pp. 224–30.
264. Liu S T.(2003) The total cost bounds of the transportation problem with varying demand and supply. Omega , vol. 31, pp. 247 – 51.
265. Michalawicz Z. (1996) Genetic Algorithms + Data structure= Evaluation Programs. Springer Verlog, Berlin.
266. Sakawa M. (2002) Genetic Algorithms and fuzzy multiobjective optimisation. Kluwer AcademicPublishers,

267. P. P. Zouein, A.M.Asce; H. Harmanani; and A. Hajar, Genetic Algorithm for Solving Site Layout Problem with Unequal-Size and Constrained Facilities, Journal Of Computing In Civil Engineering / April 2002,pp.143–151.
268. Aleksandra B Djurisic, (1998) Elite Genetic Algorithms with Adaptive Mutations for Solving Continuous Optimization Problems – Application to Modeling of the Optical Constants of Solids, Optics Communications, Vol. 151, pp.147–159.
269. B. Sareni, L. Krahenbuhl and A. Nicolas (1998), Niching Genetic Algorithms for Optimization in Electronmagnetics, IEEE Transcations on Magnetics, Vol. 34, No. 5, pp.2984–2987.
270. Heng Li and Peter Love (1997), Using Improved Genetic Algorithms to Facilitate Time-Cost Optimization, Journal of Construction Engineering and Management, Vol. 123, No. 3, pp.233–237.
271. Norman F. Foster and George S. Dulikravich, Three-Dimensional Aerodynamic Shape Optimization Using Genetic and Gradient Search Algorithms, Journal of Spacecraft and Rockets, Vol. 34, No. 1, pp.36–42.
272. S-Y. Chen, J. Situ, B. Mobasher and S. D. Rajan (1996), Use of Genetic Algorithms for the Automated Design of Residential Steel Roof Trusses, Advances in Structural Optimization-Proceedings of the First U.S.-Japan Joint Seminar on Structural Optimization, ASCE, New York.
273. S-Y. Chen (1997), Using Genetic Algorithms for the Optimal Design of Structural Systems, Dissertation for Doctor of Philosophy, Department of Civil Engineering, Arizona State University.
274. K. F. Pal (1995), Genetic Algorithm with Local Search, Biological Cybernetics, Vol. 73, pp.335- 341.
275. S. D. Rajan (1995), Sizing, Shape, and Topology Design Optimization of Trusses Using Genetic Algorithm, Journal of Structural Engineering, ASCE, Vol.121, No. 10, pp.1480–1487.
276. G. Olsen and G. N. Vanderplaats (1989), A Method for Nonlinear Optimization with Discrete Variables, AIAA Journal, Vol. 27, No. 11, pp.1584–1589.
277. D. E. Grierson and W. H. Lee (1984), Optimal Synthesis of Frameworks Using Standard Sections, Journal of Structural Mechanics, Vol. 12, No. 3, pp.335–370.
278. D. E. Grierson and G. E. Cameron (1984), Computer Automated Synthesis of Building Frameworks, Canadian Journal of Civil Engineering, Vol. 11, No. 4, pp.863–874.
279. M. P. Bendsoe and G. Strang (1988), Generating Optimal Topologies in Structural Design Using a Homogenization Method, Computer Methods in Applied Mechanics and Engineering, Vol. 71, pp.197–224.
280. Katsuyuki Suzuki and Noboru Kikuchi (1991), A Homogenization Method for Shape and Topology Optimization, Computer Methods in Applied Mechanics and Engineering, Vol. 93, pp.291–318.
281. S. Sankaranarayanan, R. T. Haftka and R. K. Kapania (1994), Truss Topology Optimization with Simultaneous Analysis and Design, AIAA Journal, Vol. 32, No. 2, pp.410–424.
282. Laetitia Jourdan. Clarisse Dhaenens. El-Ghazali Talbi, A Genetic Algorithm for Feature Selection in Data-Mining for Genetics, MIC'2001 - 4th Metaheuristics International Conference, Porto, Portugal, July 16–20, 2001,pp.29–33
283. C. Bates Congdon. A comparison of genetic algorithm and other machine learning systems on a complex classification task from common disease research. PhD thesis, University of Michigan, 1995.
284. C. Emmanouilidis, A. Hunter, and J. MacIntyre. A multiobjective evolutionary setting for feature selection and a commonality-based crossover operator. In Congress on Evolutionary Computing 2000, volume 2, pages 309–316. CEC, 2000.
285. J. Horn, D.E. Goldberg, and K. Deb. Implicit niching in a learning classi.er system : Nature's way. Evolutionary Computation, 2(1):37–66, 1994.
286. N. Monmarch'e, M. Slimane, and G. Venturini. Antclass : discovery of cluster in numeric data by an hybridization of an ant colony with the kmeans algorithm. Technical Report 213, Ecole d'Ing'enieurs en Informatique pour l'Industrie (E3i), Universit'e de Tours, Jan. 1999.

287. M. Pei, E.D. Goodman, and W.F. Punch. Feature extraction using genetic algorithms. Technical report, Michigan State University : GARAGe, June 1997.
288. M. Pei, E.D. Goodman, W.F. Punch, and Y. Ding. Genetic algorithms for classi.cation and feature extraction. In Annual Meeting : Classi.cation Society of North America, June 1995.
289. M. Pei, M. Goodman, and W.F. Punch. Pattern discovery from data using genetic algorithm. In Proc of the .rst Paci.c-Asia Conference on Knowledge Discovery and Data Mining, Feb. 1997.
290. Grimbleby, J.B.: "Automatic Analogue Network Synthesis using Genetic Algorithms", IEE/IEEE International Conference on Genetic Algorithms in Engineering Systems: Innovations and Applications (GALESIA '95), Sheffield, 12–14 September 1995, IEE Conference Publication No.414, pp. 53–58.
291. Grimbleby, J.B.: "Automatic Synthesis of Active Electronic Networks using Genetic Algorithms", IEE/IEEE International Conference on Genetic Algorithms in Engineering Systems: Innovations and Applications (GALESIA '97), Strathclyde, 2–4 September 1997, IEE Conference Publication No. 446, pp. 103–107.
292. Koza, J.R., Bennett, F.H., Andre, D. and Keane, M.A.: "Automated WYWIWYG Design for Both Topology and Component Values of Electrical Circuits using Genetic Programming", Genetic Programming 1996: Proceedings of the First Annual Conference, 28–31 July 1996, MIT Press, pp. 123–131.
293. Koza, J.R., Bennett F.H., Andre, D. and Keane, M.A.: "Evolutionary Design of Analog Electrical Circuits using Genetic Programming", Proceedings of Adaptive Computing in Design and Manufacture Conference, Plymouth, April 21–23 1998.
294. Nielsen, I.R.: "A C-T Filter Compiler – From Specification to Layout", Analog Integrated Circuits and Signal Processing, 1995, vol. 7, pp. 21–33.
295. M. Sonka, V. Hlavac and R. Boyle, Image processing, analysis and machine vision , Chapman and Hall, 1993.
296. R.M. Haralick, "Statistical and structural approaches to texture", Proc. IEEE, 67, 1979, pp. 786 - 804.
297. K. Delibasis and P.E. Undrill, "Anatomical object recognition using deformable geometric models", Image and Vision Computing, 12, 1994, pp. 423–433.
298. K. Delibasis Undrill P.E. and G.G. Cameron, "Genetic Algorithms applied to fourier descriptor based geometric models for anatomical object recognition in medical images", Comp. Vis. and Image Underst., 66 ,3, 1997, pp 286–300.
299. K. Delibasis and P.E. Undrill. Genetic algorithm implementation of stack filter design for image restoration, IEE Proc. Vision, Image and Signal Processing, 143, 1996, pp. 177 - 183.
300. M. Kass, A. Witkin and D. Terzopoulos, "Snakes: Active contour models", Intl. J. Comp. Vis., Vol. 1,No. 4, 1988, pp. 321–331.
301. Delibassis K, Undrill PE and Cameron GG, (1997) Designing Texture Filters with Genetic Algorithms : an application to Medical Images, Signal Processing, 57, 1, 19–33.
302. Y.S. Choi, R. Krishnapuram. A Robust Approach to Image Enhancement Based on Fuzzy Logic. IEEE Transactions on Image Processing, 6(6), 1997.
303. M-P. Dubuisson, A.K. Jain. A modified Hausdorff distance for object matching. In: Proceedings of the 12th IAPR Int. Conf. on Pattern Recognition, 1: 566–568, 1994.
304. J.C. Dunn. A fuzzy relative of the ISODATA process and its use in detecting compact wellseparated clusters. Journal of Cybernetics, 3: 32–57, 1973.
305. J.C. Dunn. Well-separated clusters and optimal fuzzy partitions. Journal of Cybernetics, 4: 95– 104, 1974.
306. P.D. Gader. Fuzzy Spatial Relations Based on Fuzzy Morphology. IEEE, 1997.
307. R.C. Gonzalez, R.E. Woods. Digital Image Processing. Second edition. Prentice-Hall, New Jersey, 2002.
308. K-P. Han, K-W. Song, E-Y. Chung, S-J. Cho, Y-H. Ha. Stereo matching using genetic algorithm with adaptive chromosomes. Pattern Recognition, 34: 1729–1740, 2001.
309. J. Liu, Y-H. Yang. Multiresolution Color Image Segmentation. IEEE Transactions on Pattern Analysis and Machine Intelligence, 16(7), 1994.

310. K.M. Passino, S. Yurkovich. Fuzzy Control, Addison-Wesley, California, 1998.
311. M.R. Rezaee, P.M.J. van der Zwet, B.P.F. Lelieveldt, R.J. van der Geest, J.H.C. Reiber. A Multiresolution Image Segmentation Technique Based on Pyramidal Segmentation and Fuzzy Clustering. IEEE Transactions on Image Processing, 9(7), 2000.
312. W. Rucklidge. Efficient visual recognition using the Hausdorff distance. In: Lecture Notes in Computer Science, 1173, 1996.
313. D.B. Russakoff, T. Rohlfing, C.R. Maurer Jr. Fuzzy segmentation of X-ray fluoroscopy image. Medical Imaging 2002: Image Processing. Proceedings of SPIE 2684, 2002.
314. F. Russo. Edge Detection in Noisy Images Using Fuzzy Reasoning. IEEE Transactions on Instrumentation and Measurement, 47(5), 1998.
315. R. Schallkoff. Pattern Recognition – Statistical, structural and neural approaches, John Wiley & Sons, Inc., New York, 1992.
316. M. Sonka, V. Hlavac, R. Boyle. Image Processing, Analysis, and Machine Vision. Second edition. Brooks/Cole Publishing Company, USA, 1999.
317. W-B. Tao, J-W. Tian, J. Liu. Image segmentation by three-level thresholding based on maximum fuzzy entropy and genetic algorithm. Pattern Recognition Letters, 24: 3069–3078, 2003.
318. Y.A. Tolias, S.M. Panas. Image Segmentation by a Fuzzy Clustering Algorithm Using Adaptive Spatially Constrained Membership Functions. IEEE Transactions on Systems, Man, and Cybernetics–Part A: Systems and Humans, 28(3), 1998.
319. Y. Yokoo, M. Hagiwara. Human Faces Detection Method using Genetic Algorithm. In: Proceeding of IEEE Int. Conf. on Evolutionary Computation, 113–118, 1996.
320. H. Wu, Q. Chen, M. Yachida. Face Detection From Color Images Using a Fuzzy Pattern Matching Method. IEEE Transactions on Pattern Analysis and Machine Intelligence, 21(6), 1999.
321. L.A. Zadeh. Fuzzy Logic. IEEE Computer, 21(4): 83–93, 1988.
322. Kazunori Otobe, Kei Tanaka And Masayuki Hirafuji," Knowledge Acquisition on Image Processing based On Genetic Algorithms," Proceeding of the IASTED International Conference Signal and Image Processing October 28–31, 1998, Las Vegas, Nevada – USA,pp.
323. George Karkavitsas and Maria Rangoussi," Object localization in medical images using genetic algorithms," Transactions on Engineering, Computing And Technology V2 December 2004 ISSN1305-5313
324. Brodatz, P. "A Photographic Album for Arts and Design," Dover Publishing Co., Toronto, Canada, 1966.
325. De Jong, K. "Learning with Genetic Algorithms : An overview," Machine Learning Vol. 3, Kluwer Academic publishers, 1988.
326. Devijver, P., and Kittler, J. "Pattern Recognition: A Statistical Approach," Prentice Hall, 1982.
327. Grefenstette, John J. Technical Report CS-83-11, Computer Science Dept., Vanderbilt Univ., 1984.
328. Ichino, M., and Sklansky, J.. "Optimum Feature selection by zero-one Integer Programming," IEEE Transactions on Systems, Man, and Cybernetics, Vol. 14, No. 5, 1984.
329. Michalski, R.S., Mozetic, I., Hong, J.R., and Lavrac, N.. "The Multi-purpose Incremental Learning System AQ15 and its Testing Application to Three Medical Domains, AAAI, 1986.
330. Vafaie, H., and De Jong, K.A., "Improving the performance of a Rule Induction System Using Genetic Algorithms," Proceedings of the First International Workshop on Multistrategy Learning, Harpers Ferry, W. Virginia, USA, 1991.
331. Qiang Huang, K. Yokoi, S. Kajita, K. Kaneko, H. Arai, N. Koyachi, and K. Tanie, "Planning walking patterns for a biped robot," IEEE Transactions on Robotics and Automation, vol. 17, no. 3, pp. 280–289, June 2001.
332. Jacky Baltes and Yuming Lin, "Path-tracking control of non-holonomic car-like robots using reinforcement learning," in RoboCup-99: Robot Soccer World Cup III, Manuela Veloso, Enrico Pagello, and Hiroaki Kitano, Eds., New York, 2000, pp. 162–173, Springer.
333. E. Uchibe, N. Nakamura, and M. Asada, "Cooperative behaviour acquisisition in a multiple mobile robot environment by co-evolution," in RoboCup-98: Robot Soccer World Cup II, Minoru Asada and Hiroaki Kitano, Eds. 1998, pp. 273–285, Springer Verlag.

334. T. C. Chin and X. M. Qi, "Integrated genetic algorithms based optimal fuzzy logic controller design," in Proceedings of the Fourth International Conference on Control, Automation, Robotics and Vision, 1996, pp. 563–567.
335. J. R. Koza, Genetic Programming: On the Programming of Computers by Means of Natural Selection, The MIT Press, 1992.
336. A. Patel, D. Davis, C.Guthrie, D. Tukand Tai Nguyen and J. Williams, Optimizing Cyclic-Steam Oil Production With Genetic Algorithms, SPE Western Regional Meeting, Irvine, California, 30 March–1 April, 2005.
337. M. Naghshineh and M. Schwartz, "Distributed call admission control in mobile/wireless networks" in Proc. PIMRC'95, Toronto, Canada, Sept. 1995
338. Der-Rong Din And Shian-Shyong Tseng," Genetic Algorithms for Optimal Design of the Two-Level Wireless ATM Network," Proc. Natl. Sci. Counc. ROC(A) Vol. 25, No. 3, 2001. pp. 151–162
339. Mitsuo Gen, Runwei Cheng, Genetic Algorithms and Engineering Optimization, John Wiley and Sons, Inc, New York, 1999.
340. K.Miettinen, P.Neittanmaki, M.M.Makela, J.Periaux, Evolutionary Algorithms in Engineering and Computer Science, John Wiley and Sons, Ltd, New York, 1999.
341. Practical Handbook of Genetic Algorithms- Applications Volume I, Edited by Lance Chambers, CRC Press, Inc. New York, 1995.
342. S.N.Sivanandam, S.Sumathi, S.N.Deepa, Introduction to Fuzzy Logic using MATLAB, Springer-Verlag Berlin Heidelberg, 2007.
343. S.N.Sivanandam, S.Sumathi, S.N.Deepa, Introduction to Neural Networks using MATLAB 6.0, Tata Mc-Graw Hill Publishing Company Ltd, NewDelhi, 2006.
344. Marco Dorigo, Mauro Birattari, Thomas Stutzle, "Ant Colony Optimization – Artificial Ants as a Computational Intelligence Technique", IRIDIA – technical report series, Technical Report No: TR/IRIDIA/2006-023, September 2006.
345. Marco Dorigo, Mauro Birattari, Thomas Stutzle, "Ant Colony Optimization – Artificial Ants as a Computational Intelligence Technique", IEEE Computational Intelligence Magazine, November 2006.
346. Venu G.Gudise, Ganesh. K. Venayagamoorthy, "FPGA Placement and Routing Using Particle Swarm Optimization", Proceedings of the IEEE Computer Society Annual Symposium on VLSI Emerging Trends in VLSI Systems Design (ISVLSI'04), 2004.
347. J.Kennedy, R.Eberhart, "Particle Swarm Optimization", From Proc. IEEE Int'l. Conf. on Neural Networks (Perth,Australia), IEEE Service Center, Piscataway, NJ, IV:1942–1948, 1995.

Web Bibliography

348. http://website.lineone.net/~kanta/publications/haploidGP.ps.
349. http://www.soe.rutgers.edu/ie/research/working_paper/paper%2005-011.pdf
350. http://www.rci.rutgers.edu/~coit/RESS_2007.pdf
351. http://iris.gmu.edu/~khoffman/papers/newcomb1.html
352. http://www.isps2005.dz/proceedings/papers/7-202.pdf
353. http://riot.ieor.berkeley.edu/~vinhun/index.html
354. http://garage.cse.msu.edu/papers/GARAGe97-04-01.pdf
355. http://www.genetic-programming.org/hc2005/JPT_cyclic_steam_with_genetic_algorithms.pdf
356. http://www.kecl.ntt.co.jp/as/members/yamada/unicom.pdf
357. http://www.geocities.com/jamwer2002/rep1.pdf
358. http://ipdps.cc.gatech.edu/2000/biosp3/18000605.pdf
359. http://www.sas.el.utwente.nl/home/gerez/cgi-bin/sabih/bonsma-msc.pdf?sendfile=bonsma-msc.pdf

360. http://ww1.ucmss.com/books/LFS/CSREA2006/GCA4489.pdf
361. http://www.ici.ro/camo/journal/vol8/v8a7.pdf
362. http://www-rcf.usc.edu/~maged/publications/GAinventoryrouting.pdf
363. http://vishnu.bbn.com/papers/conus.pdf
364. http://engr.louisiana.edu/asee/proceedings/VC4.pdf
365. http://osiris.tuwien.ac.at/~wgarn/VehicleRouting/Braysy.pdf
366. http://www.iasi.rm.cnr.it/ewgt/16conference/ID89.pdf
367. http://www.heinz.cmu.edu/wpapers/retrievePDF?id=2005-15
368. http://www.elec.reading.ac.uk/people/J.Grimbleby/PDF/cec99.pdf
369. http://www.pcs.cnu.edu/~riedl/research/publications/papers/Eunice1998.pdf
370. http://www.ucalgary.ca/~blais/Sahabi2006.pdf
371. http://www.smeal.psu.edu/ebrc/publications/res_papers/1999_09.pdf
372. http://www.enformatika.org/ijci/v2/v2-4-36.pdf
373. http://wifo1.bwl.uni-mannheim.de/fileadmin/files/publications/working_paper_1999_6.pdf
374. http://www.eng.auburn.edu/users/aesmith/postscript/berna.pdf
375. http://nr.stpi.org.tw/ejournal/ProceedingA/v25n3/151-162.pdf
376. http://courses.civil.ualberta.ca/cive605/GetPDFServlet_filetypepdfidJCEMD4000128000005000418000001idtypecvips.pdf
377. http://www.csc.byblos.lau.edu.lb/research/papers/jcce2002.pdf
378. http://www.ip-cc.org.uk/did/articles/miles-paper5.pdf
379. http://www.ias.ac.in/sadhana/Pdf2004Dec/Pe1229.pdf
380. http://ci.uofl.edu/rork/knowledge/publications/min_iri01.pdf
381. http://www2.lifl.fr/OPAC/Publications/Download/2001/2001_MIC_JourdanDhaenensTalbi_GeneticAlgorithm.pdf
382. http://www.cse.msstate.edu/~bridges/papers/annie2001.pdf
383. http://www-staff.it.uts.edu.au/~lbcao/publication/DM2004.pdf
384. http://www.kddresearch.org/Publications/Conference/HWWY1.pdf
385. http://arxiv.org/ftp/cs/papers/0412/0412087.pdf
386. http://www.enformatika.org/data/v2/v2-4.pdf
387. http://www.model.job.affrc.go.jp/Papers/Otobe/281077.pdf
388. http://www-cs.ccny.cuny.edu/~gertner/Students/Master/Maslov/orlando_paper_2001.PDF
389. http://www.biomed.abdn.ac.uk/Abstracts/A00033/
390. http://web.cecs.pdx.edu/~payel/fp100-ghosh.pdf
391. http://cs.gmu.edu/~eclab/papers/TAI92.pdf
392. http://eldar.mathstat.uoguelph.ca/dashlock/eprints/classify.pdf
393. http://www.recherche.enac.fr/opti/papers/articles/ieee.pdf
394. http://www.massey.ac.nz/~mgwalker/publications/walker02comparison.pdf
395. http://www.sunist.org
396. http://www.doc.ic.ac.uk/~nd/surprise_96/journal/vol4/tcw2/report.html#Introduction
397. http://dspace.nitrkl.ac.in/dspace/bitstream/2080/372/1/Nandapk-CIT-2003.pdf
398. http://tracer.lcc.uma.es/tws/cEA/documents/cant98.pdf
399. http://www.cad.zju.edu.cn/home/yqz/projects/gagpu/icnc05.pdf
400. http://www.cimms.ou.edu/~lakshman/Papers/ga/node8.html
401. http://www.itu.dk/~sathi/papers/IJCES.pdf
402. http://www.itu.dk/~sathi/papers/WSC6.pdf
403. http://www.ieindia.org/publish/cp/0503/may03cp5.pdf
404. http://www.ijicic.org/fic04-14.pdf
405. http://www.ijcsns.org/04_journal/200601/200601A28.pdf
406. http://www.nsti.org/publ/ICCN2002/272.pdf
407. http://neo.lcc.uma.es/cEA-web/documents/vrp.pdf
408. http://www.lania.mx/~ccoello/EMOO/nebro06.pdf.gz
409. http://ls11-www.cs.uni-dortmund.de/people/rudolph/publications/papers/gal95.pdf
410. http://www.genetic-programming.org/gp4chapter1.pdf
411. http://www.genetic-programming.com/gpanimatedtutorial.html
412. http://www.mathworks.com

413. http://www.paper.edu.cn
414. http://iridia.ulb.ac.be/~mdorigo/ACO/RealAnts.html
415. www.swarmintelligence.org
416. http://en.wikipedia.org/wiki/Swarm_intelligence
417. http://www.engr.iupui.edu/~shi/Coference/psopap4.html

Printing: Krips bv, Meppel
Binding: Stürtz, Würzburg